ATLAS OF NEOTROPICAL LEPIDOPTERA

Volume 2

Checklist: Part 1

ATLAS OF

NEOTROPICAL LEPIDOPTERA

Checklist: Part 1

Micropterigoidea – Immoidea

Edited by

J. B. HEPPNER

Center for Arthropod Systematics
Florida State Collection of Arthropods
Gainesville, Florida

1984 **DR W. JUNK PUBLISHERS**
a member of the KLUWER ACADEMIC PUBLISHERS GROUP
THE HAGUE / BOSTON / LANCASTER

Distributors

for the United States and Canada: Kluwer Academic Publishers, 190 Old Derby
Street, Hingham, MA 02043, USA
for the UK and Ireland: Kluwer Academic Publishers, MTP Press Limited,
Falcon House, Queen Square, Lancaster LA1 1RN, England
for all other countries: Kluwer Academic Publishers Group, Distribution Center,
P.O. Box 322, 3300 AH Dordrecht, The Netherlands

Library of Congress Cataloging in Publication Data

```
Main entry under title:

Atlas of neotropical lepidoptera.

    Bibliography: v. 2, p.
    Includes index.
    Contents:          -- 2. Checklist.
    1. Lepidoptera--Latin America--Collected works.
2. Insects--Latin America--Collected works.  I. Heppner,
John B.
QL553.A1A85  1984        595.78'098        84-7139
```

ISBN-13: 978-94-009-6535-5   e-ISBN-13: 978-94-009-6533-1
DOI: 10.1007/978-94-009-6533-1   (v. 2)

ISBN 90-6193-038-3 (this volume)
ISBN 90-6193-900-3 (series)

Cover design: Max Velthuijs

Copyright

To Latin American students and researchers
of Lepidoptera

CONTENTS

Preface ix

    Atlas of Neotropical Lepidoptera: General Plan x

    Checklist of Neotropical Lepidoptera: General Plan xi

    Contributing Authors xii

Introduction xv

    Acknowledgments xvi

    Transferred Taxa xvi

    Abbreviations xvii

    Outline of Classification of the Lepidoptera xix

    Bibliography xxiv

Checklist of Neotropical Lepidoptera: Part 1 1

    Generic Synopsis 3

    1.   Micropterigidae – *N. P. Kristensen and E. S. Nielsen* 16

    2.   Heterobathmiidae – *N. P. Kristensen and E. S. Nielsen* 16

    3.   Neopseustidae – *D. R. Davis* 16

    4.   Neotheoridae – *N. P. Kristensen* 16

    5.   Hepialidae – *G. S. Robinson and E. S. Nielsen* 16

    6.   Palaeosetidae – *G. S. Robinson and E. S. Nielsen* 17

    7.   Nepticulidae – *D. R. Davis* 17

    8.   Opostegidae – *D. R. Davis* 18

    9.   Tischeriidae – *D. R. Davis* 18

    10.   Incurvariidae – *D. R. Davis* 18

    11.   Cecidosidae – *D. R. Davis* 18

    12.   Adelidae – *D. R. Davis* 18

    13.   Prodoxidae – *D. R. Davis* 18

    14.   Heliozelidae – *J. B. Heppner* 18

    15.   Tineidae – *D. R. Davis* 19

    16.   Psychidae – *D. R. Davis* 24

    17.   Arrhenophanidae – *D. R. Davis* 25

    18.   Amphitheridae – *J. B. Heppner* 25

    19.   Lyonetiidae – *D. R. Davis and S. E. Miller* 25

    20.   Gracillariidae – *D. R. Davis and S. E. Miller* 25

    21.   Oecophoridae – *V. O. Becker* 27

    22.   Elachistidae – *V. O. Becker* 41

    23.   Blastobasidae – *V. O. Becker* 41

    24.   Coleophoridae – *V. O. Becker* 42

    25.   Momphidae – *V. O. Becker* 42

    26.   Agonoxenidae – *V. O. Becker* 43

    27.   Cosmopterigidae – *V. O. Becker* 43

    28.   Scythrididae – *V. O. Becker* 44

    29.   Gelechiidae – *V. O. Becker* 44

    30.   Copromorphidae – *J. B. Heppner* 53

    31.   Alucitidae – *S. E. Miller* 54

    32.   Carposinidae – *J. B. Heppner* 54

    33.   Epermeniidae – *J. B. Heppner* 54

    34.   Glyphipterigidae – *J. B. Heppner* 54

    35.   Plutellidae – *J. B. Heppner* 55

    36.   Yponomeutidae – *J. B. Heppner* 55

    37.   Argyresthiidae – *J. B. Heppner* 56

    38.   Douglasiidae – *J. B. Heppner* 56

    39.   Acrolepiidae – *J. B. Heppner* 56

    40.   Heliodinidae – *J. B. Heppner* 56

    41.   Immidae – *J. B. Heppner* 57

Notes 59

Index to Species 61

Index to Genera 101

PREFACE

This checklist does not supercede any other because it is the first of its kind for the Neotropical region. It represents the culmination of several years preparation by the contributing authors and editor, following the initiation of the "Atlas" project in the autumn of 1978. The collaborators to the project continue their work, either for "Atlas" fascicles or for the more immediate completion of the remaining five parts of the checklist.

The goal of the project is to provide a foundation work on the Neotropical Lepidoptera, illustrated to the extent possible and encompassing a detailed text covering the basic data known for each described species (original citation, synonymy, diagnosis, range, flight period and hosts). New species will be described as each author finds useful, but the primary purpose is initially to document all the species already known. Efforts to document the Lepidoptera faunas of individual countries in Latin America are relatively illogical, since in most cases upwards of 80% of the entire Neotropical fauna is represented in each country having tropical forest and, thus, such works would have to deal with most of the continental fauna. The "Atlas" project covers the entire region to begin with but as a cooperative project among all New World nations.

The "Atlas of Neotropical Lepidoptera" will encompass 125 volumes in total, of which 115 volumes will comprise the illustrated descriptive text, one volume will be a general introductory volume on the Lepidoptera of the neotropics and an illustrated generic synopsis, six volumes will form the checklist, two volumes will be devoted to a butterfly and a moth bibliography for the Neotropical species, and a final volume will be a general index. The project is expected to take about 20 years to complete by present and future collaborating authors. Upon the completion of the entire project a revised checklist will be issued.

The present checklist part covers the first portion of the Microlepidoptera, the first 41 families presently known to occur in the Neotropical region, encompassed by the area from the Mexican-United States border south to Patagonia. The remaining Microlepidoptera, including Pyraloidea, will be treated in Part 2 of the checklist. Names are included that were published up to the end of 1983. New taxonomic data has been included, in terms of new synonymies and new combinations, to the extent known. Much of the new data or revised taxon arrangements reflect current and on-going research by the contributing authors, details of which will be published in revisionary treatments. Where appropriate, explanations are provided by running numbered notes, to be found at the end of the volume before the terminal index. The language of the "Atlas" is English so as to have a wider international audience for the publications.

Authorship is indicated for each family, both in the table of contents and in the main checklist. Individual authors are responsible for their parts in terms of omissions and errors of spelling and dates. The format of the checklist, the higher classification adopted, editing, preparation of introductory sections and the index, and preparation of the manuscripts for printing, are the responsibility of the editor.

To the contributing authors, to those nations already pledged to the financial support of the "Atlas" project, and to the publisher and other patrons, goes the great credit of a broad and long-range outlook for this endeavor.

Gainesville   J. B. Heppner

December 30, 1983   Editor

ATLAS OF NEOTROPICAL LEPIDOPTERA: GENERAL PLAN

Volume 1	Introduction	
Vol. 2–7	Checklist: 6 pts.	
Vol. 8	Fasc. 1–14	Primitive moths
Vol. 9–10	Fasc. 15–20	Tineoidea
Vol. 11–14	Fasc. 21–29	Gelechioidea
Vol. 15	Fasc. 30–41	Copromorphoidea, Yponomeutoidea, Immoidea
Vol. 16–38	Fasc. 42–45	Pyraloidea, Pterophoroidea
Vol. 39–40	Fasc. 46–50	Sesioidea, Zygaenoidea
Vol. 41–42	Fasc. 51–55	Cossoidea, Castnioidea
Vol. 43	Fasc. 56	Tortricoidea
Vol. 44–71	Fasc. 57–61	Papilionoidea
Vol. 72–91	Fasc. 62–66	Geometroidea
Vol. 92–95	Fasc. 67–73	Bombycoidea
Vol. 96	Fasc. 74	Sphingoidea
Vol. 97–122	Fasc. 75–79	Noctuoidea
Vol. 123–124	Bibliography	
Vol. 125	General Index	

CHECKLIST OF NEOTROPICAL LEPIDOPTERA: GENERAL PLAN

Part 1   Micropterigoidea

Heterobathmioidea

Neopseustoidea

Hepialoidea

Nepticuloidea

Incurvarioidea

Tineoidea

Gelechioidea

Copromorphoidea

Yponomeutoidea

Immoidea

Part 2   Pyraloidea

Pterophoroidea

Sesioidea

Zygaenoidea

Cossoidea

Castnioidea

Tortricoidea

Part 3   Papilionoidea

Part 4   Geometroidea

Bombycoidea

Sphingoidea

Part 5   Noctuoidea

Part 6   General Index

CONTRIBUTING AUTHORS

Andrés O. Angulo   Universidad de Concepción, Concepcion, Chile
(Noctuidae, in part)

Vitor O. Becker   Centro de Pesquisa Agropecuaria dos Cerrados,   EMBRAPA, Planaltina,
D.F., Brazil
(Gelechioidea; Zygaenoidea, in part)

Keith S. Brown, Jr.   Universidade Estadual de Campinas, Campinas, São Paulo, Brazil
(Nymphalidae, in part; Papilionidae, in part)

Richard L. Brown   Mississippi State University, Starkville, Mississippi, USA
(Tortricidae, in part)

Curtis J. Callaghan   c/o Museo Nacional, Rio de Janeiro, Brazil
(Lycaenidae, in part)

†Robert H. Carcasson   c/o Provincial Museum, Victoria, British Columbia, Canada
(Sphingidae)

Mirna M. Casagrande   Universidade Federal do Paraná, Curitiba, Paraná, Brazil
(Nymphalidae, in part)

Charles V. Covell, Jr.   University of Louisville, Louisville, Kentucky, USA
(Geometridae, in part)

Donald R. Davis   Smithsonian Institution, Washington, D.C., USA
(Primitive moths, in part; Tineoidea, in part; Epipyropidae)

Julian P. Donahue   Los Angeles County Museum of Natural History, Los Angeles, California,
USA
(Cossidae)

W. Donald Duckworth   Smithsonian Institution, Washington, D.C., USA
(Sesiidae, in part)

Thomas D. Eichlin   California Department of Food and Agriculture, Sacramento, California,
USA
(Sesiidae, in part)

Douglas C. Ferguson   Systematic Entomology Laboratory, USDA, Washington, D.C., USA
(Lymantriidae)

William D. Field   Smithsonian Institution, Washington, D.C., USA
(Bibliography, in part)

John G. Franclemont   Cornell University, Ithaca, New York, USA
(Bombycoidea, in part)

Hugh A. Freeman   Garland, Texas, USA
(Hesperiidae, in part)

David F. Hardwick   Biosystematics Research Institute, Agriculture Canada, Ottawa, Ontario,
Canada
(Noctuidae: Heliothinae)

John B. Heppner   Center for Arthropod Systematics, Florida State Collection of Arthropods,
FDACS, Gainesville, Florida, USA
(Heliozelidae, Amphitheridae; Copromorphoidea; Yponomeutoidea; Immoidea; Sesioidea, in part; Tortricidae, in part; Hyblaeidae; Geometroidea,
in part; Bombycoidea, in part; Noctuoidea, in part)

Charles L. Hogue	Los Angeles County Museum of Natural History, Los Angeles, California, USA (Noctuidae, in part)
Rienk de Jong	Rijksmuseum van Natuurlijke Historie, Leiden, Netherlands (Hesperiidae, in part)
Niels P. Kristensen	Zoologisk Museum, University of Copenhagen, Copenhagen, Denmark (Micropterigidae and Heterobathmiidae, in part; Neotheoridae)
J. Donald Lafontaine	Biosystematics Research Institute, Agriculture Canada, Ottawa, Ontario, Canada (Noctuidae, in part)
Gerardo Lamas M.	Museo de Historia Natural "Javier Prado," Lima, Peru (Papilionoidea, in part; bibliography, in part)
Claude Lemaire	c/o Museum National d'Histoire Naturelle, Paris, France (Saturniidae)
Olaf H. H. Mielke	Universidade Federal do Paraná, Curitiba, Paraná, Brazil (Hesperiidae, in part)
Jacqueline Y. Miller	Allyn Museum of Entomology, Sarasota, Florida, USA (Castniidae)
Lee D. Miller	Allyn Museum of Entomology, Sarasota, Florida, USA (Papilionoidea, in part)
Scott E. Miller	Museum of Comparative Zoology, Harvard University, Cambridge, Massachusetts, USA (Lyonetiidae; Gracillariidae, in part; Alucitidae; Pterophoridae; Dalceridae)
Eugene G. Munroe	Biosystematics Research Institute, Agriculture Canada, Ottawa, Ontario, Canada (Pyralidae, in part)
Stanley S. Nicolay	Virginia Beach, Virginia (Lycaenidae, in part)
E. Schmidt Nielsen	Division of Entomology, CSIRO, Canberra City, A.C.T., Australia (Micropterigidae, Heterobathmiidae, Hepialidae, and Palaeosetidae, in part)
Jerry A. Powell	University of California, Berkeley, California, USA (Tortricidae, in part)
Józef Razowski	Institute of Systematic Zoology, Polish Academy of Sciences, Krakow, Poland (Tortricidae, in part)
Gaden S. Robinson	British Museum (Natural History), London, England, U.K. (Hepialidae and Palaeosetidae, in part)
Jay C. Shaffer	George Mason University, Fairfax, Virginia, USA (Pyralidae, in part)
Gerhard Tarmann	Tiroler Landeskundliches Museum, Innsbruck, Austria (Zygaenidae)
Paul Thiaucourt	c/o Museum National d'Histoire Naturelle, Paris, France (Notodontidae)
Allan Watson	British Museum (Natural History), London, England, U.K. (Drepanidae; Arctiidae, in part)

Paul E. S. Whalley   British Museum (Natural History), London, England, U.K.
                                          (Thyrididae)

Chris Wilkinson   Vrije Universiteit, Amsterdam, Netherlands
                                          (Nepticulidae, in part)

INTRODUCTION

The present checklist part lists the described taxa of the Neotropical region for the first 41 families currently known to occur within the area chosen for this project. Inasmuch as northern Mexico, although faunistically a part of the Nearctic region, is not normally included in Nearctic publications, the "Neotropical" region for the "Atlas" project covers the area from the Mexican-United States border south to Patagonia, including the West Indies and all oceanic islands claimed by Latin American nations. Bermuda is also included. Hawaii (USA) and Easter Id. (Chile) are not included, since their faunas are Melanesian. The 4189 species listed herein involve all those having a type-locality within this geographical area, plus those northern species having distributions extending into Mexico. Also included are a few species which have been collected in the border area but for which no verified Mexican records are known; these species are indicated by an asterisk (*).

Checklist Format

The format of the checklist is rather straightforward and follows standard entomological checklist style in listing valid species under valid generic names, with synonyms listed by date under each category. Author date follows each name, with parentheses added to indicate names not in their original generic combinations. The type-locality is noted for each name by nation only, except Brazil, Mexico, and the United States, where the states are also indicated (abbreviations used are listed on the pages covering abbreviations below). Localities noted from derived information are in brackets. All names adopted as current valid names for genera, species and subspecies are noted in boldface type. All synonyms are listed in Roman type face (this is likewise followed in the indeces). A synopsis of all included genera, including synonyms, is presented prior to the actual checklist.

Classification

Higher categories follow the arrangement proposed in a new book on the classification of Lepidoptera families (Heppner, 1984b). The new arrangement follows recent classifications, like that of Common (1970), but changes have been adopted following more recent studies. Synonyms of higher category names are not listed. All current names below the family level are listed in a phylogenetic arrangement, beginning with the taxa thought to be most primitive within each category to those thought to be most advanced. Exceptions to a phylogenetic arrangement involve only those groups which are so little known that an alphabetical arrangement is the only rational approach at this time. In Tineidae an alphabetical arrangement has also been chosen by the author, since the subfamilies require extensive revision.

Nomenclature

All described Neotropical species known to the authors for the first 41 families are included in the checklist, including all infraspecific names, emendations and misspellings. Rejected names are also listed in synonymy, as are major misidentifications often causing confusion in the literature. Various abbreviations are used for these and other situations, as noted in the list of descriptors below. Dates of authorship of names are as accurate as is known and those of certain older works follow the dates noted in Heppner (1982c) and various opinions of the Commission on Zoological Nomenclature. Date ranges and dates derived from other information are in brackets. The most recent date in a date range is always taken as the date of record in cases of priority. Original spellings of taxa are used thoughout, except for the deletion of diacritic marks which are not allowed by Article 27 of the Zoological Code. Terminal notes give information on particular aspects of certain names and are consecutively numbered throughout the checklist.

New synonyms and new combinations are clearly indicated by boldface type for the appropriate abbreviations (e.g., n. syn.). A few names are transferred in the checklist from one family to another and these names are listed below. One family, the Palaeosetidae, is listed for the first time for the Neotropical fauna.

Indexing

All families listed are given a reference number, as also indicated in the table of contents. This reference number is also used in the index to allow species names to be found in the text. Species are numbered separately for each family to allow easier future expansion of numbers. Each specific name is indexed in the terminal index by reference to its family number and its species number within that family; e.g., 29-200 refers to species 200 in family 29, which is Gelechiidae. Once readers become familiar with families of interest to them then later reference to index entries will easily identify specific names of these families merely by the family number. These number codes will also later be used in "Atlas" fascicles, other than for species that are transferred because of future revisionary studies. The generic index has the names indexed by page number.

ACKNOWLEDGMENTS

Many persons have been involved in the initial planning of the project, not least of which are the authors collaborating on the "Atlas," but W. D. Field (Smithsonian Institution, Washington) and E. L. Todd (Systematic Entomology Laboratory, USDA, Washington), deserve special mention for their encouragement. Illness has, unfortunately, prevented Dr. Todd from making his planned contributions in the Noctuidae fascicle of the "Atlas."

Mr. Wil R. Peters, the publisher for Dr. W. Junk Publishers B.V., The Hague, Netherlands, has been instrumental in fostering the "Atlas" project. His patience during these beginning years since 1978 has been crucial to the continued success of the project.

The nations thus far in support of the "Atlas" project are especially thanked for their foresight. The "Atlas," as a cooperative project, is the most logical and least expensive way for each Latin American nation to provide the basic documentation of native Lepidoptera, as noted previously, due to the generally widespread distributions of many Neotropical species. For the Netherlands and their New World interests, the Stichting voor Wetenschappelijk Orderzoek van de Troppen, The Hague, should be especially thanked here for providing already since 1981 the basic funding which in particular allowed the indexing of this first checklist volume to be completed on time (grant W87-173).

The Lepidoptera Research Foundation, Santa Barbara, California, has recently agreed to handle financial matters for the "Atlas" project by receiving donations from supporting nations, other organizations, and individuals.

A final word on one of the collaborators. Dr. Robert H. Carcasson, who passed away in September, 1982, was completing the checklist manuscript for the Neotropical Sphingidae as one of his last works, to be published in Part 4 of the Neotropical checklist. This work will be published with Dr. Carcasson as author as he expected.

TRANSFERRED TAXA

Mallomus Blanchard, 1854 – from Hepialidae to Geometridae
Mallomus ciliatus Blanchard, 1854 – from Hepialidae to Geometridae (see Checklist, Part 4)

"Gaea" lilloi Köhler, 1941 – from Sesiidae to Gelechiidae

Pexicnemidia Möschler, 1890 – from Tineidae to Yponomeutidae
Pexicnemidia mirella Möschler, 1890 – from Tineidae to Yponomeutidae

"Dalaca" perkeo Pfitzner, 1914 – from Hepialidae to Cossidae (see Checklist, Part 2)

Syncerastis Meyrick, 1931 – from Tineidae to Yponomeutidae
Syncerastis ptisanopa Meyrick, 1931 – from Tineidae to Yponomeutidae

Sericostola Meyrick, 1927 – from Yponomeutidae to Glyphipterigidae
Sericostola rhodanopa Meyrick, 1927 – from Yponomeutidae to Glyphipterigidae

Elachista rubella Blanchard, 1852 – from Elachistidae to Heliodinidae (**Heliodines**)

ABBREVIATIONS

Descriptors

ab.	aberration
auct.	of authors (= misdetermination)
aut. f.	autumn form
emend.	emendation (incorrect emendation)
extralim.	extralimital (not occurring in the Neotropical region; usually involving an extralimital species having Neotropical subspecies)
f.	form
gynand.	gynandromorph
hyb.	hybrid
incorr. spell.	incorrect spelling (use of diacritic marks not allowed by ICZN)
misid.	misidentified
mispl.	misplaced (generic placement uncertain or incorrect)
missp.	misspelling
n. comb.	new combination
n. syn.	new synonymy
nom. dub.	nomen dubium (= identity uncertain)
nom. nud.	nomen nudum (= undescribed name)
nom. oblit.	nomen oblitum (= forgotten name; unused senior synonym)
preocc.	preoccupied name
redesc.	redescribed name (a second original description)
repl. name	replacement name (for a preoccupied name)
rev. stat.	revised status
[sic]	original misspelling (= lapsus calami)
spr. f.	spring form
sum. f.	summer form
suppr.	suppressed (or in a rejected work according to ICZN)
unassoc.	unassociated female or male
unavail.	unavailable name (by ICZN article)
uncert. stat.	status uncertain
var.	variety
wint. f.	winter form
*	prefix for extralimital taxa (also for species from the United States border area not yet collected in northern Mexico but which are presumed to occur in Mexico as well)
/	= "and" (used to separate two or more localities where a lectotype has not yet been designated to restrict the type-locality)
ICZN	International Commission on Zoological Nomenclature
teste	= verified

Abbreviations for States of Brazil

Acre	Ac	Paraiba	Pb
Alagoas	Al	Paraná	Pr
Amapa	Ap	Pernambuco	Pe
Amazonas	Am	Piaui	Pi
Bahia	Ba	Rio de Janeiro	RJ
Ceara	Ce	Rio Grande do Norte	RN
Distrito Federal	DF	Rio Grande do Sul	RS
Espirito Santo	ES	Rondonia	Rd
Goiás	Go	Roraima	Rr
Maranhao	Ma	Santa Catarina	SC
Mato Grosso	MT	Sao Paulo	SP
Minas Gerais	MG	Sergipe	Se
Pará	Pa		

Abbreviations for States of Mexico

Aguascalientes	Agu	Morelos	Mor
Baja California Norte	BCN	Nayarit	Nay
Baja California Sur	BCS	Nuevo Leon	NL
Campeche	Cam	Oaxaca	Oax
Chiapas	Chia	Puebla	Pue
Chihuahua	Chih	Queretaro	Que
Coahuila	Coa	Quintana Roo	QR
Colima	Col	San Luis Potosi	SLP
Distrito Federal	DF	Sinaloa	Sin
Durango	Dur	Sonora	Son
Guanajuato	Gua	Tabasco	Tab
Guerrero	Gue	Tamaulipas	Tam
Hidalgo	Hid	Tlaxcala	Tla
Jalisco	Jal	Veracruz	Ver
Mexico	Mex	Yucatan	Yuc
Michoacan	Mic	Zacatecas	Zac

Abbreviations for States of the United States

Alabama	Al	Montana	Mt
Alaska	Ak	Nebraska	Nb
Arizona	Az	Nevada	Nv
Arkansas	Ar	New Hampshire	NH
California	Ca	New Jersey	NJ
Colorado	Co	New Mexico	NM
Connecticut	Ct	New York	NY
Delaware	De	North Carolina	NC
District of Columbia	DC	North Dakota	ND
Florida	Fl	Ohio	Oh
Georgia	Ga	Oklahoma	Ok
Hawaii	Hi	Oregon	Or
Idaho	Id	Pennsylvania	Pa
Illinois	Il	Rhode Island	RI
Indiana	In	South Carolina	SC
Iowa	Ia	South Dakota	SD
Kansas	Ks	Tennessee	Tn
Kentucky	Ky	Texas	Tx
Louisiana	La	Utah	Ut
Maine	Me	Vermont	Vt
Maryland	Md	Virginia	Va
Massachusetts	Ma	Washington	Wa
Michigan	Mi	West Virginia	WV
Minnesota	Mn	Wisconsin	Wi
Mississippi	Ms	Wyoming	Wy
Missouri	Mo		

OUTLINE OF CLASSIFICATION OF THE LEPIDOPTERA

The list of families, subfamilies, and associated higher categories, presented here shows in outline form the classification adopted for the "Atlas of Neotropical Lepidoptera." The classification follows that of a book recently completed by the editor (Heppner, 1984b). This classification is similar to that of Common (1970) but includes the most recent changes based on current understandings of various specialists (e. g., Brock, 1971; Common, 1975; Davis, 1975b, 1978a, 1978b; Davis and Nielsen, 1980; Heppner, 1977, 1982a, 1982b; Hodges, 1978; Hodges, et al., 1983; Kristensen, 1978a, 1978b; Kristensen and Nielsen, 1979, 1980; Nielsen and Davis, 1981; Powell, 1980; Razowski, 1974; Zagulajev, 1973; Zimmerman, 1978), as well as such innovations as the placement of Amphitheridae in Tineoidea, the placement of Pyraloidea after Immoidea, and the rearrangement of families within the Cossoidea and Zygaenoidea. Specific documentation for these latter changes will be found in Heppner (1984b). Refinements may be made in the future as new information becomes available to possibly alter our concepts of Lepidoptera phylogeny. The families known to occur in the Neotropical region are numbered in the checklist (as also in the Table of Contents) and these numbers are used in the index to refer species names to their position in the checklist. In the outline the unnumbered families are extralimital and not known to occur in the Neotropical region. The number following each Neotropical family for the first 41 families indicates the species total listed in the checklist for that family; the numbers for families 42 to 79 involve the estimated numbers of described species for the Neotropical region as noted in Heppner (1984b) (some species totals for families do not correspond to the number indicated in Heppner (1984b) due to the inclusion herein of Nearctic species that overlap the border region).

Under Tineidae recognized subfamilies are listed but since extensive revision is needed to redefine the limits of these subfamilies, possibly with major synonymies, the genera and species have been listed in the checklist entirely alphabetically for this family.

Some major recent literature references that revise the Neotropical fauna for the families covered in this first part of the checklist, or otherwise form the basis of current taxonomic decisions by the authors, are as follows:

Micropterigidae (Kristensen and Nielsen, 1979, 1982)
Heterobathmiidae (Kristensen and Nielsen, 1979, 1983)
Neopseustidae (Davis, 1975b; Davis and Nielsen, 1980)
Neotheoridae (Kristensen, 1978a)
Hepialidae Robinson, 1977; Robinson and Nielsen, 1983)
Palaeosetidae (Robinson and Nielsen, 1983)
Nepticulidae (Newton and Wilkinson, 1982; Scoble, 1982, 1983; Wilkinson and Newton, 1981)
Opostegidae (Davis, 1984; Eyer, 1964)
Tischeriidae (Braun, 1972; Hering, 1926)
Incurvariidae (Nielsen and Davis, 1981)
Cecidosidae (Becker, 1977)
Adelidae (Nielsen, 1980; Pastrana, 1961; Razowski and Wojtusiak, 1978)
Prodoxidae (Davis, 1967)

Heliozelidae (Wojtusiak, 1978)
Tineidae (Davis, 1978b; Dietz, 1905; Hasbrouck, 1964; Petersen, 1957-58; Robinson, 1976, 1978; Zagulajev, 1960-79, 1973)
Psychidae (Davis, 1964, 1975a; Vázquez, 1941-[44])
Arrhenophanidae (Bradley, 1951)
Amphitheridae (Moriuti, 1978)
Lyonetiidae (Braun, 1963; Kuroko, 1964)
Gracillariidae (Ely, 1918; Opler and Davis, 1981; Vári, 1961)
Oecophoridae (Becker, 1982; Clarke, 1971, 1978; Duckworth, 1969, 1970, 1971; Hodges, 1974; Kasy, 1973; Powell, 1973)
Elachistidae (Braun, 1948; Traugott-Olsen and Nielsen, 1977)
Blastobasidae (Dietz, 1910; Gozmány, 1963)
Coleophoridae (Hodges, 1966b; Sattler and Tremewan, 1974, 1978; Toll, 1962)
Momphidae (Hodges, 1966b; Riedl, 1969)
Agonoxenidae (Clarke, 1962)
Cosmopterigidae (Hodges, 1978)
Scythrididae (Jäckh, 1977-78; Passerin d'Entrèves, 1977)
Gelechiidae (Busck, 1903b, 1939; Hodges, 1966a; Povolný, 1967a, 1967b; Sattler, 1960, 1973)
Copromorphidae (Clarke, 1955; Heppner, 1984a, 1984c)
Alucitidae (Buszko, 1977)
Carposinidae (Davis, 1969)
Epermeniidae (Gaedike, 1977)
Glyphipterigidae (Diakonoff, 1984; Heppner, 1981a, 1981b, 1982a, 1984d; Toll, 1956)
Plutellidae (Busck, 1903a; Moriuti, 1977)
Yponomeutidae (Moriuti, 1977)
Argyresthiidae (Moriuti, 1977)
Douglasiidae (Gaedike, 1974)
Acrolepiidae (Davis, 1980; Gaedike, 1970)
Heliodinidae (Gaedike, 1967; Meyrick, 1914)
Immidae (Heppner, 1982d)

Classification Outline

LEPIDOPTERA

Suborder Zeugloptera

Micropterigoidea

1. **Micropterigidae** (2)

Suborder Aglossata

Agathiphagoidea

* Agathiphagidae

Suborder Heterobathmiina

Heterobathmioidea

2. **Heterobathmiidae** (2)

Suborder Glossata

Infraorder Dacnonypha

Eriocranioidea

*   Eriocraniidae
*   Acanthopteroctetidae
*   Lophocoronidae

Infraorder Neopseustina

Neopseustoidea

3.   **Neopseustidae** (4)

Infraorder Exoporia

Mnesarchaeoidea

*   Mnesarchaeidae

Hepialoidea

4.   **Neotheoridae** (1)
*   Anomosetidae
*   Prototheoridae
5.   **Hepialidae** (131)
6.   **Palaeosetidae** (1)

Infraorder Heteroneura

Division Monotrysia

Nepticuloidea

7.   **Nepticulidae** (13)
      Pectinivalvinae
      Nepticulinae
8.   **Opostegidae** (16)
9.   **Tischeriidae** (8)

Incurvarioidea

10. **Incurvariidae** (4)
11. **Cecidosidae** (4)
12. **Adelidae** (17)
13. **Prodoxidae** (6)
14. **Heliozelidae** (15)

Division Ditrysia

Section Tineina

Tineoidea

15. **Tineidae** (456)
      Nemapogoninae
      Scardiinae
      Euplocaminae
      Tinissinae
      Meessiinae
      Teichobiinae
      Dryadaulinae
      Acrolophinae
      Setomorphinae
      Phthoropoeinae

      Myrmecozelinae
      Siloscinae
      Tineinae
      Erechthiinae
      Hieroxestinae
*   Eriocottidae
      Eriocottinae
      Compsocteninae
16. **Psychidae** (73)
      Penestoglossinae
      Taleporiinae
      Psychinae
      Oiketicinae
*   Pseudarbelidae
17. **Arrhenophanidae** (7)
*   Ochsenheimeriidae
18   **Amphitheridae** (1)
19. **Lyonetiidae** (19)
      Bedelliinae
      Lyonetiinae
      Cemiostominae
      Bucculatriginae
20. **Gracillariidae** (147)
      Gracillariinae
      Lithocolletinae
      Phyllocnistinae

Gelechioidea

21. **Oecophoridae** (1733)
      Depressariinae
      Ethmiinae
      Peleopodinae
      Autostichinae
      Xyloryctinae
      Stenomatinae
      Oecophorinae
      Hypertrophinae
      Chimabachinae
      Deuterogoniinae
22. **Elachistidae** (7)
      Coelopoetinae
      Elachistinae
*   Pterolonchidae
23. **Blastobasidae** (114)
      Symmocinae
      Blastobasinae
24. **Coleophoridae** (32)
      Coleophorinae
      Batrachedrinae
25. **Momphidae** (29)
26. **Agonoxenidae** (48)
      Agonoxeninae
      Blastodacninae
27. **Cosmopterigidae** (110)
      Antequerinae
      Cosmopteriginae
      Chrysopeleiinae
28. **Scythrididae** (12)
29. **Gelechiidae** (835)
      Anomologinae
      Gelechiinae
      Anacampsinae
      Chelariinae
      Dichomeridinae

Lecithocerinae
Physoptilinae

Copromorphoidea

30. **Copromorphidae** (12)
31. **Alucitidae** (24)
32. **Carposinidae** (7)
33. **Epermeniidae** (3)
Epermeniinae
Ochromolopinae
34. **Glyphipterigidae** (75)

Yponomeutoidea

35. **Plutellidae** (27)
Orthotaeliinae
Galacticinae
Plutellinae
Scythropinae
Praydinae
36. **Yponomeutidae** (90)
Attevinae
Saridoscelinae
Yponomeutinae
Cedestinae
37. **Argyresthiidae** (7)
38. **Douglasiidae** (1)
39. **Acrolepiidae** (21)
40. **Heliodinidae** (39)
Schreckensteiniinae
Heliodininae

Immoidea

41. **Immidae** (36)

Pyraloidea

42. **Hyblaeidae** [2]
43. **Thyrididae** [183]
Argyrotypinae
Thyridinae
Siculodinae
Striglininae
* Lathrotelidae
44. **Pyralidae** [3350]
Midilinae
Scopariinae
Nymphulinae
Odontiinae
Glaphyriinae
Evergestinae
Pyraustinae
Schoenobiinae
Cybalomiinae
Crambinae
Pyralinae
Chrysauginae
Epipaschiinae
Galleriinae
Phycitinae
Peoriinae

Pterophoroidea

* Tineodidae
* Oxychirotidae
45. **Pterophoridae** [135]
Agdistinae
Platyptiliinae
Pterophorinae

Sesioidea

46. **Brachodidae** [13]
Brachodinae
Phycodinae
47. **Sesiidae** [193]
Tinthiinae
Paranthreninae
Sesiinae
48. **Choreutidae** [91]
Millieriinae
Brenthiinae
Choreutinae

Zygaenoidea

* Heterogynidae
49. **Zygaenidae** [131]
Zygaeninae
Phaudinae
Charideinae
Chalcosiinae
Anomoeotinae
Himantopterinae
Procridinae
* Somabrachyidae
50. **Megalopygidae** [251]

Section Cossina

Cossoidea

51. **Cossidae** [166]
Pseudocossinae
Hypoptinae
Cossulinae
Cossinae
Zeuzerinae
* Dudgeoneidae
* Metarbelidae
* Cyclotornidae
52. **Epipyropidae** [4]
53. **Dalceridae** [80]
Acraginae
Dalcerinae
54. **Limacodidae** [260]
* Chrysopolomidae
Ectropinae
Chrysopolominae

Castnioidea

55. **Castniidae** [160]

            Synemoninae
            Neocastniinae
            Castniinae

Tortricoidea

    56. **Tortricidae** [676]
            Chlidanotinae
            Ceracinae
            Tortricinae
            Cochylinae
            Olethreutinae

Papilionoidea

    57. **Hesperiidae** [1786]
            Megathyminae
            Coeliadinae
            Pyrrhopyginae
            Pyrginae
            Trapezitinae
            Heteropterinae
            Hesperiinae
    58. **Papilionidae** [119]
            Baroniinae
            Parnassiinae
            Papilioninae
    59. **Pieridae** [275]
            Pseudopontiinae
            Dismorphiinae
            Pierinae
            Coliadinae
    60. **Lycaenidae** [2575]
            Poritiinae
            Miletinae
            Curetinae
            Lycaeninae
            Riodininae
    61. **Nymphalidae** [2798]
            Danainae
            Ithomiinae
            Acraeinae
            Heliconiinae
            Nymphalinae
            Calinaginae
            Libytheinae
            Apaturinae
            Amathusiinae
            Morphinae
            Brassolinae
            Satyrinae

Geometroidea

    Series Calliduliformes

    *   Callidulidae
    *   Pterothysanidae
            Pterothysaninae
            Hibrildinae
    *   Ratardidae

    Series Uraniformes

    62. **Sematuridae** [35]

    63. **Uraniidae** [6]
            Microniinae
            Uraniinae
    *   Epicopeiidae
    *   Apoprogonidae
    64. **Epiplemidae** [230]

Series Geometriformes

    65. **Geometridae** [7804]
            Archiearinae
            Oenochrominae
            Ennominae
            Geometrinae
            Sterrhinae
            Larentiinae

Series Drepaniformes

    *   Axiidae
    66. **Thyatiridae** [5]
    *   Cyclidiidae
    *   Drepanidae
            Drepaninae
            Oretinae
            Nidarinae

Bombycoidea

Series Bombyciformes

    *   Carthaeidae
    67. **Eupterotidae** [3]
            Striphnopteryginae
            Janinae
            Eupterotinae
    68. **Apatelodidae** [245]
            Apatelodinae
            Epiinae
    *   Bombycidae
    69. **Mimallonidae** [250]
    *   Anthelidae
            Munychryiinae
            Anthelinae
    70. **Lasiocampidae** [707]
            Chondrosteginae
            Chionopsychinae
            Archaeopachinae
            Gonometinae
            Macromphaliinae
            Gastropachinae
            Lasiocampinae

Series Saturniiformes

    *   Endromidae
    *   Lemoniidae
            Lemoniinae
            Sabaliinae
    *   Brahmaeidae
    71. **Oxytenidae** [60]
    72. **Cercophanidae** [30]
            Cercophaninae
            Janiodinae
    73. **Saturniidae** [809]

Arsenurinae
Ceratocampinae
Hemileucinae
Agliinae
Ludiinae
Salassinae
Saturniinae

Sphingoidea

74. **Sphingidae** [397]
Sphinginae
Macroglossinae

Noctuoidea

75. **Notodontidae** [1650]
Notodontinae
Dilobinae
Thaumetopoeinae
76. **Dioptidae** [500]
*   Thyretidae
77. **Lymantriidae** [180]
78. **Arctiidae** [6300]
Aganainae
Pericopinae
Lithosiinae
Arctiinae
Ctenuchinae
79. **Noctuidae** [8516]
Hypenodinae
Hypeninae
Herminiinae
Rivulinae
Ophiderinae
Catocalinae
Plusiinae
Euteliinae
Stictopterinae
Sarrothripinae
Chloephorinae
Nolinae
Acontiinae
Pantheinae
Acronictinae
Agaristinae
Cocytiinae
Amphipyrinae
Cuculliinae
Hadeninae
Noctuinae
Heliothinae

BIBLIOGRAPHY

Becker, V. O.
  1977. The taxonomic position of the Cecidosidae Brèthes (Lepidoptera). Polski Pismo Ent., 47: 79-86.
  1982. Stenomine moths of the Neotropical genus *Timocratica* (Oecophoridae). Bull. Br. Mus. (Nat. Hist.), Ent., 45:211-306.
Bradley, J. D.
  1951. Notes on the family Arrhenophanidae (Lepidoptera: Heteroneura), with special reference to the morphology of the genitalia, and descriptions of one new genus and two new species. Ent. (London), 84:178-185.
Braun, A. F.
  1948. Elachistidae of North America (Microlepidoptera). Mem. Amer. Ent. Soc., 13:1-110, 26 pl.
  1963. The genus *Bucculatrix* in America north of Mexico (Microlepidoptera). Mem. Amer. Ent. Soc., 18:1-208, 45 pl.
  1972. Tischeriidae of America north of Mexico (Microlepidoptera). Mem. Amer. Ent. Soc., 28:1-148.
Brock, J. P.
  1971. A contribution towards an understanding of the morphology and phylogeny of the ditrysian Lepidoptera. J. Nat. Hist., 5:29-102.
Busck, A.
  1903a. Notes on the *Cerostoma* group of Yponomeutidae, with description of new North American species. J. New York Ent. Soc., 9:45-59.
  1903b. A revision of the American moths of the family Gelechiidae, with descriptions of new species. Proc. U. S. Natl. Mus., 25:767-938, pl. 28-32.
  1939. Restriction of the genus *Gelechia* (Lepidoptera: Gelechiidae), with descriptions of new genera. Proc. U. S. Natl. Mus., 86:563-593, pl. 59-71.
Buszko, J.
  1977. Alucitidae. In, Klucze do oznaczania owadów Polski, 27 (Lepidoptera), 38:1-18. Warsaw. [In Polish].
Clarke, J. F. G.
  1955. Copromorphidae. In, Catalogue of the type specimens of Microlepidoptera in the British Museum (Natural History) described by Edward Meyrick, 2:509-531. London: Br. Mus. (Nat. Hist.).
  1962. The genus *Homoeoprepes* Walsingham (Lepidoptera: Blastodacnidae). Neotropical Microlepidoptera, I and II. Proc. U. S. Natl. Mus., 113:373-382.
  1971. Neotropical Microlepidoptera, XIX: notes on and new species of Oecophoridae (Lepidoptera). Smithson. Contr. Zool., 95:1-39.
  1978. Neotropical Microlepidoptera, XXI: new genera and species of Oecophoridae from Chile. Smithson. Contr. Zool., 273:1-79.
Common, I. F. B.
  1970. Lepidoptera (moths and butterflies). In, The insects of Australia: a textbook for students and research workers. Pp. 765-866. Canberra: Melbourne Univ. Press.
  1975. Evolution and classification of the Lepidoptera. Ann. Rev. Ent., 20:183-203.
Davis, D. R.
  1964. Bagworm moths of the Western Hemisphere (Lepidoptera: Psychidae). Bull. U. S. Natl. Mus., 244:1-233.
  1967. A revision of the moths of the subfamily Prodoxinae (Lepidoptera: Incurvariidae). Bull. U. S. Natl. Mus., 255:1-170.
  1969. A revision of the American moths of the family Carposinidae (Lepidoptera: Carposinoidea). Bull. U. S. Natl. Mus., 289:1-105.
  1975a. A review of the West Indian moths of the family Psychidae with descriptions of new taxa and immature stages. Smithson. Contr. Zool., 188:1-66.
  1975b. Systematics and zoogeography of the family Neopseustidae with the proposal of a new superfamily (Lepidoptera: Neopseustoidea). Smithson. Contr. Zool., 210:1-45.
  1978a. A revision of the North American moths of the superfamily Eriocranioidea with the proposal of a new family, Acanthopteroctetidae (Lepidoptera). Smithson. Contr. Zool., 251:1-131.
  1978b. The North American moths of the genera *Phaeoses, Opogona,* and *Oinophila,* with a discussion of their supergeneric affinities (Lepidoptera: Tineidae). Smithson. Contr. Zool., 282:1-39.
  1980. A redescription of *"Micropteryx" selectella* Walker with a discussion concerning its family affinities (Acrolepiidae). J. Lepid. Soc., 32 187-190.
  1984. A preliminary review of the family Opostegidae, with descriptions of new taxa and immature stages. Smithson. Contr. Zool., [in press].
Davis, D. R., and E. S. Nielsen
  1980. Description of a new genus and two new species of Neopseustidae from South America, with discussion of phylogeny and biological observations (Lepidoptera: Neopseustoidea). Steenstrupia (Copenhagen), 6:253-289.
Diakonoff, A. N.
  1984. Glyphipterigidae, sensu lato. In, H. G. Amsel, et al. (eds.), Microlepidoptera Palearctica, volume 7. Vienna: G. Fromme. [In Press].
Dietz, W. G.
  1905. Revision of the genera and species of the tineid subfamilies Amydriinae and Tineinae inhabiting North America. Trans. Amer. Ent. Soc., 31:1-95, pl. 1-6.
  1910. Revision of the Blastobasidae of North America. Trans. Amer. Ent. Soc., 36:1-72, pl. 1-4.
Duckworth, W. D.
  1969. Bredin-Archbold-Smithsonian biological survey of Dominica: West Indian Stenomidae (Lepidoptera: Gelechioidea). Smithson. Contr. Zool., 4:1-21.

1970. Neotropical Microlepidoptera, XVIII: revision
       of the genus *Peleopoda* (Lepidoptera: Oecopho-
       ridae). Smithson. Contr. Zool., 48:1-30.
1971. Neotropical Microlepidoptera, XX: revision of
       the genus *Setiostoma* (Lepidoptera: Stenomidae).
       Smithson. Contr. Zool., 106:1-45.
Ely, C. R.
1918. A revision of the North American Gracilarii-
       dae [sic] from the standpoint of venation. Proc.
       Ent. Soc. Wash., 19:29-77. (1917)
Eyer, J. R.
1964. A pictorial key to the North American moths
       of the family Opostegidae. J. Lepid. Soc., 17:
       237-242.
Gaedike, R.
1967. Zur systematischen Stellung einiger Gattungen
       der Heliodinidae Schreckensteiniidae sowie Re-
       vision der paläarktischen Arten der Gattung *Pan-
       calia* Curtis, 1830 (Lepidoptera). Beitr. Ent.
       (Berlin), 17:363-374.
1970. Revision der paläarktischen Acrolepiidae (Le-
       pidoptera). Ent. Abh. (Dresden), 38:1-54.
1974. Revision der paläarktischen Douglasiidae (Le-
       pidoptera). Acta Faun. Ent. Mus. Natl. Prague,
       15:79-101.
1977. Revision der nearktischen und neotropischen
       Epermeniidae (Lepidoptera). Beitr. Ent. (Berlin),
       27:301-312.
Gozmány, L. A.
1963. The family Symmocidae and the descriptions
       of new taxa mainly from the Near East (Lepido-
       ptera). Acta Zool. Acad. Sci. Hung. (Budapest),
       9:67-134.
Hasbrouck, F. F.
1964. Moths of the family Acrolophidae in America
       north of Mexico (Microlepidoptera). Proc. U. S.
       Natl. Mus., 114:487-706.
Heppner, J. B.
1977. The status of the Glyphipterigidae and a reas-
       sessment of relationships in yponomeutoid fam-
       ilies and ditrysian superfamilies. J. Lepid. Soc.,
       31:124-134.
1981a. Revision of the new genus *Diploschizia* (Lep-
       idoptera: Glyphipterigidae) for North America.
       Fla. Ent., 64:309-336.
1981b. *Neomachlotica*, a new genus of Glyphipterig-
       idae (Lepidoptera). Proc. Ent. Soc. Wash., 83:
       479-488.
1982a. Synopsis of the Glyphipterigidae (Lepidopte-
       ra: Copromorphoidea). Proc. Ent. Soc. Wash.,
       84:38-66.
1982b. A world catalog of genera associated with the
       Glyphipterigidae auctorum (Lepidoptera).   J.
       New York Ent. Soc., 89:220-294.
1982c. Dates of selected Lepidoptera literature for
       the Western Hemisphere fauna. J. Lepid. Soc.,
       36:87-111.
1982d. Review of the family Immidae with a world
       checklist (Lepidoptera: Immoidea).   Entomo-
       graphy (Sacramento), 1:257-279.
1984a. Revision of the Oriental and Nearctic genus
       Ellabella (Lepidoptera: Copromorphoidea).   J.
       Res. Lepid. (Santa Barbara), 22: [in press].
1984b. Lepidoptera family classification. A guide to
       the higher categories, world diversity and liter-

ature resources of the butterflies and moths.
       Medford, New Jersey: Plexus Publ. [In press].
1984c. Revision of the genus *Lotisma* (Lepidoptera:
       Copromorphidae). J. Res. Lepid., 23: [in press].
1984d. Sedge moths of North America (Lepidoptera:
       Glyphipterigidae). Gainesville: Flora and Fauna
       Press. (ca. 185 pp.)
Hering, O.
1926. Die Blattminierer-Gattung *Tischeria* in ihren
       paläarktischen Arten. Ent. Jahrb. (Leipzig), 35:
       99-106, 1 pl.
Hodges, R. W.
1966a. Revision of Nearctic Gelechiidae, I. The
       Lita group (Lepidoptera: Gelechioidea). Proc.
       U. S. Natl. Mus., 119:1-66, pl. 1-31.
1966b. Review of the New World species of *Batra-
       chedra* with descriptions of three new genera
       (Lepidoptera: Gelechioidea). Trans. Amer. Ent.
       Soc., 92:585-651.
1974. Gelechioidea. Oecophoridae. In, R. B. Domi-
       nick, et al. (eds.), The moths of America north
       of Mexico. Fasc. 6.2. London: E. W. Classey.
       142 pp., 8 pl.
1978. Gelechioidea. Cosmopterigidae. In, R. B. Do-
       minick, et al. (eds.), The moths of America
       north of Mexico.   Fasc. 6.1. London: E. W.
       Classey. 166 pp., 6 pl.
Hodges, R. W., et al. (eds.).
1983. Check list of the Lepidoptera of America north
       of Mexico. London: E. W. Classey. 284 pp.
Jäckh, E.
1977-78. Bearbeitung der Gattung *Scythris* Hübner
       (Lepidoptera, Scythrididae). Dtsch. Ent. Zeit.
       (Berlin), 24:261-271 (1977); 25:71-89 (1978).
Kasy, F.
1973. Beitrag zur Kenntnis der Familie Stathmopo-
       didae Meyrick (Lepidoptera, Gelechioidea). Tij-
       ds. Ent. (Amsterdam), 116:227-299.
Kuroko, H.
1964. Revisional studies on the family Lyonetiidae
       of Japan (Lepidoptera). Esakia (Fukuoka), 4:1-
       61, pl. 1-17.
Kristensen, N. P.
1978a. A new familia of Hepialoidea from South
       America, with remarks on the phylogeny of the
       subordo Exoporia (Lepidoptera). Ent. German-
       ica (Stuttgart), 4:272-294.
1978b. Observations on *Anomoses hylecoetes* (Ano-
       mosetidae), with a key to hepialoid families (In-
       secta: Lepidoptera). Steenstrupia (Copenhagen),
       5:1-19.
Kristensen, N. P., and E. S. Nielsen
1980. The ventral diaphragm of primitive (non-di-
       trysian) Lepidoptera. A morphological and phy-
       logenetic study. Zeit. Zool. Syst. Evolutionsfor.
       (Frankfurt), 18:123-146.
1982. South American micropterigid moths: two
       new genera of the *Sabatinca*-group (Lepidopte-
       ra: Micropterigidae). Ent. Scand., 13:513-529.
1983. The *Heterobathmia* life history elucidated: im-
       mature stages contradict assignment to subor-
       der Zeugloptera (Insecta: Lepidoptera).   Zeit.
       Zool. Syst. Evolutionsfor. (Frankfurt), 21:101-
       124.
Meyrick, E.

1914. Lepidoptera Heterocera. Fam. Heliodinidae. Genera Insectorum (Brussels), 165:1-29, 2 pl.

Moriuti, S.
1977. Yponomeutoidea s. lat. (Insecta: Lepidoptera). In, Fauna Japonica. Tokyo: Keigaku Publ. 327 pp., 95 pl.
1978. Amphitheridae (Lepidoptera): four new species from Asia, Telethera blepharacma Meyrick new to Japan and Formosa, and Sphenograptis Meyrick transferred to the family. Bull. Univ. Osaka Pref., (B) 30:1-17.

Newton, P. J., and C. Wilkinson
1982. A taxonomic revision of the North American species of Stigmella (Lepidoptera: Nepticulidae). Syst. Ent. (London), 7:367-463.

Nielsen, E. S.
1980. A cladistic analysis of the Holarctic genera of adelid moths (Lepidoptera: Incurvaroidea [sic]). Ent. Scand., 11:161-178.

Nielsen, E. S., and D. R. Davis.
1981. A revision of the Neotropical Incurvariidae s. str., with the description of two new genera and two new species (Lepidoptera: Incurvarioidea). Steenstrupia (Copenhagen), 7:25-57.

Opler, P. A., and D. R. Davis
1981. The leafmining moths of the genus Cameraria associated with Fagaceae in California (Lepidoptera: Gracillariidae). Smithson. Contr. Zool., 333:1-58.

Passerin d'Entrèves, P.
1977. Revisione degli scitrididi (Lepidoptera, Scythrididae) palearctici. Boll. Istit. Mus. Zool. Univ. Torino (Turin), 5:57-76.

Pastrana, J. A.
1961. La familia Adelidae (Lep.) en la República Argentina. Physis (Buenos Aires), 12:191-201, 3 pl.

Petersen, G.
1957-58. Die Genitalien der paläarktischen Tineiden (Lepidoptera: Tineidae). Beitr. Ent. (Berlin), 7: 55-176, 338-379, 557-595 (1957); 8:111-118, 398-430 (1958).

Povolný, D.
1967a. Genitalia of some Nearctic and Neotropic members of the tribe Gnorimoschemini (Lep., Gel.). Acta Ent. Mus. Natl. Prague, 37:51-127.
1967b. Die stammgeschichtlichen Beziehungen der Tribus Gnorimoschemini im Weltrahmen (Lep., Gel.). Acta Ent. Mus. Natl. Prague, 37:161-232.

Powell, J. A.
1973. A systematic monograph of New World ethmiid moths (Lepidoptera: Gelechioidea). Smithson. Contr. Zool., 120:1-302.
1980. Evolution of larval food preferences in Microlepidoptera. Ann. Rev. Ent., 25:133-159.

Razowski, J.
1974. Phylogeny and classification of Lepidoptera. Acta. Zool. Cracoviensis (Krakow), 19:1-18.

Razowski, J., and J. Wojtusiak
1978. Family-group taxa of Adeloidea (Lepidoptera). Polski Pismo Ent., 48:3-18.

Riedl, T.
1969. Matériaux pour la connaissance des Momphidae paléarctiques (Lepidoptera). Part IX. Revue des Momphidae européenes et compris quelques espèces d'Afrique du Nord et du Proche-Orient. Polski Pismo Ent., 39:635-919. [Momphidae sensu Riedl = Cosmopterigidae, Agonoxenidae and Momphidae].

Robinson, G. S.
1976. A taxonomic revision of the Tinissinae of the World (Lepidoptera: Tineidae). Bull. Br. Mus. (Nat. Hist.), Ent., 32:253-300.
1977. A taxonomic revision of the genus Callipielus Butler (Lepidoptera: Hepialidae). Bull. Br. Mus. (Nat. Hist.), Ent., 35:103-121, 6 pl.
1979. Clothes-moths of the Tinea pellionella complex: a revision of the world's species (Lepidoptera: Tineidae). Bull. Br. Mus. (Nat. Hist.), Ent., 38:57-128.

Robinson, G. S., and E. S. Nielsen
1983. Ghost-moths of southern South America. Entomonog. (Klampenborg), 4:1-192.

Sattler, K.
1960. Generische Gruppierung der europäischen Arten der Sammelgattung Gelechia (Lepidoptera, Gelechiidae) auf Grund der Untersuchungen der männlichen und weiblichen Genitalarmaturen. Dtsch. Ent. Zeit. (Berlin), 7:10-118.
1973. A catalogue of the family-group and genus-group names of the Gelechiidae, Holcopogonidae, Lecithoceridae, and Symmocidae (Lepidoptera). Bull. Br. Mus. (Nat. Hist.), Ent., 28:153-282.

Sattler, K., and W. G. Tremewan
1974. A catalogue of the family-group and genus-group names of the Coleophoridae (Lepidoptera). Bull. Br. Mus. (Nat. Hist.), Ent., 30:184-214.
1978. A supplementary catalogue of the family-group and genus-group names of the Coleophoridae (Lepidoptera). Bull. Br. Mus. (Nat. Hist.), Ent., 37:73-96.

Scoble, M. J.
1982. A pectinifer in the Nepticulidae (Lepidoptera) and its phylogenetic implications. Ann. Transvaal Mus., 33:123-129.
1983. A revised cladistic classification of the Nepticulidae (Lepidoptera) with descriptions of new taxa mainly from South Africa. Transvaal Mus. Monog., 2:1-105.

Toll, S.
1956. Glyphipterygidae. In, Klucze do oznaczania owadów Polski, 27 (Lepidoptera), 39:1-36 Warsaw. [In Polish].
1962. Materialien zur Kenntnis der paläarktischen Arten der Familie Coleophoridae (Lepidoptera). Polski Pismo Ent., 7:577-720, pl. 1-133.

Traugott-Olsen, E., and E. S. Nielsen
1977. The Elachistidae (Lepidoptera) of Fennoscandia and Denmark. In, Fauna Ent. Scand., 6:1-299.

Vári, L.
1961. Lithocolletidae. In, South African Lepidoptera. Volume 1. Transvaal Mus. Mem., 12:1-238, 23 pl.

Vázquez Garcia, L.
1941-[44]. Estudio monográfico de los Psychidae de Mexico. Anal. Inst. Biol. Univ. Mexico, 12:295-310, 3 pl. (1941); 13:257-300, 4 pl. (1942 [1944]).

Wilkinson, C., and P. J. Newton
    1981. The microlepidopteran genus *Ectoedemia* Busck (Nepticulidae) in North America. Tijds. Ent. (Amsterdam), 124-27-92, 93-110.
Wojtusiak, J.
    1978. Heliozelidae. In, Klucze do oznaczania owadów Polski, 27 (Lepidoptera), 94:1-18. Warsaw. [In Polish].
Zagulajev, A. K.
    1960-79. Tineidae. In, Fauna USSR, Lepidoptera, 4(3):1-267 (1960); 4(2):1-424 (1964); 4(4):1-126 (1973); 4(6):1-408 (1979). Leningrad: Akad. Nauk USSR. [In Russian].
    1973. On phylogeny in the superfamily Tineoidea (Lepidoptera). Trudy Vsesoy. Ent. Obshch. (Moscow), 56:170-183. [In Russian].
Zimmerman, E. C.
    1958-78. Lepidoptera. In, Insects of Hawaii, 7:1-542 (1958); 8:1-456 (1958); 9:1-1876 (1978). Honolulu: Univ. Hawaii Press.

Checklist of Neotropical Lepidoptera

Part 1

LEPIDOPTERA

Suborder Zeugloptera

Micropterigoidea

1. MICROPTERIGIDAE

Hypomartyria Kristensen & Nielsen, 1982
Squamicornia Kristensen & Nielsen, 1982

Suborder Heterobathmiina

Heterobathmioidea

2. HETEROBATHMIIDAE

Heterobathmia Kristensen & Nielsen, 1979

Suborder Glossata

Infraorder Neopseustina

Neopseustoidea

3. NEOPSEUSTIDAE

Synempora Davis & Nielsen, 1980
Apoplania Davis, 1975

Infraorder Exoporia

Hepialoidea

4. NEOTHEORIDAE

Neotheora Kristensen, 1978

5. HEPIALIDAE

Dalaca Walker, 1856
    Dolaca Druce, 1901, missp.
    Huapina Bryk, 1945
    Maculella Viette, 1950
    Toenga Tindale, 1954
Callipielus Butler, 1882
    Stachyocera Ureta, 1957
Blanchardina Viette, 1950
Calada Nielsen & Robinson, 1983
Puermytrans Viette, 1951
Parapielus Viette, 1949
    Lossbergiana Viette, 1950
Andeabatis Nielsen & Robinson, 1983
Druceiella Viette, 1949
Trichophassus Le Cerf, 1919
Phassus Walker, 1856
Schausiana Viette, 1949
Aplatissa Viette, 1953

Pfitzneriana Viette, 1952
    Pseudophassus Pfitzner, 1938, preocc.
Aepytus Herrich-Schäffer, [1858]
  Subgenus **Hampsoniella** Viette, 1949
  Subgenus **Pseudodalaca** Viette, 1949
  Subgenus **Gymelloxes** Viette, 1952
  Subgenus **Alloaepytus** Viette, 1952
  Subgenus **Aepytus** Herrich-Schäffer, [1858]
  Subgenus **Thiastyx** Viette, 1951
  Subgenus **Schaefferiana** Viette, 1950
  Subgenus **Paragorgopis** Viette, 1951
  Subgenus **Hepialyxodes** Viette, 1951
  Subgenus **Xytrops** Viette, 1950
  Subgenus **Cibyra** Walker, 1856
  Subgenus **Lamelliformia** Viette, 1951
  Subgenus **Tricladia** R. Felder, 1874
    Pseudophassus R. Felder, 1874
    Parana Viette, 1949
  Subgenus **Pseudophilaenia** Viette, 1950
  Subgenus **Philaenia** Kirby, 1892
  Subgenus **Yleuxas** Viette, 1951
Phialuse Viette, 1961
Roseala Viette, 1950
"Dalaca" auct. (not Walker, 1856)
Pfitzneriella Viette, 1950

6. PALAEOSETIDAE

Osrhoes Druce, 1900

Infraorder Heteroneura

Division Monotrysia

Nepticuloidea

7. NEPTICULIDAE

Nepticulinae

Tribe Nepticulini

Stigmella Schrank, 1802
    Nepticula Heyden, 1843
    Stigmalla Herrich-Schäffer, 1853, missp.
    Dysnepticula Börner, 1925
    Johanssonia Borkowski, 1972

Tribe Trifurculini

Enteucha Meyrick, 1915

8. OPOSTEGIDAE

Opostega Zeller, 1839

9. TISCHERIIDAE

Tischeria Zeller, 1839
        Evexia Gistel, 1847
        Philodoxa Gistel, 1848
        Coptotriche Walsingham, 1890
        Tisheria Busck [1903], missp.

Incurvarioidea

10. INCURVARIIDAE

Basileura Nielsen & Davis, 1981
Simacauda Nielsen & Davis, 1981

11. CECIDOSIDAE

Cecidoses Curtis, 1835
        Clistoses Kieffer & Jörgensen, 1910
        Eucecidoses Brèthes, 1916
Oliera Brèthes, 1916
Ridiaschinia Brèthes, 1916
Dicranoses Kieffer & Jörgensen, 1910

12. ADELIDAE

Ceromitia Zeller, 1854
        Trichorrhabda Meyrick, 1912, **n. syn.**
Adela Latreille, 1796
        Capillaria Haworth, 1828
        Metallitis Sodoffsky, 1837
        Aedilis Gistel, 1873
        Dicte Chambers, 1873

13. PRODOXIDAE

Tegeticula Zeller, 1873
        Pronuba Riley, 1872, preocc.
        Promiba Kirbey, 1874, missp.
        Thia H. Edwards, 1888, preocc.
        Thelethia Dyar, 1893
        Valentina Coolidge, 1909, preocc.
Parategeticula Davis, 1967
Prodoxus Riley, 1880
Prodoxoides Nielsen & Davis, 1984

14. HELIOZELIDAE

Antispila Hübner, [1825]
Coptodisca Walsingham, 1895
        Aspidisca Clemens, 1860, preocc.
Heliozela Herrich-Schäffer, 1853
        Dyselachista Spuler, 1910
Lamprozela Meyrick, 1916
Monachozela Meyrick, 1931
Phanerozela Meyrick, 1921

Division Ditrysia

Section Tineina

Tineoidea

15. TINEIDAE

Acrolophus Poey, 1832
        Anaphora Clemens, 1859
        Daulia Walker, 1863, preocc.
        Derchis Walker, 1863
        Eddara Walker, 1863, preocc.
        Hibita Walker, 1863
        Tirasia Walker, 1863
        Bazira Walker, 1864
        Phlongia Walker, 1864
        Urbara Walker, 1864
        Tachasara Walker, 1865
        Eutheca Grote, 1881, preocc.
        Eulepiste Walsingham, 1882
        Pseudoconchylis Walsingham, 1884
        Homonymus Walsingham, 1887
        Neolophus Walsingham, 1887
        Hypoclopus Walsingham, 1887
        Thysanoscelis Walsingham, 1887
        Thysanoskelis Walsingham, 1887
        Ankistrophorus Walsingham, 1887
        Caenogenes Walsingham, 1887
        Felderia Walsingham, 1887
        Ortholophus Walsingham, 1887
        Pseudanophora Walsingham, 1887
        Sapinella Kirby, 1892
        Atopocera Walsingham, 1897
        Pilanophora Walsingham, 1897
        Thysanosedes Druce, 1901, missp.
        Hypocolypus Dyar, [1903], missp.
        Anophora Kearfott, 1903, missp.
        Brachysymbola Meyrick, 1912
        Orothyntis Meyrick, 1913
        Apoclisis Walsingham, 1914
        Psephocrita Meyrick, 1919
        Xerocausta Meyrick, 1928, **n. syn.**
        Anaphorina Strand, 1932
Amydria Clemens, 1859
        Casape Walker, 1864
        Amadria Chambers, 1875, missp.
        Amadrya Chambers, 1875, missp.
        Amydrya Kearfott, 1903, missp.
        Neomeristis Meyrick, 1919, **n. syn.**
Antipolistes Forbes, 1933
Atticonviva Busck, 1934
Augolychna Meyrick, 1922
Axiagasta Meyrick, 1930
Barymochtha Meyrick, 1922
Basanasca Meyrick, 1922
Brithyceros Meyrick, 1932
Bythocrates Meyrick, 1919
Choropleca Durrant, 1914
        Cyane Chambers, 1873, preocc.
        Diachalastis Meyrick, 1920
Clepticodes Meyrick, 1927
Clinograptis Meyrick, 1932
Cnismorectis Meyrick, 1936
Colpocrita Meyrick, 1930
Comodica Meyrick, 1880
Compsocrita Meyrick, 1922
Cranaodes Meyrick, 1919
Crepidochares Meyrick, 1922
Cryphiotechna Meyrick, 1932
Dasmophora Meyrick, 1919
Demobrotis Meyrick, 1892
Diataga Walsingham, 1914

Dorata Busck, 1904
    Pterolonche Walsingham, 1889, preocc.
    Dorota Kearfott, 1907, missp.
Drastea Walsingham, 1914
Dyotopasta Busck, 1907
    Pseudoxylesthia Walsingham, 1907
    Dystopasta McDunnough, 1939, emend.
Dysoptus Walsingham, 1914
Ectinocampa Silvestri 1944
Ephedroxena Meyrick, 1919
Episyrta Meyrick, 1929
Erechthias Meyrick, 1880
    Ereunetis Meyrick, 1880
    Decadarchis Meyrick, 1886
    Nesoxena Meyrick, 1929
    Amphisyncentris Meyrick, 1933
    Caryolestis Meyrick, 1934
    Triadogona Meyrick, 1937
    Anemerarcha Meyrick, 1937
    Erechtias Ghesquière, 1940
    Tinexotaxa Gozmány, 1968
    Neodecadarchis Zimmerman, 1978
    Lepidobregma Zimmerman, 1978
    Pantheus Zimmerman, 1978
Exoncotis Meyrick, 1919
Homodoxus Walsingham, 1914
Homosetia Clemens, 1863
Homostinea Dietz, 1905
    Homotinea Meyrick, 1932, emend.
Hormantris Meyrick, 1927
Hybroma Clemens, 1862
Infurcitinea Spuler, 1910
    Atinea Amsel, 1954
    Microtinea Amsel, 1954
Isocorypha Dietz, 1905
    Socorypha Busck, 1909, missp.
Leptochersa Meyrick, 1919
Lepyrotica Meyrick, 1921
Leucophasma Walsingham, 1897
Lindera Blanchard, 1852
    Palpula Blanchard, 1852, preocc.
    Safra Walker, 1864, preocc.
    Chresotes Butler, 1881
    Paraneura Dietz, 1905
    Cervitinea Amsel, 1956
Lipomerinx Walsingham, 1914
Lithopsaestis Meyrick, 1932
Mea Busck, 1906
    Progona Dietz, 1905, preocc.
Monopis Hübner, [1825]
    Blabophanes Zeller, 1852
    Hyalospila Herrich-Schäffer, 1853
    Rhitia Walker, 1864
    Eusynopa Lower, 1903
Myrmecozela Zeller, 1852
    Psecadioides Butler, 1881
    Promasia Chrètien, 1905
    Protolophe Rebel, 1915
Mythoplastis Meyrick, 1919
Nemapogon Schrank, 1802
    Diaphthirusa Hübner, [1825]
Neomeristis Meyrick, 1919
Niditinea Petersen, 1957
    Tineidia Zagulajev, 1960
Oenoe Chambers, 1874

Oinophila Stephens, 1848
    Oenophila Zeller, 1853, missp.

Opogona Zeller, 1853
    Lozostoma Stainton, 1859
    Conchyliospila Wallengren, 1861
    Cachura Walker, 1864
    Dendroneura Walsingham, 1892
    Hieroxestis Meyrick, 1892
    Exala Meyrick, 1912
Opsodoca Meyrick, 1919
Otochares Meyrick, 1919
Pachydyta Meyrick, 1922
    Pachdyta Meyrick, 1922, missp.
Palaephatus Butler, 1883
Panthytarcha Meyrick, 1922
Pedaliotis Meyrick, 1930
Perilicmetis Meyrick, 1932
Phaeoses Forbes, 1922
Phereoeca Hinton & Bradley, 1956
Polypsecta Meyrick, 1930
Pompostolella Fletcher, 1940
    Pompostola Meyrick, 1927, preocc.
Praeacedes Amsel, 1954
    Titaenoses Hinton & Bradley, 1956
Proboloptila Meyrick, 1921
Protodarcia Forbes, 1931
Ptilopsaltis Meyrick, 1935
Scardia Treitschke, 1830
    Agarica Sodoffsky, 1837
    Morophaga Herrich-Schäffer, 1853
    Atabyria Snellen, 1884
    Osphretica Meyrick, 1910
    Microscardia Amsel, 1952
Setiarcha Meyrick, 1932
Setomorpha Zeller, 1852
    Epilegis Dietz, 1905
    Apotomia Dietz, 1905
    Semiota Dietz, 1905
    Trisyntopa Lower, 1918
Syncraternis Meyrick, 1922
Syrmologa Meyrick, 1919
Taeniodictys Forbes, 1933
Tetrapalpus Davis, 1972
Tinea Linnaeus, 1758
    Autoses Hübner, [1825]
    Acedes Hübner, [1825]
    Edosa Walker, 1866
    Chrysoryctis Meyrick, 1886
    Dystinea Börner, 1925
Tiquadra Walker, 1863
    Oscella Walker, 1864
    Manchana Walker, 1866
    Ventia Walker, 1866
    Acureuta Zeller, 1877
Trichophaga Ragonot, 1894
Trierostola Meyrick, 1932
Xylesthia Clemens, 1859
    Xylestia Dyar, [1903], missp.
Xystrologa Meyrick, 1919
    Achanodes Meyrick, 1922
    Syrrhoaula Meyrick, 1932
Zonochares Meyrick, 1922
Zymologa Meyrick, 1919

16. PSYCHIDAE

Penestoglossinae

Amiantastis Meyrick, 1932
  Autocnaptis Meyrick, 1935
Plumana Busck, 1911
  Homilostola Meyrick, 1917
Perisceptis Meyrick, 1931
Pterogyne Davis, 1975

Psychinae

Epichnopterix Hübner, [1825]
  Epichnopteryx Heylaerts, 1875, missp.
Paucivena Davis, 1975
Prochalia Barnes & McDunnough, 1913
Metaxypsyche Davis, 1975
Naevipenna Davis, 1964
Cryptothelea Duncan, 1841
  Platoeceticus Packard, 1869
  Cryptotheles Costa Lima, 1945, missp.
Dendropsyche F. M. Jones, 1926
Lumacra Davis, 1964
Curtorama Davis, 1964
Astala Davis, 1964

Oiketicinae

Thantopsyche Butler, 1882
Animula Herrich-Schäffer, 1856
 Subgenus **Animula** Herrich-Schäffer, 1856
 Subgenus **Antipenna** Davis, 1964
Biopsyche Dyar, 1905
Oiketicus Guilding, 1827
  Oeketicus Lefebre, 1842, missp.
  Oeceticus Herrich-Schäffer, 1858, missp.
  Oiketikus Seitz, 1919, missp.
  Oicocestis Köhler, 1924, missp.
  Oiceticus Lahille, 1926, missp.
 Subgenus **Paraoiketicus** Davis, 1964
 Subgenus **Oiketicus** Guilding, 1827
Thyridopteryx Stephens, 1834
  Hymenopsyche Grote, 1865

17. ARRHENOPHANIDAE

Arrhenophanes Walsingham, 1913
Cnissostages Zeller, 1863
Ecpathophanes Bradley, 1951
Harmaclona Busck, 1914
  Ptychoxena Meyrick, 1916

18. AMPHITHERIDAE

Dasycarea Zeller, 1877

19. LYONETIIDAE

Bedelliinae

Euprora Busck, 1906
Philonome Chambers, 1874
  Phillonome Chambers, 1875, missp.
  Eurynome Chambers, 1875, preocc.
  Busckia Dyar, [1903]

Bedellia Stainton, 1849

Lyonetiinae

Lyonetia Hübner, [1825]
  Argyromiges Curtis, 1829
  Argyromis Stephens, 1829
  Lyoneta Matsumura, 1931, missp.

Cemiostominae

Leucoptera Hübner, [1825]
  Cemiostoma Zeller, 1848
Perileucoptera Silvestri, 1943
Compsoschema Walsingham, 1897
Hapalothyma Meyrick, 1920
Thomictis Meyrick, 1920

Bucculatriginae

Bucculatrix Zeller, 1848
  Ceroclastis Zeller, 1848
  Bacculatrix Flint, Noble, & Shaw, 1978, missp.

20. GRACILLARIIDAE

Gracillariinae

Caloptilia Hübner, [1825]
  Poeciloptilia Hübner, [1825]
  Ornix Treitschke, 1833
  Coriscium Zeller, 1839
  Antiolopha Meyrick, 1894
Eucosmophora Walsingham, 1897
Neurostrota Ely, 1918
  Neurostrata Ely, 1918, missp.
Parectopa Clemens, 1860
  Euspilopteryx Zeller, 1847, emend., misid.
  Euspilapteryx Spuler, 1910, missp.
Neurobathra Ely, 1918
Penica Walsingham, 1914
Parornix Spuler, 1910
Chilocampyla Busck, 1900
Dialectica Walsingham, 1897
  Eutrichocnemis Spuler, 1910
Acrocercops Wallengren, 1881
Cuphodes Meyrick, 1897
  Phrixosceles Meyrick, 1908
Spanioptila Walsingham, 1897
Leucanthiza Clemens, 1859
Marmara Clemens, 1863
  Aesyle Chambers, 1875

Lithocolletinae

Cremastobombycia Braun, 1908
Phyllonorycter Hübner, [1825]
  Lithocolletis Hübner, [1825]
  Eucestis Hübner, [1825]
  Lithcolletes Matsumura, 1907, missp.
  Phyllorycter Walsingham, 1914, missp.
Cameraria Chapman, 1902
Porphyrosela Braun, 1908

Phyllocnistinae

Phyllocnistis Zeller, 1848
      Phylloenistis Chambers, 1875, missp.
      Phylloetis Chambers, 1876, missp.
      Phyllocnitis Busck, 1900, missp.

Gelechioidea

21. OECOPHORIDAE

Depressariinae

Tribe Depressariini

Exaeretia Stainton, 1849
      Depressariodes Turati, 1924
      Martyrhilda Clarke, 1941
Talitha Clarke, 1978

Tribe Amphisbatini

Afdera Clarke, 1978
Ancipita Busck, 1914
Anthrinacia Walsingham, 1911
Auxotricha Meyrick, 1931
Chariphylla Meyrick, 1921
Comotechna Meyrick, 1920
Compsistis Meyrick, 1888
Coptotelia Zeller, 1863
      Hyphypena Warren, 1889
Costoma Busck, 1914
      Phalarotarsa Meyrick, 1924
Cryptolechia Zeller, 1852
Doina Clarke, 1978
Doshia Clarke, 1978
Ectaga Walsingham, 1912
Erithyma Meyrick, 1914
Eupragia Walsingham, 1911
Filinota Busck, 1911
      Lupercalia Busck, 1912
      Mnesichara Walsingham, 1912
Gnathotona Meyrick, 1931
Gonada Busck, 1911
Gonionota Zeller, 1877
Habrophylax Meyrick, 1931
Hamadera Busck, 1914
Hastamea Fletcher, 1929, repl. name
      Hasta Busck, 1911, preocc.
Himotica Meyrick, 1912
Hypercallia Stephens, 1834
      Agriocoma Zeller, 1877
      Brachyplatea Zeller, 1877
      Eumimographe Dognin, 1905
Idiocrates Meyrick, 1909
Lucyna Clarke, 1978
Machimia Clemens, 1860
Maesara Clarke, 1968
Melaneulia Butler, 1883
Muna Clarke, 1978
Nedenia Clarke, 1978
Nematochares Meyrick, 1931
Osmarina Clarke, 1978
Palinorsa Meyrick, 1924
Perzelia Clarke, 1978

Philtronoma Meyrick, 1914
Pholcobates Meyrick, 1931
Phytomimia Walsingham, 1912
Pisinidea Butler, 1883
Profilinota Clarke, 1973
Pseudocentris Meyrick, 1921
Psilocorsis Clemens, 1860
      Paepia Walker, 1864
      Hagno Chambers, 1872
Psittacastis Meyrick, 1909
      Necedes Walsingham, 1912
Rhindoma Busck, 1914
Scoliographa Meyrick, 1916
Taruda Walker, 1864
      Ecliptoloma Zeller, 1877
Trycherodes Meyrick, 1914
      Teratomorpha Walsingham, 1912, preocc.
Zemiocrita Meyrick, 1933

Ethmiinae

Erysiptila Meyrick, 1914
      Cyrictodes Meyrick, 1926
Ethmia Hübner, [1819]
      Psecadia Hübner, [1825]
      Anesychia Hübner, [1825]
      Disthymnia Hübner, [1825]
      Melanoleuca Stephens, 1829
      Aedia Duponchel, 1836
      Chalybe Duponchel, 1836
      Azinis Walker, 1863
      Tamarrha Walker, 1864
      Ceratophysetis Meyrick, 1887
      Theoxenia Walsingham, 1887
      Babaiaxa Busck, 1902
      Wiltshireia Amsel, 1949
Macrocirca Meyrick, 1931

Peleopodinae

Durrantia Busck, 1908
      Dolidiria Busck, 1912
Peleopoda Zeller, 1877
      Theatria Walsingham, 1912
Pseuderotis Clarke, 1956
Schistonoea Forbes, 1931

Stenomatinae

Anadasmus Walsingham, 1897
Anapatris Meyrick, 1932
Antaeotricha Zeller, 1854
      Mesoptycha Zeller, 1854
      Brachiloma Clemens, 1863
      Harpalyce Chambers, 1874
      Ide Chambers, 1880
      Antoeotricha Walsingham, 1881, missp.
      Aedemoses Walsingham, 1912
      Athleta Walsingham, 1912
      Prasolithites Meyrick, 1912
      Aphanoxena Meyrick, 1915
      Psephomeres Meyrick, 1916
      Eumiturga Meyrick, 1925
Baeonoma Meyrick, 1916
Catarata Walsingham, 1912

Cerconota Meyrick, 1915
  Pomphocrita Meyrick, 1930
  Cerconata Busck, 1935, missp.
Chlamydastis Meyrick, 1916
  Ptilogenes Meyrick, 1917
Diastoma Möschler, 1882
Energia Walsingham, 1912
Falculina Zeller, 1877
Gonioterma Walsingham, 1897
Hyalopseustis Meyrick, 1925
Lethata Duckworth, 1964
Loxotoma Zeller, 1854
Menesta Clemens, 1860
Mothonica Walsingham, 1912
Mysaromima Meyrick, 1926
Orphnolechia Meyrick, 1909
Parascaeas Meyrick, 1936
Paraspastis Meyrick, 1915
Petalothyrsa Meyrick, 1931
Petasanthes Meyrick, 1925
Phelotropa Meyrick, 1915
Promenesta Busck, 1914
Rectiostoma Becker, 1982, repl. name
  Setiostoma Zeller, 1875, preocc.
Rhodonassa Meyrick, 1915
Stenoma Zeller, 1839
  Auxocrossa Zeller, 1854
  Auxocrassa Walsingham, 1882, missp.
Thioscelis Meyrick, 1909
Timocratica Meyrick, 1912
  Lychnocrates Meyrick, 1926
Zetesima Walsingham, 1912

Oecophorinae

Tribe Oecophorini

Acartophila Meyrick, 1932
Aliciana Clarke, 1978
Altiura Clarke, 1978
Alynda Clarke, 1978
Aniuta Clarke, 1978
Arctopoda Butler, 1883
  Polypseustis Dognin, 1908
Atelosticha Meyrick, 1883
  Crypsynarthra Lower, 1901
Atha Clarke, 1978
Atopotorna Meyrick, 1932
Borkhausenia Hübner, [1825]
  Amaurosetia Stephens, [1835]
  Pseudatemelia Rebel, 1910
Brymblia Hodges, 1974
Callistenoma Butler, 1883
Cecidolechia Kieffer & Jörgensen, 1910
  Cecidolechia Strand, 1911, redesc.
Chezala Walker, 1864
Corita Clarke, 1978
Decantha Busck, 1914
Deia Clarke, 1978
Despina Clarke, 1978
Dinotropa Meyrick, 1916
Dita Clarke, 1978
Doliotechna Meyrick, 1914
Dysgnorima Zeller, 1877
Endrosis Hübner, [1825]

Eomichla Meyrick, 1916
  Astoxena Meyrick, 1930
Eraina Clarke, 1978
Exosphrantis Meyrick, 1931
Halimarmara Meyrick, 1931
Harpella Schrank, 1802
Heliostibes Zeller, 1874
Hofmannophila Spuler, 1910
Hyperskeles Butler, 1883
Ifeda Hodges, 1966
Inga Busck, 1908
  Doxa Walsingham, 1912
  Lysigrapha Meyrick, 1914
  Pelomimas Meyrick, 1924
  Orsimacha Meyrick, 1914
  Siderograptis Meyrick, 1920
  Phanerodoxa Meyrick, 1921
  Epimoryctis Meyrick, 1930
  Horomeristis Meyrick, 1931
  Agriotorna Meyrick, 1931
Irenia Clarke, 1978
Lelita Clarke, 1978
Lygronoma Meyrick, 1913
Macarocosma Meyrick, 1931
Mathildana Clarke, 1941
Melochrysis Meyrick, 1916
Odonna Clarke, 1982
Pachyphoenix Butler, 1883
  Tyriomorpha Meyrick, 1918
  Mattea Duckworth, 1966
Philomusaea Meyrick, 1931
  Philomusea Clarke, 1978, missp.
Pseudoecophora Staudinger, 1899
Pycnotarsa Meyrick, 1920
Pyramidobela Braun, 1923
  Idioptila Meyrick, 1927
Retha Clarke, 1978
Revonda Clarke, 1978
Stasixena Meyrick, 1930
Struthoscelis Meyrick, 1913
Teresita Clarke, 1978
Thaumatolita Walsingham, 1912
Theama E. M. Hering, 1958
Utilia Clarke, 1978
Zymrina Clarke, 1978

Tribe Stathmopodini

Arauzona Walker, [1865]
Snellenia Walsingham, 1889
Stathmopoda Herrich-Schäffer, 1853
  Boocara Butler, 1880
  Placostola Meyrick, 1887
  Erineda Busck, 1909
  Agrioscelis Meyrick, 1913
  Kakivoria Nagano, 1916
Tinaegeria Walker, 1856

Tribe Pleurotini

Pleurota Hübner, [1825]
  Eupleuris Hübner, [1825]
  Holoscolia Zeller, 1839
  Protasis Herrich-Schäffer, 1853

22. ELACHISTIDAE

Elachistinae

Dicranoctetes Braun, 1918
        Donacivola Busck, [1934]
Elachista Treitschke, 1833
        Aphelosetia Stephens, 1834
        Cycnodia Herrich-Schäffer, 1853
        Phigalia Chambers, 1875, preocc.
        Atachia Wocke, 1876
        Neaera Chambers, 1880, preocc.
        Hecista Wallengren, 1881
        Aphigalia Dyar, 1903
        Irenicodes Meyrick, 1919
        Euproteodes Viette, 1954

23. BLASTOBASIDAE

Symmocinae

Eupolella Fletcher, 1940, repl. name
        Eupolis Meyrick, 1923, preocc.
Glyphidocera Walsingham, 1892
        Harpagandra Meyrick, 1918
Oecia Walsingham, 1897
        Macroceras Staudinger, 1876, preocc.
Sceptea Walsingham, 1911
        Sceptia McDunnough, 1939, missp.
Symmoca Hübner, [1825]
        Simoca Weiler, 1877, missp.
        Parasymmoca Rebel, 1903
        Asarista Meyrick, 1935
        Conquassata Gozmány, 1957
        Symmoletria Gozmány, 1963

Blastobasinae

Tribe Blastobasini

Auximobasis Walsingham, 1892
Blastobasis Zeller, 1855
        Epistetus Walsingham, 1894
        Protesis Walsingham, 1908
Holcocera Clemens, 1863
        Catacrypsis Walsingham, 1907
        Cynotes Walsingham, 1907
        Prosodica Walsingham, 1907
Iconisma Walsingham, 1897
Metallocrates Meyrick, 1930
Valentinia Walsingham, 1907

Tribe Pigritiini

Pigritia Clemens, 1860

24. COLEOPHORIDAE

Coleophorinae

Coleophora Hübner, [1822]
        Eupista Hübner, [1825]
        Apista Hübner, [1825]
        Haploptilia Hübner, [1825]
        Porrectaria Haworth, 1828

        Damophila Curtis, 1832
        Astyages Stephens, 1834
        Metallosetia Stephens, 1834
        Casas Wallengren, 1881
        Casigneta Wallengren, 1881
        Corythangela Meyrick, 1897
        Casignetella Strand, 1928
        Calaritania Mariani, 1943
        Heringiella Börner, 1944
        Tolleophora Căpuse, 1971
        Ionescumia Căpuse, 1971
        Stollia Căpuse, 1971
        Razowskia Căpuse, 1971
        Orghidania Căpuse, 1971
        Frederickoenigia Căpuse, 1971
        Suireia Căpuse, 1971
        Zagulajevia Căpuse, 1971
        Amseliphora Căpuse, 1971
        Nemesia Căpuse, 1971
        Zangheriphora Căpuse, 1971
        Bourgogneja Căpuse, 1971
        Aureliania Căpuse, 1971
        Bacescuia Căpuse, 1971
        Klinzigedia Căpuse, 1971
        Vladdelia Căpuse, 1971
        Klimeschja Căpuse, 1971
        Glaseria Căpuse, 1971
        Valvulongia Căpuse, 1971
        Falkovitshia Căpuse, 1972
        Helopharea Falkovitsh, 1972
        Cricotechna Falkovitsh, 1972
        Plegmidia Falkovitsh, 1972
        Agapalsa Falkovitsh, 1972
        Phylloschema Falkovitsh, 1972
        Bima Falkovitsh, 1972
        Systrophoeca Falkovitsh, 1972
        Aporiptura Falkovitsh, 1972
        Symphypoda Falkovitsh, 1972
        Oedicaula Falkovitsh, 1972
        Argyractinia Falkovitsh, 1972
        Chnoocera Falkovitsh, 1972
        Orthographis Falkovitsh, 1972
        Phagolamia Falkovitsh, 1972
        Monotemachia Falkovitsh, 1972
        Corethropoea Falkovitsh, 1972
        Characia Falkovitsh, 1972
        Perygra Falkovitsh, 1972
        Perygridia Falkovitsh, 1972
        Luzulina Falkovitsh, 1972
        Carpochena Falkovitsh, 1972
        Abaraschia Căpuse, 1973
        Amselghia Căpuse, 1973
        Ardania Căpuse, 1973
        Ascleriductia Căpuse, 1973
        Baraschia Căpuse, 1973
        Benanderpia Căpuse, 1973
        Calcomarginia Căpuse, 1973
        Caleophora Căpuse, 1973, missp.
        Corothropoea Căpuse, 1973, missp.
        Cornulivalvulia Căpuse, 1973
        Dumitrescumia Căpuse, 1973
        Ecebalia Căpuse, 1973
        Globulia Căpuse, 1973
        Hamuliella Căpuse, 1973
        Helvalbia Căpuse, 1973

Ionnemesia Căpuse, 1973
Kasyfia Căpuse, 1973
Kuznetzovvlia Căpuse, 1973
Latisacculia Căpuse, 1973
Longibacillia Căpuse, 1973
Lucidaesia Căpuse, 1973
Lvaria Căpuse, 1973
Membrania Căpuse, 1973
Metapista Căpuse, 1973
Multicoloria Căpuse, 1973
Neugenvia Căpuse, 1973
Nosyrislia Căpuse, 1973
Ortohgraphis Căpuse, 1973, missp.
Oudejansia Căpuse, 1973
Paravalvulia Căpuse, 1973
Patzakia Căpuse, 1973
Postvinculia Căpuse, 1973
Proglaseria Căpuse, 1973
Quadratia Căpuse, 1973
Rhamnia Căpuse, 1973
Sacculia Căpuse, 1973
Scleriductia Căpuse, 1973
Tollsia Căpuse, 1973
Tuberculia Căpuse, 1973
Ulna Căpuse, 1973
Ductispira Căpuse, 1973
Klimeschjosefia Căpuse, 1975
Tocasta Busck, 1912

Batrachedrinae

Amblytenes Meyrick, 1930
Batrachedra Herrich-Schäffer, 1853
    Eustaintonia Spuler, 1910
Chedra Hodges, 1966

25. MOMPHIDAE

Anchimompha Clarke, 1965
Mompha Hübner, [1825]
    Laverna Curtis, 1839
    Lophoptilus Sircom, 1848
    Cyphophora Herrich-Schäffer, 1853
    Psacaphora Herrich-Schäffer, 1853
    Anybia Stainton, 1854
    Wilsonia Clemens, 1864
    Leucophryne Chambers, 1875
Synallagma Busck, 1907

26. AGONOXENIDAE

Blastodacninae

Homoeoprepes Walsingham, 1909
    Homeoprepes Hodges, 1978, missp.
Microcolona Meyrick, 1897
Nanodacna Clarke, 1964
Nicanthes Meyrick, 1928
Pammeces Zeller, 1863
Panclintis Meyrick, 1929
Prochola Meyrick, 1915
Syntetrernis Meyrick, 1922
Zaratha Walker, 1864

27. COSMOPTERIGIDAE

Antequerinae

Antequera Clarke, 1941
Euclemensia Grote, 1878, repl. name
    Hamadryas Clemens, 1864, preocc.

Cosmopteriginae

Anoncia Clarke, 1941
Aphanosara Forbes, 1931
Cosmopterix Hübner, [1825]
    Cosmopteryx Zeller, 1839, emend.
Dromiaulis Meyrick, 1922
Ecballogonia Walsingham, 1912
Harpograptis Meyrick, 1925
Labdia Walker, 1864
Limnaecia Stainton, 1851
    Lymnaecia Kimball, 1965, missp.
Moriloma Busck, 1912
Pyroderces Herrich-Schäffer, 1853
    Anatrachyntis Meyrick, 1915
    Sathrobrota Hodges, 1962
"Scaeosopha" auct., (not Meyrick, 1914)
Sematoptis Meyrick, 1931
Teladoma Busck, 1932
Triclonella Busck, 1901
    Anorcota Meyrick, 1920
    Pharmacoptis Meyrick, 1932
Urangela Busck, 1912

Chrysopeleiinae

Ascalenia Wocke, 1876
Eritarbes Walsingham, 1909
Ithome Chambers, 1875
    Eriphia Chambers, 1875, preocc.
Leptozestis Meyrick, 1924
Obithome Hodges, 1964
Perimede Chambers, 1874
Periploca Braun, 1919
Stilbosis Clemens, 1860
    Aeaea Chambers, 1874
    Amaurogramma Braun, 1919
Walshia Clemens, 1864

28. SCYTHRIDIDAE

Scythris Hübner, [1825]
    Butalis Treitschke, 1833, preocc.
    Arotrura Walsingham, 1888
    Colinita Busck, 1907

29. GELECHIIDAE

Anomologinae

Isophrictis Meyrick, 1917
Megacraspedus Zeller, 1839
    Neda Chambers, 1874, preocc.
    Pycnobathra Lower, 1901
    Autoneda Busck, 1903, repl. name
    Toxoceras Chrétien, 1915
    Megacraspedas Barnes & McDunnough, 1917,
        missp.

Monochroa Heinemann, 1870
        Catabrachmia Rebel, 1909
Nealyda Dietz, 1900

Gelechiinae

Aristotelia Hübner, [1825]
        Ergatis Heinemann, 1870, preocc.
        Isochasta Meyrick, 1886
        Eucatoptus Walsingham, 1897
Arla Clarke, 1942
Aroga Busck, 1914
        Aruga Janse, 1958, missp.
Arogalea Walsingham, 1910
Barticeja Povolný, 1967
Calliprora Meyrick, 1914
Chionodes Hübner, [1825]
        Chionoda Hübner, [1826], missp.
Coleotechnites Chambers, 1880
        Evagora Clemens, 1860, preocc.
        Eidothea Chambers, 1873, preocc.
        Eidothoa Chambers, 1873, missp.
        Coleotechnistes Riley, 1891, missp.
        Eucordylea Dietz, 1900
        Pulicalvaria T. N. Freeman, 1963
Deltophora Janse, 1950
Ephysteris Meyrick, 1908
        Microcraspedus Janse, 1958
        Ephystereris Janse, 1960, missp.
        Echinoglossa Clarke, 1965
Euchionodes Clarke, 1950
Eudactylota Walsingham, 1911
Evippe Chambers, 1873
        Phaetusa Chambers, 1875, preocc.
        Tholerostola Meyrick, 1917
Exoteleia Wallengren, 1881
        Paralechia Busck, 1903
        Heringia Spuler, 1910, preocc.
        Heringiola Strand, 1917, repl. name
Faculta Busck, 1939
Fascista Busck, 1939
Friseria Busck, 1939
Gelechia Hübner, [1825]
        Guenea Bruand, 1850
        Galechia Desmarest, [1857], missp.
        Gelschia Nowicki, 1865, missp.
        Cirrha Chambers, 1872
        Oeseis Chambers, 1875
        Gelecia Watt, 1920, missp.
Gnorimoschema Busck, 1900
        Gnorimochema Dyar, [1903], missp.
        Tuta Kieffer & Jörgensen, 1910
        Gonorimoschema Deurs, 1954, missp.
        Gnorrimoschema Hartig, 1964, missp.
        Lerupsia Riedl, 1965
        Neoschema Povolný, 1967
        Larupsia Soffner, 1967, missp.
Hapalosaris Meyrick, 1917
Keiferia Busck, 1939
Lita Treitschke, 1833
Locharca Meyrick, 1923
Neodactylota Busck, 1903
Orsotricha Meyrick, 1914
Phthorimaea Meyrick, 1902
        Phtyorimaea Turner, 1919, missp.

        Phthorimoea Povolný & Zakopal, 1951, missp.
        Pthorimaea Issiki, 1957, missp.
        Phthorimea Diakonoff, [1968], missp.
        Phtorimea Oei-Dharma, 1969, missp.
Polyhymno Chambers, 1874
        Copocercia Zeller, 1877
        Oegoconiodes Matsumura, 1931
        Oegoconides Neave, 1940, missp.
Pseudarla Clarke, 1965
Ptycerata Ely, 1910
        Scrobipalpopsis Povolný, 1967
  Subgenus **Ptycerata** Ely, 1910
  Subgenus **Scrobischema** Povolný, 1979
Recurvaria Haworth, 1828
        Telea Stephens, 1834, preocc.
        Aphanaula Meyrick, 1895
        Hinnebergia Spuler, 1910
        Microlechia Turati, 1924
Schistophila Chrétien, 1899
Scrobipalpula Povolný, 1964
  Subgenus **Eurysacca** Povolný, 1967
  Subgenus **Magnifascia** Povolný, 1967
  Subgenus **Scrobipalpula** Povolný, 1964
Sophronia Hübner, [1825]
Stegasta Meyrick, 1904
Stomopteryx Heinemann, 1870
        Inotica Meyrick, 1913
        Instica Sharp, 1915, missp.
        Acraeologa Meyrick, 1921
        Stomopterix Turati, 1922, missp.
        Stromopteryx Pierce & Metcalfe, 1935, missp.
Symmetrischema Povolný, 1967
Taygete Chambers, 1873
Telphusa Chambers, 1872
        Adrasteia Chambers, 1872
        Adrastia Kirby, 1874, emend.
        Geniadophora Walsingham, 1897
        Telephusa Beirne, 1938, missp.
Thiotricha Meyrick, 1886
        Reuttia O. Hofmann, 1898
        Thiotrica Inoue, 1954, missp.
        Thiothricha Hartig, 1956, missp.
Tildenia Povolný, 1967

Anacampsinae

Anacampsis Curtis, 1827
        Anacompsis Desmarest, [1857], missp.
        Tachyptilia Heinemann, 1870
        Tuchyptilia Kirby, 1871, missp.
        Tachiptilia Chambers, 1878, missp.
        Tachyptilix Hartmann, 1880, missp.
        Agriastis Meyrick, 1914
        Agriastsi Busck, 1919, missp.
        Tachoptilia Daltry, 1926, missp.
        Trachyphilia Le Marchand, 1947, missp.
        Trachyptilia Le Marchand, 1947, missp.
Battaristis Meyrick, 1914
        Duvita Busck, 1916
Compsolechia Meyrick, 1918
Holophysis Walsingham, 1910
        Hoplophysis McDunnough, 1939, missp.
Strobisia Clemens, 1860
        Systasiota Walsingham, 1910
Untomia Busck, 1906

Chelariinae

Crasimorpha Meyrick, 1923
Epilechia Busck, 1939
Haplochela Meyrick, 1923
Hypatima Hübner, [1825]
      Chelaria Haworth, 1828
      Hypatina Stephens, 1835, missp.
      Allocota Meyrick, 1904, preocc.
      Cynestomorpha Meyrick, 1904
      Deuteroptila Meyrick, 1904
      Cymatomorpha Meyrick, 1904
      Semodictis Meyrick, 1909
      Allocotaniana Strand, 1913, repl. name
      Episacta Turner, 1919
      Cellaria Neave, 1939, missp.
      Cheleria Lhomme, [1948], missp.
Pectinophora Busck, 1917
Pessograptis Meyrick, 1923
Porpodryas Meyrick, 1920
Prostomeus Busck, 1903
Semophylas Meyrick, 1932
Sitotroga Heinemann, 1870
      Silotroga Kirby, 1871, missp.
      Nesolechia Meyrick, 1921
      Syngenomictis Meyrick, 1927
      Sitotrogus Matsumura, 1931, missp.
      Sitotrega Borg, 1932, missp.
      Sititroga Costa Lima, 1945, missp.

Dichomeridinae

Acompsia Hübner, [1825]
      Acampsia Westwood, 1840, missp.
      Accompsia Bruand, 1850, missp.
      Brachycrossata Heinemann, 1870
      Brachicrossata Hartmann, 1880, missp.
Anorthosia Clemens, 1860
      Sagaritis Chambers, 1872, preocc.
      Anorthodisca Gaede, 1937, missp.
Brachmia Hübner, [1825]
      Braclunia Stephens, 1834, missp.
      Cladodes Heinemann, 1870, preocc.
      Ceratophora Heinemann, 1870, preocc.
      Eudodacles Snellen, 1889, repl. name
      Aulacomima Meyrick, 1904
      Apethistis Meyrick, 1908
Brachyacma Meyrick, 1886
      Lathontogenus Walsingham, 1897
      Paraspistes Meyrick, 1905
      Lipatia Busck, 1910
      Paraspistis Busck, 1914, missp.
      Brachyaema Povolný, 1964, missp.
      Lathontogonus Diakonoff, [1968], missp.
      Brachiacma Common, 1970, missp.
      Lathontogenes Hodges, 1983, missp.
Cotyloscia Meyrick, 1923
Cymotricha Meyrick, 1923
      Oxysactis Meyrick, 1923
Dichomeris Hübner, [1818]
      Oxybelia Hübner, [1825]
      Rhinosia Treitschke, 1833
      Rhobonda Walker, 1864, preocc.
      Carna Walker, 1864, repl. name
      Gaeza Walker, 1864

      Eurysara Turner, 1919
      Euryzancla Turner, 1919
      Macrozancla Turner, 1919
      Brochometis Meyrick, 1923
Eunebristis Meyrick, 1923
Hapalonoma Meyrick, 1914
Ilingiotis Meyrick, 1914
      Sirogenes Meyrick, 1923
Myrophila Meyrick, 1923
Onebala Walker, 1864
      Helcystogramma Zeller, 1877
      Dectobathra Meyrick, 1904
Pachysaris Meyrick, 1914
Paranoea Walsingham, 1911
Plocamosaris Meyrick, 1912
      Noeza Walker, 1866, preocc.
      Neochrista Meyrick, 1923
Prophoraula Meyrick, 1922
Semiomeris Meyrick, 1923
Teuchophanes Meyrick, 1914
Trichotaphe Clemens, 1860
      Begoe Chambers, 1862
      Malacotricha Zeller, 1873
      Tricotaphe Riley, 1891, missp.
      Malacotriche Busck, 1903, missp.
      Malachotriche Busck, 1903, missp.
Vazugada Walker, 1864

Lecithocerinae

Deoclona Busck, 1903
      Proclesis Walsingham, 1911
      Lioclepta Meyrick, 1922
      Deoclana Fletcher, 1929, missp.

Unplaced Genera

Acrophiletes Meyrick, 1932
Adullamitis Meyrick, 1932
      Adullanitis Gaede, 1937, missp.
Aerotypia Walsingham, 1911
Agathactis Meyrick, 1929
Ageliarches Meyrick, 1923
Alsodryas Meyrick, 1914
Anomoxena Meyrick, 1917
Anthinora Meyrick, 1914
Anthistarcha Meyrick, 1925
      Antistarcha Costa Lima, 1945, missp.
Apotactis Meyrick, 1918
Apothetoeca Meyrick, 1922
Arotromima Meyrick, 1929
Beltheca Busck, 1914
      Anterethista Meyrick, 1914
      Antherethista Gaede, 1937, missp.
Besciva Busck, 1914
Brachypsaltis Meyrick, 1931
Bruchiana Jörgensen, 1916
Catalexis Walsingham, 1909
Catoptristis Meyrick, 1925
Cerycangela Meyrick, 1925
Chalcomima Meyrick, 1929
Charistica Meyrick, 1925
Clistothyris Zeller, 1877
Coleostoma Meyrick, 1922
Colonanthes Meyrick, 1923

Commatica Meyrick, 1909
     Apopira Walsingham, 1911
Compsosaris Meyrick, 1914
     Gompsosaris Gaede, 1937, missp.
Copticostola Meyrick, 1929
Crambodoxa Meyrick, 1913
Darlia Clarke, 1950
Diastaltica Walsingham, 1910
Dissoptila Meyrick, 1914
Drepanoterma Walsingham, 1897
Elasiprora Meyrick, 1914
Empedaula Meyrick, 1918
Eripnura Meyrick, 1914
Eristhenodes Meyrick, 1935
Erythriastis Meyrick, 1925
Ethirostoma Meyrick, 1914
Eunomarcha Meyrick, 1923
     Atoponeura Busck, 1914, preocc.
Euzonomarcha Meyrick, 1925
Fortinea Busck, 1914
Galtica Busck, 1914
Glaucacna Forbes, 1931
     Glaucagna Gaede, 1937, missp.
Ilarches Meyrick, 1933
Iphimachaera Meyrick, 1931
Isembola Meyrick, 1926
Leistogenes Meyrick, 1927
Logisis Walsingham, 1909
Lophaeola Meyrick, 1932
Meridorma Meyrick, 1925
Metabolaea Meyrick, 1923
Metopleura Busck, 1912
Molopostola Meyrick, 1920
Oestomorpha Walsingham, 1911
Oxycryptis Meyrick, 1912
Oxylechia Meyrick, 1917
Pachygeneia Meyrick, 1923
Parastega Meyrick, 1912
Parelectroides Clarke, 1952, repl. name
     Parelectra Meyrick, 1925, preocc.
Pavolechia Busck, 1914
     Desmaucha Meyrick, 1918
Pelocnistis Meyrick, 1932
Perioristica Walsingham, 1910
Phylopatris Meyrick, 1923
Promolopica Meyrick, 1925
Ptilostonychia Walsingham, 1911
     Ptilonostychia Fletcher, 1929, missp.
Rhynchotona Meyrick, 1923
Satrapodoxa Meyrick, 1925
Sclerograptis Meyrick, 1923
Simoneura Walsingham, 1911
Scrotacta Meyrick, 1914
Stachyostoma Meyrick, 1923
Stagmaturgis Meyrick, 1923
Steremniodes Meyrick, 1923
Stereodmeta Meyrick, 1931
Stibarenches Meyrick, 1930
Symphanactis Meyrick, 1925
Synactias Meyrick, 1931
Tabernillaia Walsingham, 1911
     Tabernillaea Meyrick, 1925, emend.
Taphrosaris Meyrick, 1922
Tecia Kieffer & Jörgensen, 1910
     Fapua Kieffer & Jörgensen, 1910
     Lata Kieffer & Jörgensen, 1910

Trypsigenes Meyrick, 1914
     Thripsigenes Clarke, 1955, missp.
Thyrsomnestis Meyrick, 1929
Tocmia Walker, 1864
Trichembola Meyrick, 1918
Zelosyne Walsingham, 1911

Copromorphoidea

30. COPROMORPHIDAE

Cathelotis Meyrick, 1926
Endothamna Meyrick, 1922
Lotisma Busck, 1909
Neophylarcha Meyrick, 1926
Ordrupia Busck, 1911
     Ordupia Busck, 1911, missp.
     Ardrupia Busck, 1911, missp.
Phycomorpha Meyrick, 1914
Rhopalosetia Meyrick, 1926
Saridacma Meyrick, 1930
Syncamaris Meyrick, 1932

31. ALUCITIDAE

Alinguata Fleming, 1948
Alucita Linnaeus, 1758
     Orneodes Latreille, 1796, suppr.
     Euchiradia Hübner, [1825]
     Alucitina Heydenreich, 1851
Hexeretmis Meyrick, 1929
Paelia Walker, 1866

32. CARPOSINIDAE

Carposina Herrich-Schäffer, 1853
 Subgenus Carposina Herrich-Schäffer, 1853
 Subgenus Trepsitypa Meyrick, 1913
 Subgenus Dipremna Davis, 1969
 Subgenus Epipremna Davis, 1969
 Subgenus Hypopremna Davis, 1969
Atoposea Davis, 1969

33. EPERMENIIDAE

Ochromolopinae

Parochromolopis Gaedike, 1977

34. GLYPHIPTERIGIDAE

Cotaena Walker, [1865]
Myrsila Boisduval, [1875]
Phalerarcha Meyrick, 1913
Cronicombra Meyrick, 1920
Taeniostolella Fletcher, 1940, repl. name
     Taeniostola Meyrick, 1920, preocc.
Machlotica Meyrick, 1909
     Maclotica Busck, 1915, missp.
Neomachlotica Heppner, 1981
Trapeziophora Walsingham, 1892
Rhabdocrates Meyrick, 1931
Ussara Walker, 1864
     Setiostoma R. Felder & Rogenhofer, 1875
     Usara Busck, [1934], missp.
Sericostola Meyrick, 1927

Glyphipterix Hübner, [1825]
  Heribeia Stephens, 1829
  Aechmia Treitschke, 1833
  Aecimia Boisduval, 1836, missp.
  Glyphipteryx Zeller, 1839, emend.
  Glyphiteryx Fischer von Röslerstamm, 1841,
    missp.
  Anacampsoides Bruand, 1850, nom. oblit.
  Glypipteryx Stainton, 1854, missp.
  Glyphopteryx Herrich-Schäffer, 1854, emend.
  Glyphiptoryx Mann & Rogenhofer, 1878, missp.
  Glyphptieryx Turati, 1879, missp.
  Glyphipterys Christoph, 1882, missp.
  Glyphyteryx Hampson, 1918, missp.
  Glyphteryx Watt, 1920, missp.
Diploschizia Heppner, 1981

Yponomeutoidea

35. PLUTELLIDAE

Plutellinae

Calliathla Meyrick, 1931
Eucalliathla Clarke, 1967
Euceratia Walsingham, 1881
Eudolichura Clarke, 1965
Leuroperna Clarke, 1965
Orthenches Meyrick, 1886
Philaustera Meyrick, 1927
Plutella Schrank, 1802
  Anadetia Hübner, [1825]
  Euota Hübner, [1825]
  Cerostoma Stephens, 1834
  Creagria Sodoffsky, 1837
Thalassonympha Meyrick, 1931
Ypsolopha Latreille, 1796
  Ypsolophus Fabricius, 1798
  Cerostoma Latreille, 1802
  Hypsolopha Billberg, 1820
  Theristis Hübner, [1825]
  Harpipterix Hübner, [1825]
  Abebaea Hübner, [1825]
  Harpipteryx Treitschke, 1833, emend.
  Chaetochilus Stephens, 1834
  Harpepteryx Sodoffsky, 1837, emend.
  Hypolepia Guenée, 1845, nom. nud.
  Pteroxia Guenée, 1845, nom. nud.
  Harpopteryx Agassiz, 1846, emend.
  Hypsilophus Agassiz, 1846, emend.
  Periclymenobius Wallengren, 1880
  Credemnon Wallengren, 1880
  Trachoma Wallengren, 1880
  Pluteloptera Chambers, 1880
  Plutelloptera Walsingham, 1881, missp.
  Alapa Kieffer & Jörgensen, 1910
  Mapa Strand, 1911
  Pycnopogon Chrétien, 1922
  Credemna Forbes, 1923, missp.
  Melitonympha Meyrick, 1927
  Chalconympha Meyrick, 1931
  Credemon Moriuti, 1977, missp.

Praydinae

Atemelia Herrich-Schäffer, 1853
Prays Hübner, [1825]

36. YPONOMEUTIDAE

Attevinae

Atteva Walker, 1854
  Peciloptera Clemens, 1860, preocc.
  Amblothridia Wallengren, 1861
  Corinea Walker, 1863
  Oeta Grote, 1865
  Carthara Walker, 1866, preocc.
  Synadia Walker, 1866, repl. name
  Scintilla Guenée, 1879, preocc.
  Syblis Guenée, 1879
Lactura Walker, 1854
  Dianasa Walker, 1854
  Mieza Walker, 1854
  Sarbena Walker, 1865, preocc.
  Themiscyra Walker, 1865
  Cyptasia Walker, 1866
  Buxeta Walker, 1866
  Enaemia Zeller, 1872
  Pseudotalara Druce, 1885
  Pseudocaprima Walsingham, 1900
  Epidictica Turner, 1903
  Hedycharis Turner, 1903
  Eriopyrrha Meyrick, 1913
Pygmocrates Meyrick, 1932

Yponomeutinae

Anchimacheta Walsingham, 1914
Ditrigonophora Walsingham, 1897
Euarne Möschler, 1890
Ithutomus Butler, 1883
  Ithytomus Meyrick, 1914, emend.
Spiladarcha Meyrick, 1913
Syncerastis Meyrick, 1931
Teinoptila Sauber, 1902
Toecorhychia Butler, 1883
Urodus Herrich-Schäffer, 1854
  Trichostibas Zeller, 1863
  Pexicnemidia Möschler, 1890, **n. syn.**
  Paratiquadra Walsingham, 1897
Xyrosaris Meyrick, 1907
  Xyrosaria Kearfott, [1903], missp.
Yponomeuta Latreille, 1796
  Hyphantes Hübner, [1806], suppr.
  Erminea Haworth, [1811]
  Hyponomeuta Billberg, 1820, emend.
  Coenyphantes Hübner, [1822]
  Nygmia Hübner, [1825], preocc.
  Hyponomeuta Sodoffsky, 1837, emend.
  Hyponomenta Turner, 1898, missp.
Zelleria Stainton, 1849

37. ARGYRESTHIIDAE

Argyresthia Hübner, [1825]
  Argyrosetia Stephens, 1829
  Oligos Treitschke, 1830
  Ederesa Curtis, 1833
  Ismene Stephens, 1834

Blastotere Ratzeburg, 1840
Argyrestia MacKay, 1972, missp.

38. DOUGLASIIDAE

Protonyctia Meyrick, 1931

39. ACROLEPIIDAE

Acrolepiopsis Gaedike, 1970
Argiope Chambers, 1873, preocc.
Antispastis Meyrick, 1926

40. HELIODINIDAE

Schreckensteiniinae

Schreckensteinia Hübner, [1825]
Chrysocorys Curtis, 1833

Heliodininae

Amphiclada Meyrick, 1912
Copocentra Meyrick, 1909
Crembalastis Meyrick, 1915
Cycloplastis Clemens, 1864
Heliodines Stainton, 1854
Aetole Chambers, 1875
Aetola Frey, 1884, missp.
Heliodinides Turner, 1941, missp.
Lamprolophus Busck, 1900
Embola Walsingham, 1909
Lithariapteryx Chambers, 1876
Pseudastasia Walsingham, 1909
Scelorthus Busck, 1900
Thrasydoxa Meyrick, 1912

Immoidea

41. IMMIDAE

Moca Walker, 1863
Adricara Walker, 1863
Alicadra Walker, [1866]
Jobula Walker, 1866
Callartona Hampson, [1893]
Loxotrochis Meyrick, 1906
Imma Walker, [1859]
Pingrassa Walker, [1859]
Tortricomorpha C. Felder, 1861
Topaza Walker, 1864
Vinzela Walker, [1866]
Thylacopleura Meyrick, 1886
Davendra Moore, 1887
Pseudotortrix Turner, 1900

LEPIDOPTERA

Suborder Zeugloptera

Micropterigoidea

1. MICROPTERIGIDAE

by N. P. Kristensen & E. S. Nielsen

HYPOMARTYRIA Kristensen & Nielsen, 1982

1	**micropteroides** Kristensen & Nielsen, 1982	Chile

SQUAMICORNIA Kristensen & Nielsen, 1982

2	**aequatoriella** Kristensen & Nielsen, 1982	Ecuador

Suborder Heterobathmiina

Heterobathmioidea

2. HETEROBATHMIIDAE

by N. P. Kristensen & E. S. Nielsen

HETEROBATHMIA Kristensen & Nielsen, 1979

1	**diffusa** Kristensen & Nielsen, 1979	Argentina
2	**pseuderiocrania** Kristensen & Nielsen, 1979	Argentina

Suborder Glossata

Infraorder Neopseustina

Neopseustoidea

3. NEOPSEUSTIDAE

by D. R. Davis

SYNEMPORA Davis & Nielsen, 1980

1	**andesae** Davis & Nielsen, 1980	Argentina

APOPLANIA Davis, 1975

2	**penai** Davis & Nielsen, 1980	Chile
3	**valdiviae** Davis & Nielsen, 1984	Chile
4	**chilensis** Davis, 1975	Chile

Infraorder Exoporia

Hepialoidea

4. NEOTHEORIDAE

by N. P. Kristensen

NEOTHEORA Kristensen, 1978

1	**chiloides** Kristensen, 1978	Brazil (MT)

5. HEPIALIDAE

by G. S. Robinson & E. S. Nielsen

DALACA Walker, 1856
Dolaca Druce, 1901, missp.
Huapina Bryk, 1945
Maculella Viette, 1950
Toenga Tindale, 1954

1	**crocatus** (Ureta, 1956) (Hepialus)	Chile
2	**chiliensis** (Viette, 1950) (Maculella)	Chile
	chilensis (Viette, 1950) (Maculella), missp.	
3	**pallens** (Blanchard, 1852) (Hepialus)	Chile
	hemileuca Butler, 1882	Chile
	marmorata Butler, 1882	Chile
	subfervens Butler, 1882	Chile
	violacea Butler, 1882	Chile
	dimidiatus (Berg, 1882) (Aepytus)	Chile
	noctuides Pfitzner, 1914	Chile
	parviguttata (Bryk, 1945) (Huapina)	Argentina
	pseudodimiata (Paclt, 1953) (Lossbergiana)	Argentina
	oceanica (Tindale, 1954) (Toenga)	"Cook Is." [Chile?]
4	**quadricornis** Nielsen & Robinson, 1983	Argentina
5	**nigricornis** Walker, 1856	Chile
6	**patriciae** Nielsen & Robinson, 1983	Argentina

7	**laminata** Nielsen & Robinson, 1983	Chile
8	**fuscus** (Mabille, 1885) (Hepialus)	Chile
9	**postvariabilis** Nielsen & Robinson, 1983	Argentina
10	**variabilis** (Viette, 1950) (Maculella)	Chile

CALLIPIELUS Butler, 1882
Stachyocera Ureta, 1957

11	**arenosus** Butler, 1882	Chile
	antarcticus (Staudinger, 1899) (Hepialus),	Argentina
	preocc. (not Wallengren, 1860)	
	staudingeri Wagner, 1911, repl. name	Argentina
	leukogramma Bryk, 1945	Argentina
	chiliensis Viette, 1950	Chile
12	**digitata** Robinson, 1977	Chile
	brunnescens Robinson, 1977	Chile
	castilloi Robinson, 1977	Chile
13	**salasi** Robinson, 1977	Chile
14	**perforata** Nielsen & Robinson, 1983	Argentina
15	**gentilii** Nielsen & Robinson, 1983	Argentina
16	**fumosa** Nielsen & Robinson, 1983	Chile
17	**argentata** Ureta, 1957	Chile
18	**krahmeri** Nielsen & Robinson, 1983	Chile
19	**izquierdoi** (Ureta, 1957) (Stachyocera)	Chile
20	**vulgaris** Nielsen & Robinson, 1983	Argentina

BLANCHARDINA Viette, 1950

21	**venosus** (Blanchard, 1852) (Hepialus)	Chile

CALADA Nielsen & Robinson, 1983

22	**fuegensis** Nielsen & Robinson, 1983	Argentina
23	**migueli** Nielsen & Robinson, 1983	Argentina

PUERMYTRANS Viette, 1951

24	**chiliensis** Viette, 1951	Chile

PARAPIELUS Viette, 1949
Lossbergiana Viette, 1950

25	**luteicornis** (Berg, 1882) (Pielus)	Argentina
	f. popperi (Pfitzner, 1938) (Pielus)	[Argentina?]
26	**oberthuri** (Viette, 1950) (Lossbergiana)	Chile
27	**heimlichi** (Ureta, 1956) (Hepialus)	Chile
28	**reedi** (Ureta, 1957) (Hepialus)	Chile

ANDEABATIS Nielsen & Robinson, 1983

29	**chilensis** (Ureta, 1951) (Xyleutes)	Chile

DRUCEIELLA Viette, 1949

30	**basirubra** (Schaus, 1901) (Dalaca)	Peru
	songoensis (Pfitzner, 1914) (Pseudophassus)	Bolivia
31	**amazonensis** Viette, 1950	Brazil (Pa)
32	**metellus** (Druce, 1890) (Hepialus)	Ecuador
33	**momus** (Druce, 1890) (Hepialus)	Ecuador
	metricus (Pfitzner, 1914) (Pseudophassus),	?
	nom. nud.	
	f. metricus (Pfitzner, 1938) (Pseudophassus)	?

TRICHOPHASSUS Le Cerf, 1919

34	**giganteus** (Herrich-Schäffer, [1853]) (Epiolus [sic])	"S. Am."
	hayeki (Foetterle, 1903) (Phassus)	Brazil (RJ)

PHASSUS Walker, 1856

35	**triangularis** H. Edwards, 1885	Mexico (Ver)
	f. triangularides Pfitzner, 1938	Mexico
36	**huebneri** (Geyer, [1838]) (Pharmacis)	Mexico
	argentiferus Walker, 1856	Mexico (Oax)
	pedipogon Strand, 1916	Costa Rica
37	**basirei** Schaus, 1890	Mexico (Ver)
38	**n-signatus** Weymer, 1907	Guatemala
39	**phalerus** Druce, 1887	Mexico (Ver)
40	**marcius** Druce, 1892	Mexico (Dur)
41	**exclamationis** Pfitzner, 1938	?
42	**aurigenus** Pfitzner, 1914	Costa Rica
43	**championi** Druce, 1887	Guatemala
44	**pharus** (Druce, 1887) (Hepialus)	Guatemala
45	**rosulentus** Weymer, 1907	Mexico (Ver)
46	**eldorado** Pfitzner, 1906	Venezuela
47	**pretiosus** (Herrich-Schäffer, [1856]) (Epialus [sic])	Brazil
	plusia (Herrich-Schäffer, [1856]) (Epialus [sic]),	Brazil
	unavail.	
48	**agrionides** Walker, 1856	Brazil
49	**tessellatus** (Herrich-Schäffer, [1854]) (Epialus [sic])	"New Holland"
50	**smithi** Druce, 1889	Mexico (Ver)
51	**costaricensis** Druce, 1887	Costa Rica
52	**absyrtus** Schaus, 1892	Brazil (RJ)
53	**guianensis** Schaus, 1940	Guyana

54	**chrysodidyma** Dyar, 1915	Mexico (Mex)
55	**transversus** Walker, 1856	Brazil (RJ)

SCHAUSIANA Viette, 1949

56	**trojesa** (Schaus, 1901) (Phassus)	Mexico (Jal)

APLATISSA Viette, 1953

57	**michaelis** (Pfitzner, 1914) (Dalaca)	Brazil (Am)/Peru
	michaeli (Pfitzner, 1937) (Dalaca), missp.	
58	**strangoides** Viette, 1953	Brazil (Am)

PFITZNERIANA Viette, 1952
  Pseudophassus Pfitzner, 1938, preocc. (Pfitzner,
  1914 [Hepialidae])

59	**olivescens** (Pfitzner, 1914) (Dalaca)	
	a) **olivescens** (Pfitzner, 1914) (Dalaca)	Colombia
	b) **boliviensis** Viette, 1961	Bolivia
60	**vogli** Viette, 1952	Venezuela
61	**allura** Viette, 1961	Bolivia
62	**prosopus** (Druce, 1901) (Hepialus)	Colombia

AEPYTUS Herrich-Schäffer, [1858]
  Subgenus **Hampsoniella** Viette, 1949

63	**equatorialis** Viette, 1949	Ecuador
64	**serta** (Schaus, 1894) (Dalaca)	Mexico (Ver)
65	**assa** (Druce, 1887) (Dalaca)	Guatemala

  Subgenus **Pseudodalaca** Viette, 1949

66	**mexicanensis** (Viette, 1952) (Pseudodalaca)	Mexico (Ver)
67	**gugelmanni** Viette, 1949	Mexico (Ver)

  Subgenus **Gymelloxes** Viette, 1952

68	**terea** (Schaus, 1892) (Dalaca)	Mexico ("Paso San Juan")
	muysca (Pfitzner, 1914) (Dalaca)	Panama
69	**trilinearis** (Pfitzner, 1914) (Dalaca)	Colombia
	trilinearides (Pfitzner, 1937) (Dalaca), missp.	
70	**paropus** (Druce, 1890) (Hepialus)	Ecuador

  Subgenus **Alloaepytus** Viette, 1952

71	**tesselloides** (Schaus, 1901) (Dalaca)	Paraguay
	coscinophora (Pfitzner, 1914) (Dalaca)	Brazil (MT)

  Subgenus **Aepytus** Herrich-Schäffer, [1858]

72	**jeanneli** (Viette, 1950) (Schaefferiana)	Brazil (Pr)
73	**biedermanni** (Viette, 1950) (Schaefferiana)	Brazil (MG)
74	**exclamans** (Herrich-Schäffer, [1854]) (Epialus [sic])	Brazil
75	**forsteri** Viette, 1961	Bolivia
76	**munona** Schaus, 1929	Brazil (SC)
77	**petropolisiensis** Viette, 1951	Brazil (RJ)
78	**helga** Schaus, 1929	Brazil (SC)
79	**zischkai** Viette, 1961	Bolivia
80	**danieli** Viette, 1961	Argentina

  Subgenus **Thiastyx** Viette, 1951

81	**catharinae** (Viette, 1951) (Thiastyx)	Brazil (SC)

  Subgenus **Schaefferiana** Viette, 1950

82	**epigramma** (Herrich-Schäffer, [1854]) (Epialus [sic])	Brazil
83	**simplex** (Viette, 1955) (Schaefferiana)	Brazil (MG)

  Subgenus **Paragorgopis** Viette, 1951

84	**foetterlei** (Viette, 1951) (Paragorgopis)	Brazil (RJ)
85	**oreas** (Schaus, 1892) (Dalaca)	Brazil (RJ)
86	**spitzi** (Viette, 1955) (Paragorgopis)	Brazil (SP)
87	**jordani** (Viette, 1955) (Paragorgopis)	Brazil (MG)
88	**pittionii** (Viette, 1955) (Paragorgopis)	Brazil (RJ)
89	**schausi** (Viette, 1951) (Paragorgopis)	Brazil (SP)
90	**nigrovenosalis** (Viette, 1955) (Paragorgopis)	Brazil (MG)

  Subgenus **Hepialyxodes** Viette, 1951

91	**rileyi** (Viette, 1951) (Hepialyxodes)	Brazil (SP)

  Subgenus **Xytrops** Viette, 1950

92	**dorita** (Schaus, 1901) (Cibyra)	Brazil (Pr)
93	**monoargenteus** Viette, 1950	Brazil (Pr)
94	**yungas** (Viette, 1961) (Xytrops)	Bolivia
95	**pluriargenteus** (Viette, 1955) (Xytrops)	Brazil (SP)
96	**verresi** Schaus, 1929	Brazil (SC)

  Subgenus **Cibyra** Walker, 1856

97	**poltrona** (Schaus, 1901) (Cibyra)	Brazil (Pr)
98	**ferruginosa** (Walker, 1856) (Cibyra)	Brazil
	ferruginea (Kirby, 1892) (Cibyra), missp.	
	dormita (Schaus, 1901) (Cibyra)	Brazil (RJ)

  Subgenus **Lamelliformia** Viette, 1951

99	**sladeni** (Hampson, 1903) (Dalaca)	Brazil (MT)
	tupi (Pfitzner, 1914) (Cibyra)	Brazil (RJ)
100	**prytanes** (Schaus, 1892) (Dalaca)	Brazil (RJ)

  Subgenus **Tricladia** R. Felder, 1874
    Pseudophassus Pfitzner, 1914
    Parana Viette, 1949

101	**mahagoniatus** (Pfitzner, 1914) (Pseudophassus)	Bolivia
102	**umbrifera** (R. Felder, 1874) (Tricladia)	Brazil
103	**philiponi** Viette, 1949	Brazil (Pa)

  Subgenus **Pseudophilaenia** Viette, 1950

104	**omagua** (Pfitzner, 1937) (Philaenia)	Peru

  Subgenus **Philaenia** Kirby, 1892

105	**guyanensis** Viette, 1950	French Guiana
106	**lagopus** (Möschler, 1877) (Pharmacis)	Surinam
107	**thisbe** (Druce, 1901) (Dolaca [sic])	Colombia
	f. hemichrysea (Pfitzner, 1937) (Dalaca)	Colombia
108	**indicata** (Strand, 1912) (Dalaca)	Ecuador
109	**brasiliensis** (Viette, 1951) (Philaenia)	Brazil (RJ)
110	**saguanmachica** (Pfitzner, 1914) (Dalaca)	Colombia
111	**fasslii** (Pfitzner, 1914) (Dalaca)	Colombia

  Subgenus **Yleuxas** Viette, 1951

112	**brunnea** (Schaus, 1901) (Cibyra)	Venezuela
	bradleyi (Viette, 1951) (Yleuxas)	Peru

PHIALUSE Viette, 1961

113	**palmar** Viette, 1961	Bolivia

ROSEALA Viette, 1950

114	**bourgognei** Viette, 1950	Brazil (RJ)

"DALACA" auct., not Walker, 1856 [1]

115	**chiriquensis** Pfitzner, 1914	Panama
116	**cocama** Pfitzner, 1914	Peru
	nannophyes Pfitzner, 1914	Ecuador
117	**cuprifera** Pfitzner, 1914	Peru
118	**guarani** Pfitzner, 1914	Brazil (SC)
119	**katharinae** Pfitzner, 1914	Brazil (SC)
120	**manoa** Pfitzner, 1914	Colombia
121	**niepelti** Pfitzner, 1914	Ecuador
122	**obliquestrigata** Strand, 1912	Peru
123	**stigmatica** Pfitzner, 1937	Paraguay
124	**tapuja** Pfitzner, 1914	Colombia
125	**usaque** Pfitzner, 1914	Colombia
126	**vibicata** Pfitzner, 1914	Ecuador
127	**mummia** Schaus, 1892	Brazil (RJ)
	mummea Pfitzner, 1937, missp.	

PFITZNERIELLA Viette, 1950

128	**lucicola** (Maassen, 1890) (Triodia)	Ecuador
129	**monticola** (Maassen, 1890) (Triodia)	Ecuador
130	**similis** (Zukowsky, 1954) (Triodia)	Peru
131	**remota** (Pfitzner, 1906) (Hepialus)	Peru

6. PALAEOSETIDAE[2]

by G. S. Robinson & E. S. Nielsen

OSRHOES Druce, 1900

1	**coronta** Druce, 1900	Colombia

Infraorder Heteroneura

Division Monotrysia

Nepticuloidea

7. NEPTICULIDAE

by D. R. Davis

Nepticulinae[3]

Tribe Nepticulini[4]

STIGMELLA Schrank, 1802
  Nepticula Heyden, 1843
  Stigmalla Herrich-Schäffer, 1853, missp.
  Dysnepticula Börner, 1925
  Johanssonia Borkowski, 1972

1	**aerifica** (Meyrick, 1915) (Nepticula), **n. comb.**	Peru
2	**andina** (Meyrick, 1915) (Nepticula), **n. comb.**	Peru
3	**costalimai** (Bourquin, 1962) (Nepticula), **n. comb.**	Argentina
4	**cuprata** (Meyrick, 1915) (Nepticula), **n. comb.**	Peru
5	**epicosma** (Meyrick, 1915) (Nepticula), **n. comb.**	Peru
6	**eurydesma** (Meyrick, 1915) (Nepticula), **n. comb.**	Guyana
7	**gossypii** (Forbes, 1930) (Nepticula), **n. comb.**	Puerto Rico
8	**guittonae** (Bourquin, 1962) (Nepticula), **n. comb.**	Argentina
9	**hylomaga** (Meyrick, 1931) (Nepticula), **n. comb.**	Argentina
10	**johannis** (Zeller, 1877) (Nepticula), **n. comb.**	Colombia
11	**molybditis** (Zeller, 1877) (Nepticula), **n. comb.**	Colombia
12	**olyritis** (Meyrick, 1915) (Nepticula), **n. comb.**	Peru

Tribe Trifurculini

ENTEUCHA Meyrick, 1915

13	**cyanochlora** Meyrick, 1915	Guyana

8. OPOSTEGIDAE

by D. R. Davis

OPOSTEGA Zeller, 1839

1	**abrupta** Walsingham, 1897	Virgin Is. (St. Thomas)
2	**accessoriella** Frey & Boll, 1876	USA (Tx)
3	**acidata** Meyrick, 1915	Ecuador
4	**adusta** Walsingham, 1897	Virgin Is. (St. Thomas)
5	**congruens** Walsingham, 1914	Mexico (Gue)
6	**elachista** Walsingham, 1914	Mexico (Gue)
7	**microlepta** Meyrick, 1915	Ecuador/Guyana
8	**monosperma** Meyrick, 1931	Brazil (MG)
9	**paromias** Meyrick, 1915	Peru
10	**perdigna** Walsingham, 1914	Mexico (Gue)
11	**pexa** Meyrick, 1920	Brazil (Pa)
12	**pontifex** Meyrick, 1915	Colombia
13	**protomochla** Meyrick, 1935	Argentina
14	**pumila** Walsingham, 1914	Mexico (Tab)
15	**sacculata** Meyrick, 1915	Ecuador
16	**saltatrix** Walsingham, 1897	Virgin Is. (St. Thomas)

9. TISCHERIIDAE

by D. R. Davis

TISCHERIA Zeller, 1839
  Evexia Gistel, 1847
  Philodoxa Gistel, 1848
  Coptotriche Walsingham, 1890
  Tisheria Busck, [1903], missp.

1	**capnota** Meyrick, 1915	Peru
2	**deliquescens** Meyrick, 1915	Guyana
3	**elongata** Walsingham, 1914	Mexico (Gue)
4	**ephaptis** Meyrick, 1915	Peru
5	**koehleri** Bourquin, 1962	Argentina
6	**plagifera** Meyrick, 1915	Ecuador
7	**pulverea** Walsingham, 1897	Virgin Is. (St. Thomas)
8	**unicolor** Walsingham, 1897	Virgin Is. (St. Croix)

Incurvarioidea

10. INCURVARIIDAE

by D. R. Davis

BASILEURA Nielsen & Davis, 1981

1	**elongata** Nielsen & Davis, 1981	Argentina

SIMACAUDA Nielsen & Davis, 1981

2	**dicommatias** (Meyrick, 1931) (Lampronia)	Chile
3	**heliocephala** (Meyrick, 1931) (Lampronia)	Chile
4	**virescens** Nielsen & Davis, 1981	Argentina

11. CECIDOSIDAE

by D. R. Davis

CECIDOSES Curtis, 1835
  Clistoses Kieffer & Jörgensen, 1910
  Eucecidoses Brèthes, 1916

1	**eremita** Curtis, 1835	Uruguay
	artifex (Kieffer & Jörgensen, 1910) (Clistoses)	Argentina
2	**minutanus** Brèthes, 1916	Argentina

OLIERA Brèthes, 1916

3	**argentinana** Brèthes, 1916	Argentina

RIDIASCHINIA Brèthes, 1916

4	**congregatella** Brèthes, 1916	Argentina

DICRANOSES Kieffer & Jörgensen, 1910

5	**capsulifex** Kieffer & Jörgensen, 1910	Argentina

12. ADELIDAE

by D. R. Davis

CEROMITIA Zeller, 1854
  Trichorrhabda Meyrick, 1912, **n. syn.**

1	**chionocrossa** Meyrick, 1921	Brazil (Pa)
2	**eremarcha** Meyrick, 1914	Paraguay
3	**exalbata** Meyrick, 1921	Brazil (Pa)
4	**fasciolata** (Butler, 1883) (Nemophora)	Chile
5	**ilyodes** Meyrick, 1931	Argentina
6	**laminensis** Pastrana, 1961	Argentina
7	**lizeri** Pastrana, 1961	Argentina
8	**ochrodyta** Meyrick, 1921	Brazil (Am)
9	**phaeoceros** Meyrick, 1921	Brazil (Am)
10	**pucaraensis** Pastrana, 1961	Argentina
11	**schajovskoii** Pastrana, 1961	Argentina
12	**sciographa** Meyrick, 1921	Brazil (Pa)
13	**viscida** Meyrick, 1921	Brazil (Pa)

ADELA Latreille, 1796
  Capillaria Haworth, 1828
  Metallitis Sodoffsky, 1837
  Aedilis Gistel, 1873
  Dicte Chambers, 1873

14	**aethiops** R. Felder & Rogenhofer, 1875	"Australia" [C. Am.]
15	**astrella** Walsingham, 1915	Mexico (Gue)
16	**lithopola** Walsingham, 1915	Mexico (Ver)
17	**ridingsella** Clemens, 1864	USA (Va)

13. PRODOXIDAE

by D. R. Davis

TEGETICULA Zeller, 1873
  Pronuba Riley, 1872, preocc. (Thompson, 1860
    [Coleoptera])
  Promiba Kirby, 1874, missp.
  Thia H. Edwards, 1888, preocc. (Leach, 1815 [Crustacea])
  Thelethia Dyar, 1893
  Valentinia Coolidge, 1909, preocc. (Walsingham, 1907
    [Blastobasidae])

*1	**maculata** (Riley, 1881) (Pronuba)	
	a) **extranea** (H. Edwards, 1888) (Thia)	USA (Ca)
	aterrima (Trelease, 1893) (Pronuba)	USA (Ca)
2	**yuccasella** (Riley, 1872) (Pronuba)	USA (Mo)
	alba Zeller, 1873	USA (Tx)
	yuccaella (Boll, 1876) (Pronùba), missp.	
	intermedius (Riley, 1881) (Prodoxus)	USA (Tx)
	mexicana Bastida, 1962	Mexico (SLP)

PARATEGETICULA Davis, 1967

3	**pollenifera** Davis, 1967	USA (Az)

PRODOXUS Riley, 1880

4	**quinquepunctellus** (Chambers, 1875) (Hyponomeuta[sic])	USA (Tx)
	paradoxica (Chambers, 1878) (Hyponomeuta[sic])	USA (Co)
	decipiens Riley, 1880	USA (SC)
5	**intricatus** Riley, 1893	Mexico (Ver)

PRODOXOIDES Nielsen & Davis, 1984 [5]

6	**assymmetra** Nielsen & Davis, 1984	Chile

14. HELIOZELIDAE

by J. B. Heppner

ANTISPILA Hübner, [1825]

1	**chlorosema** Meyrick, 1931	Chile
2	**cyclosema** Meyrick, 1921	Brazil (Am)
3	**mesogramma** Meyrick, 1921	Peru
3.1	**nolckeni** Zeller, 1877	Colombia
4	**orthodelta** Meyrick, 1931	Brazil (Am)
5	**pentalitha** Meyrick, 1916	Guyana
6	**postscripta** Meyrick, 1921	Peru
7	**praecincta** Meyrick, 1921	Brazil (Pa)

8	**trypherantis** Meyrick, 1916	Guyana

COPTODISCA Walsingham, 1895
    Aspidisca Clemens, 1860, preocc. (Ehrenburg, 1830 [Protura])

9	**rhizophorae** Walsingham, 1897	Virgin Is. (St. Thomas)

HELIOZELA Herrich-Schäffer, 1853
    Dyselachista Spuler, 1910

10	**ahenea** Walsingham, 1897	Haiti
11	**cuprea** Walsingham, 1897	Virgin Is. (St. Thomas)

LAMPROZELA Meyrick, 1916

12	**praefulgens** Meyrick, 1916	Guyana

MONACHOZELA Meyrick, 1931

13	**neoleuca** Meyrick, 1931	Brazil (Pa)

PHANEROZELA Meyrick, 1921

14	**polydora** Meyrick, 1921	Brazil (Am)

Division Ditrysia

Section Tineina

Tineoidea

15. TINEIDAE [6]

by D. R. Davis

ACROLOPHUS Poey, 1832
    Anaphora Clemens, 1859
    Daulia Walker, 1863, preocc. (Walker, 1859 [Pyralidae])
    Derchis Walker, 1863
    Eddara Walker, 1863, preocc. (Walker, 1858 [Hemiptera])
    Hibita Walker, 1863
    Tirasia Walker, 1863
    Bazira Walker, 1864
    Phlongia Walker, 1864
    Urbara Walker, 1864
    Tachasara Walker, 1865
    Eutheca Grote, 1881, preocc. (Kiesenwetter, 1877 [Coleoptera])
    Eulepiste Walsingham, 1882
    Pseudoconchylis Walsingham, 1884
    Homonymus Walsingham, 1887
    Neolophus Walsingham, 1887
    Hypoclopus Walsingham, 1887
    Thysanoscelis Walsingham, 1887
    Thysanoskelis Walsingham, 1887, missp.
    Ankistrophorus Walsingham, 1887
    Caenogenes Walsingham, 1887
    Felderia Walsingham, 1887
    Ortholophus Walsingham, 1887
    Pseudanaphora Walsingham, 1887
    Sapinella Kirby, 1892
    Atopocera Walsingham, 1897
    Pilanaphora Walsingham, 1897
    Thysanosedes Druce, 1901, missp.
    Hypocolypus Dyar, [1903], missp.
    Anophora Kearfott, 1903, missp.
    Brachysymbola Meyrick, 1912
    Orothyntis Meyrick, 1913
    Apoclisis Walsingham, 1914
    Psephocrita Meyrick, 1919
    Xerocausta Meyrick, 1928, n. syn.
    Anaphorina Strand, 1932

1	**abdita** Walsingham, 1914	Guatemala
*2	**acornus** Hasbrouck, 1964	USA (Az)
3	**albipennis** Meyrick, 1931	Brazil (MG?)
4	**anaphorella** (Walsingham, 1892) (Amydria), n. comb.	St. Vincent
5	**anathyrsa** Meyrick, 1919	French Guiana
6	**angulatella** (Walsingham, 1897) (Morophaga), n. comb.	Dominica
7	**apertella** Busck, 1913	Mexico (Mex)
8	**arcasalis** (Walker, 1858) (Palthis)	Dom. Rep.
9	**arcturella** (Walker, 1863) (Hibita)	Brazil (Am)
10	**argentinus** Walsingham, 1887	Argentina
11	**arida** Walsingham, 1914	Mexico (Gue)
12	**arimasalis** (Walker, 1850) (Palthis)	Brazil (Pa)
*13	**arizonellus** Walsingham, 1887	USA (Az)
14	**australis** (Walsingham, 1897) (Xylesthia), n. comb.	Haiti
15	**bactra** Busck, 1914	Panama
*16	**baldufi** Hasbrouck, 1964	USA (Az)
17	**barbipalpus** Busck, 1912	Mexico (Pue)
18	**barema** Durrant, 1915	Guatemala
19	**baryspila** Meyrick, 1931	Brazil (ES)
20	**bidens** Walsingham, 1914	Mexico (Dur)

21	**bifurcata** Busck, 1914	Panama
22	**bogotensis** (Walsingham, 1887) (Anaphora)	Colombia
23	**bombaulia** Meyrick, 1922	Argentina
24	**boucardi** Druce, 1901	Mexico (Ver)
25	**bugabae** Walsingham, 1914	Panama
26	**capax** Meyrick, 1928	Colombia
27	**catagnampta** Meyrick, 1930	Brazil (Pa)
28	**ceramocha** (Meyrick, 1928) (Xerocausta), n. comb.	Colombia
29	**cervicolor** Meyrick, 1931	Argentina
30	**chloropelta** Meyrick, 1922	Brazil (Am)
31	**chonactis** Meyrick, 1932	Brazil (RJ)
32	**cleptica** Walsingham, 1914	Mexico (Gue)
*33	**cockerelli** (Dyar, 1900) (Eulepiste)	USA (NM)
34	**condita** Durrant, 1914	Mexico (Ver)
35	**contubernalis** Meyrick, 1922	Brazil (RJ)
36	**corrientis** (Walsingham, 1887) (Ankistrophorus)	Argentina
37	**corticinicolor** Strand, 1920	Costa Rica
38	**corvula** Walsingham, 1914	Mexico (Dur)
39	**corymba** Durrant, 1914	Guatemala
40	**cosmeta** Walsingham, 1914	Guatemala
41	**cossoides** R. Felder & Rogenhofer, 1875	Brazil
42	**crinifrons** Walsingham, 1914	Mexico (Ver)
43	**cyclophora** Meyrick, 1931	Paraguay
44	**damina** Walsingham, 1914	Panama
45	**diachelota** Meyrick, 1931	Argentina
46	**dictyopsamma** Meyrick, 1931	Brazil (ES)
47	**dimidiella** (Walsingham, 1892) (Felderia)	Cuba
48	**directus** Busck, 1912	Mexico (DF)
49	**doeri** (Walsingham, 1887) (Felderia)	Brazil (RJ)
50	**ductifera** Meyrick, 1927	Antigua
51	**echinon** (Druce, 1901) (Felderia)	Mexico (Ver)
	cassicordis Dyar, 1907	Mexico (DF)
	robertus Busck, 1920, n. syn.	Mexico (Mex)
52	**echinura** Meyrick, 1915	Peru
53	**ectenes** Walsingham, 1914	Mexico (Gue)
54	**empedocles** Meyrick, 1930	Brazil (RJ)
55	**emphytopa** Meyrick, 1932	Brazil (RS)
56	**erethismia** Meyrick, 1930	Brazil (Pa)
57	**euporia** Walsingham, 1914	Mexico (Gue)
58	**outclcs** Walsingham, 1914	Mexico (Ver)
59	**exigua** Meyrick, 1915	Guyana
60	**farracea** Meyrick, 1931	Brazil (ES)
61	**ferrarenella** (Walker, 1864) (Phlongia)	[S. Am.?]
62	**ferruginea** (Walsingham, 1887) (Anaphora)	Colombia
63	**fervidus** Busck, 1912	Mexico (Ver)
	antonellus (Barnes & McDunnough, 1913) (Neolophus)	USA (Tx)
64	**filicicornis** (Walsingham, 1887) (Felderia)	USA (Az)
	mexicanellus Beutenmüller, 1888	Mexico (DF)
	filicornis (Dyar, 1900) (Felderia), missp.	
65	**forreri** Walsingham, 1914	Mexico (Dur)
66	**fumida** Walsingham, 1914	Costa Rica
67	**galeata** (Walker, 1864) (Urbara)	Brazil (Am)
68	**garleppi** (Druce, 1901) (Felderia)	Bolivia
69	**giganteus** (Druce, 1901) (Ankistrophorus)	Mexico (Ver)
70	**goniocentra** Meyrick, 1923	Brazil (Am)
*71	**griseus** (Walsingham, 1887) (Hypoclopus)	
	a) griseus (Walsingham, 1887) (Hypoclopus)	USA (Az)
	leucallactis Meyrick, 1919	USA (Az)
72	**halidora** Meyrick, 1915	Guyana
73	**hamiferella** (Hübner, [1827-31]) (Pinaris)	Brazil (RJ)
74	**harmoniella** Busck, 1913	Mexico (Pue)
75	**harparsen** Forbes, 1931	Puerto Rico
76	**hedemanni** (Walsingham, 1897) (Pilanaphora)	Virgin Is. (St. Croix)
77	**hirsutevestita** (Walsingham, 1897) (Morophaga)	Jamaica
78	**hirsutus** (Walsingham, 1887) (Thysanoscelis)	Brazil (ES)
79	**horridalis** (Walker, 1863) (Derchis)	Brazil (Am)
80	**hypophaea** Meyrick, 1923	Brazil (Am)
81	**icarus** Busck, 1912	Mexico (Pue)
82	**illudens** Meyrick, 1924	Jamaica
83	**infida** Meyrick, 1912	Venezuela
84	**interfusa** Meyrick, 1927	Ecuador
85	**invida** Durrant, 1914	Panama
86	**irrisoria** Meyrick, 1924	Jamaica
87	**jalapae** Walsingham, 1914	Mexico (Ver)
	harpella Walsingham, 1914, n. syn.	Mexico (Ver)
*88	**kearfotti** (Dyar, 1903) (Eulepiste)	USA (Az)
	diversus Busck, 1912	USA (Ca)
*88.1	**klotsi** Hasbrouck, 1964	USA (Az)
89	**laetifica** Durrant, 1914	Mexico (Ver)
90	**latiberbis** Meyrick, 1931	Brazil (SP)
91	**lerodes** Durrant, 1914	Mexico (Ver)
92	**leucodocis** (Zeller, 1877) (Anaphora)	Cuba
	medioliniella (Kearfott, 1907) (Dorata)	USA (Ca)
93	**leucopogon** Walsingham, 1914	Mexico (Tab)
94	**leucotricha** Meyrick, 1931	Paraguay
95	**libitina** Druce, 1901	Guatemala
96	**linus** Druce, 1901	Mexico (Ver)
97	**lithopa** Durrant, 1914	Mexico (Dur)
*98	**macrogaster** (Walsingham, 1887) (Anaphora)	USA (Az)
	bipectinicornis Hasbrouck, 1964, n. syn. [7]	USA (Az)
	unipectinicornis Hasbrouck, 1964, n. syn.	USA (Az)
	laminicornis Hasbrouck, 1964, n. syn.	USA (Az)
*99	**macrophallus** Hasbrouck, 1964	USA (Az)

100	**macrozancla** Meyrick, 1922	Peru
101	**maculata** Walsingham, 1887	Brazil (RJ)
102	**maculisecta** Busck, 1913	Mexico (Ver)
	dentiger Walsingham, 1914, **n. syn.**	Mexico (Ver)
103	**manticodes** Meyrick, 1919	French Guiana
104	**marcida** Walsingham, 1914	Mexico (Gue)
105	**merocoma** Meyrick, 1922	Peru
106	**micromacha** Meyrick, 1932	Costa Rica
107	**mimasalis** (Walker, 1858) (Palthis)	Dom. Rep.
	indecora (Walker, 1863) (Daulia), **n. syn.**	Dom. Rep.
	languidalis (Walker, 1865) (Tachasara), **n. syn.**	Dom. Rep.
108	**minima** (Walsingham, 1887) (Anaphora)	Brazil (Ce)
109	**misema** Walsingham, 1914	Mexico (Gue)
110	**modestus** Busck, 1912	Mexico (Pue)
111	**monoctenis** Meyrick, 1931	Paraguay
112	**morbidula** Meyrick, 1931	Brazil (SP)
113	**niveipunctata** Walsingham, 1892	Cuba
114	**noctivaga** (Walsingham, 1897) (Pseudanaphora)	Grenada
115	**noctuina** (Walsingham, 1892) (Anaphora)	Cuba
116	**nubifer** Walsingham, 1914	Panama
117	**numidia** (Druce, 1901) (Anaphora)	Mexico (Ver)
118	**occultum** (Walsingham, 1897) (Atopocera)	Haiti
119	**ochracea** (Möschler, 1890) (Caenogenes)	Puerto Rico
120	**orasiusalis** (Walker, 1859) (Palthis)	[Belize]
	orasialis Walsingham, 1914, missp.	
121	**orizabae** (Dyar, 1907) (Anaphora)	Mexico (Ver)
122	**ornata** (Walsingham, 1887) (Anaphora)	Colombia?
123	**pachynta** Meyrick, 1913	Colombia
124	**pallidus** Möschler, 1881	Surinam
125	**panamae** Busck, 1914	Panama
126	**pannephela** Meyrick, 1931	Brazil (ES)
127	**particeps** Meyrick, 1913	Venezuela
128	**parvus** (Walsingham, 1897) (Hypoclopus)	Dominica
129	**pauper** Walsingham, 1914	Mexico (Ver)
130	**penumbra** Walsingham, 1914	Panama
131	**perissarcha** Meyrick, 1919	Guyana
132	**perpetua** Meyrick, 1913	Paraguay
133	**perrensella** (Walsingham, 1887) (Caenogenes)	Argentina
134	**perrensi** (Druce, 1901) (Anaphora)	Brazil (Go)
135	**phaeomalla** Meyrick, 1913	Argentina
136	**pinnifera** Meyrick, 1931	Brazil (MG?)
137	**poeyi** Walsingham, 1892	St. Vincent
*138	**popeanellus** (Clemens, 1859) (Anaphora)	USA (Tx)
	agrotipennella (Grote, 1872) (Anaphora)	USA (Al)
	scardina (Zeller, 1873) (Anaphora)	USA (Tx)
	morrisoni (Walsingham, 1887) (Anaphora)	USA (Fl)
	confusellus J. B. Smith, 1891	USA (Ga)
139	**practica** Meyrick, 1913	Brazil (MG?)
140	**pretusalis** (Walker, 1858) (Palthis)	[Belize]
	granulatella (Walker, 1863) (Tirasia)	Brazil (Am)
	praetusalis Walsingham, 1914, missp.	
	cathecta Walsingham, 1914, **n. syn.**	Mexico (Ver)
141	**prepodes** Walsingham, 1914	Mexico (Gue)
142	**pristinella** (Walker, 1863) (Tinea)	Venezuela
143	**psammophila** Meyrick, 1931	Brazil (MG?)
144	**pseudohirsutus** Hasbrouck, 1964, repl. name	USA (Ca)
	hirsutus Busck, 1912, preocc. (not Walsingham, 1887)	USA (Ca)
145	**pseudonoma** Meyrick, 1922	Brazil (Am)
146	**psoloessa** Meyrick, 1932	Brazil (SC)
147	**pumicea** Meyrick, 1913	Paraguay
148	**punctata** (Druce, 1901) (Anaphora)	Costa Rica
149	**pusilla** (Zeller, 1877) (Anaphora)	Colombia
150	**pygmaea** (Walsingham, 1887) (Felderia)	Argentina
151	**pyramellus** (Barnes & McDunnough, 1913) (Eulepiste)	USA (Nv)
152	**rastricornis** Meyrick, 1932	Virgin Is. (St. Thomas)
153	**reflexa** (Fabricius, 1793) (Bombyx)	[Virgin Is.?]
154	**ridicula** Meyrick, 1913	Surinam
155	**rupestris** (Walsingham, 1914) (Apoclisis)	Jamaica
156	**sachari** Busck, 1913	Guyana
157	**sagaritis** Meyrick, 1932	Bolivia
158	**salvini** (Druce, 1901) (Thysanosedes [sic])	Panama
	arcei (Druce, 1901) (Anaphora), **n. syn.**	Panama
	caprimulgus Walsingham, 1914, **n. syn.**	Panama
159	**sarista** Meyrick, 1913	Venezuela
160	**satyrisca** Meyrick, 1927	Colombia
161	**schistodes** Meyrick, 1913	Venezuela
162	**scopodes** Meyrick, 1913	Venezuela
163	**scotera** Walsingham, 1914	Mexico (Gue)
164	**scotina** (Walsingham, 1914) (Amydria)	Mexico (Gue)
165	**scrupulata** (Meyrick, 1913) (Orothyntis)	Colombia
166	**seminigera** Meyrick, 1913	Venezuela
167	**sepulcralis** (Meyrick, 1912) (Brachysymbola)	Argentina
*168	**serratus** Hasbrouck, 1964	USA (Az)
169	**setiacma** Meyrick, 1923	Brazil (Am)
170	**signatus** Busck, 1920	Mexico (Chia)
171	**spathista** Meyrick, 1919	Colombia
172	**spinifera** Meyrick, 1913	Paraguay
173	**subfusca** Meyrick, 1913	Argentina
174	**superstes** Walsingham, 1914	Guatemala
175	**suspensilis** Meyrick, 1913	Brazil (SC)
176	**synapta** Durrant, 1914	Mexico (Tab)
177	**tetrancyla** Meyrick, 1913	Peru
*178	**texanellus** (Chambers, 1878) (Anaphora)	USA (Tx)

	hulstellus Beutenmüller, 1887	USA (Fl)
	barnesii (Dyar, 1900) (Atopocera)	USA (Tx)
179	**thaminodes** Meyrick, 1919	Guyana
180	**tholomicta** Meyrick, 1929	Colombia
181	**torta** (Meyrick, 1922) (Orothyntis)	Peru
182	**tretus** Kaye, 1925	Trinidad
183	**triatomella** (Walsingham, 1877) (Anaphora)	Virgin Is. (St. Thomas)
184	**tricausta** Meyrick, 1913	Argentina
185	**trichosoma** Durrant, 1914	Guatemala
186	**umbratipalpis** (Walsingham, 1892) (Eulepiste)	Dom. Rep.
187	**uncigera** (Walsingham, 1897) (Anaphora)	Colombia
188	**uncispinis** Walsingham, 1914	Costa Rica
189	**underwoodi** Druce, 1901	Costa Rica
*190	**vanduzeei** Hasbrouck, 1964	USA (Tx)
191	**variabilis** (Walsingham, 1887) (Ortholophus)	USA (Az)
192	**vauriei** Hasbrouck, 1964	USA (Tx)
193	**vespertilio** Meyrick, 1931	Brazil (ES)
194	**victrix** Walsingham, 1914	Guatemala
195	**vigia** Beutelspacher, 1969	Mexico (Ver)
196	**vitellus** Poey, 1832	Cuba
197	**walsinghami** Möschler, 1890	Puerto Rico
	triformellus Forbes, 1930, **n. syn.**	Puerto Rico
	triformalis Forbes, 1930, missp.	
198	**whitelyi** (Druce, 1901) (Anaphora)	Guyana
199	**xylinella** (Walker, 1863) (Eddara)	Jamaica
200	**zanclophora** Meyrick, 1922	Peru

AMYDRIA Clemens, 1859
    Casape Walker, 1864
    Amadria Chambers, 1875, missp.
    Amadrya Chambers, 1875, missp.
    Amydrya Kearfott, 1903, missp.
    Neomeristis Meyrick, 1919, **n. syn.**

201	**abscensella** (Walker, 1863) (Tinea), **n. comb.**	Venezuela
202	**anceps** Walsingham, 1914	Mexico (Gue)
	socialis (Beutelspacher, 1977) (Acrolophus), **n. syn.**	Mexico (Jal)
*203	**apachella** Dietz, 1905	USA (Az)
204	**arizonella** Dietz, 1905	USA (Az)
205	**margoriella** Dietz, 1905	USA (Tx)
	marjorieella Dietz, 1905, missp.	
	margorieella Dietz, 1905, missp.	
	marjoriella Busck, 1906, missp.	
206	**meridionalis** Walsingham, 1914	Costa Rica
207	**muricolor** Walsingham, 1914	Mexico (Dur)
208	**obliquella** Dietz, 1905	USA (Ca)
209	**pauculella** (Walker, 1864) (Casape)	Venezuela
210	**pogonites** Walsingham, 1914	Mexico (Gue)
211	**taracta** Walsingham, 1914	Guatemala

ANTIPOLISTES Forbes, 1933

212	**anthracella** Forbes, 1933	Puerto Rico
	latebrivora (Meyrick, 1935) (Tinea), **n. syn.**	Trinidad

ATTICONVIVA Busck, 1934

213	**melichrosta** (Meyrick, 1922) (Tiquadra), **n. comb.**	Brazil (Pa)
	eidmannella Busck, 1934, **n. syn.**	Brazil (RJ)

AUGOLYCHNA Meyrick, 1922

214	**septemstrigella** (Chambers, 1878) (Tinea)	USA (Tx)

AXIAGASTA Meyrick, 1930

215	**stactogramma** Meyrick, 1930	Brazil (Pa)

BARYMOCHTHA Meyrick, 1922

216	**entherastis** Meyrick, 1922	Brazil (Am)/Guyana

BASANASCA Meyrick, 1922

217	**parcens** Meyrick, 1922	Brazil (Am)

BRITHYCEROS Meyrick, 1932

218	**dichroanthes** Meyrick, 1932	Brazil (Am)/Peru

BYTHOCRATES Meyrick, 1919

219	**drosocycla** Meyrick, 1919	Guyana

CHOROPLECA Durrant, 1914
    Cyane Chambers, 1873, preocc. (C. Felder, 1861
    [Nymphalidae])
    Diachalastis Meyrick, 1920

220	**acrodisca** Meyrick, 1917	Guyana
221	**boviceps** Walsingham, 1914	Mexico (Gue)
222	**catorthota** Meyrick, 1917	Guyana
223	**discatella** Walker, 1864	Brazil (Am)

224	**germana** Walsingham, 1914	Mexico (Gue)
225	**isodisca** Meyrick, 1917	Guyana
226	**murenula** Meyrick, 1924	Peru
227	**panscia** Meyrick, 1917	Ecuador
228	**poecilta** Walsingham, 1914	Mexico (Gue)
229	**rhombifera** Meyrick, 1917	Guyana
230	**sublimis** Meyrick, 1917	Colombia
231	**terpsichorella** (Busck, 1910) (Cyane) [8]	USA (Hi)
	tetraglossa (Meyrick, 1920) (Diachalastis)	Fiji
232	**tripudians** Meyrick, 1924	Peru
233	**zygoterma** Meyrick, 1917	Colombia/Ecuador

CLEPTICODES Meyrick, 1927

234	**hexaleuca** Meyrick, 1932	Virgin Is. (St. Thomas)

CLINOGRAPTIS Meyrick, 1932

235	**ogmodes** Meyrick, 1932	Brazil (Am)

CNISMORECTIS Meyrick, 1936

236	**choritica** Meyrick, 1936	Bolivia

COLPOCRITA Meyrick, 1930

237	**diptila** Meyrick, 1930	Brazil (Pa)

COMODICA Meyrick, 1880

238	**ordinata** Walsingham, 1914	Mexico (Tab)

COMPSOCRITA Meyrick, 1922

239	**florens** Meyrick, 1922	Brazil (Pa)

CRANAODES Meyrick, 1919

240	**prostylias** Meyrick, 1927	Colombia
241	**stereopa** Meyrick, 1919	Colombia

CREPIDOCHARES Meyrick, 1922

242	**subtigrina** Meyrick, 1922	Brazil (Am)

CRYPHIOTECHNA Meyrick, 1932

243	**ochracma** Meyrick, 1932	Peru

DASMOPHORA Meyrick, 1919

244	**xerospila** Meyrick, 1919	French Guiana

DEMOBROTIS Meyrick, 1892

245	**haliplancta** Meyrick, 1927	Bermuda

DIATAGA Walsingham, 1914

246	**compsacma** Meyrick, 1919	Guyana
247	**leptosceles** Walsingham, 1914	Mexico (Ver)

DORATA Busck, 1904
Pterolonche Walsingham, 1889, preocc. (Zeller, 1847 [Pterolonchidae])
Dorota Kearfott, 1907, missp.

248	**nigritella** Busck, 1920	Mexico (Sin)
249	**sagittella** Busck, 1913	Mexico (Pue)

DRASTEA Walsingham, 1914

250	**mexica** Walsingham, 1914	Mexico (Ver)

DYOTOPASTA Busck, 1907
Pseudoxylesthia Walsingham, 1907
Dystopasta McDunnough, 1939, emend.

251	**yumaella** Kearfott, 1907	USA (Az)
	angustella (Walsingham, 1907) (Pseudoxyles-thia)	USA (Az/Tx)

DYSOPTUS Walsingham, 1914

252	**probata** Walsingham, 1914	Guatemala
253	**tantalota** Meyrick, 1919	Guyana

ECTINOCAMPA Silvestri, 1944

254	**nasutitermina** Silvestri, 1944	Brazil (SP)

EPHEDROXENA Meyrick, 1919

255	**incisoria** Meyrick, 1919	Guyana

EPISYRTA Meyrick, 1929

256	**coniomicta** Meyrick, 1929	Colombia
257	**protonistis** Meyrick, 1930	Brazil (Pa)

ERECHTHIAS Meyrick, 1880
Ereunetis Meyrick, 1880
Decadarchis Meyrick, 1886
Nesoxena Meyrick, 1929
Amphisyncentris Meyrick, 1933
Caryolestis Meyrick, 1934
Triadogona Meyrick, 1937
Anemerarcha Meyrick, 1937
Erechtias Ghesquière, 1940, missp.
Tinexotaxa Gozmány, 1968
Neodecadarchis Zimmerman, 1978
Lepidobregma Zimmerman, 1978
Pantheus Zimmerman, 1978

258	**darwini** Robinson, 1983	Brazil (St. Paul's Rocks)
259	**minuscula** (Walsingham, 1897) (Ereunetis)	Virgin Is. (St. Croix/ St. Thomas)
260	**zebrina** (Butler, 1881) (Argyresthia)	USA (Hi)
	lanceolata (Walsingham, 1897) (Ereunetis)	Virgin Is. (St. Thomas)
	xenica (Meyrick, 1911) (Ereunetis)	Seychelles
	travestita (Gozmány, 1968) (Tinexotaxa)	Sierra Leone

EXONCOTIS Meyrick, 1919

261	**gemistis** (Meyrick, 1909) (Amydria), **n. comb.**	Bolivia
262	**resona** Meyrick, 1929	Colombia
263	**umbraticella** (Busck, 1914) (Amydria)	Panama
	increpans Meyrick, 1919, **n. syn.**	French Guiana

HOMODOXUS Walsingham, 1914

264	**aristula** Walsingham, 1914	Guatemala

HOMOSETIA Clemens, 1863

265	**anaphrictis** Meyrick, 1919	Guyana
266	**ancyropis** Meyrick, 1919	Guyana
267	**cosmopa** Meyrick, 1919	Guyana
268	**iambica** Meyrick, 1919	Colombia
269	**scandalitis** Meyrick, 1919	Guyana
270	**tephropis** Meyrick, 1919	Ecuador

HOMOSTINEA Dietz, 1905
Homotinea Meyrick, 1932, emend.

271	**chersadacta** Meyrick, 1932 (Homotinea [sic])	Brazil (Am/Pa)

HORMANTRIS Meyrick, 1927

272	**astragalopa** Meyrick, 1927	Colombia

HYBROMA Clemens, 1862

273	**anticosma** Meyrick, 1919	Colombia
274	**crocorrhoa** Meyrick, 1919	Colombia
275	**dulica** Walsingham, 1914	Mexico (Tab)
276	**paedisca** Walsingham, 1914	Mexico (Gue)
277	**pegaea** Meyrick, 1919	Guyana
278	**zacharis** Meyrick, 1919	Guyana

INFURCITINEA Spuler, 1910
Atinea Amsel, 1954
Microtinea Amsel, 1954

279	**luteela** Forbes, 1931	Puerto Rico
280	**palpella** Forbes, 1931	Puerto Rico

ISOCORYPHA Dietz, 1905
Socorypha Busck, 1909, missp.

281	**limbata** Walsingham, 1914	Mexico (Tab)

LEPTOCHERSA Meyrick, 1919

282	**diarthra** Meyrick, 1919	Guyana

LEPYROTICA Meyrick, 1921

283	**brevistrigata** (Walsingham, 1897) (Tinea), **n. comb.**	Virgin Is. (St. Croix/ St. Thomas)
284	**delotoma** (Meyrick, 1919) (Tinea), **n. comb.**	French Guiana
285	**diluticornis** (Walsingham, 1897) (Tinea), **n. comb.**	Virgin Is. (St. Thomas)
	scythropiella (Walsingham, 1897) (Tinea), **n. syn.**	Virgin Is. (St. Thomas)
286	**fragilella** (Walsingham, 1897) (Tinea)	Haiti
287	**reduplicata** (Walsingham, 1897) (Tinea)	Virgin Is. (St. Thomas)
288	**scardamyctis** Meyrick, 1921	West Indies ("Leeward Is.")

LEUCOPHASMA Walsingham, 1897

| 289 | **carmodiella** Busck, 1910 | Tobago |
| 290 | **phantasmella** Walsingham, 1897 | Grenada |

LINDERA Blanchard, 1852
    Palpula Blanchard, 1852, preocc. (Treitschke,
      1833 [Oecophoridae])
    Safra Walker, 1864, preocc. (Walker, 1863 [Pyralidae])
    Chresotes Butler, 1881
    Paraneura Dietz, 1905
    Cervitinea Amsel, 1956

291	**baliopa** Meyrick, 1917	Brazil ("Rio Tapenaya")
292	**onychias** Meyrick, 1931	Brazil (ES)
293	**venezuelensis** (Amsel, 1956) (Cervitinea)	Venezuela
294	**tessellatella** Blanchard, 1852	Chile
	variegella (Blanchard, 1852) (Palpula)	Chile
	bogotatella (Walker, 1864) (Safra)	Colombia
	simulella (Dietz, 1905) (Paraneura)	USA (Ca)
	ehrhornella (Dietz, 1905) (Paraneura)	USA (Ca)
	cruciferella (Dietz, 1905) (Paraneura)	USA (Ca)

LIPOMERINX Walsingham, 1914

| 295 | **prismatica** Walsingham, 1914 | Mexico (Ver) |
| | erebantha (Meyrick, 1919) (Trithamnora), **n. syn.** | Guyana |

LITHOPSAESTIS Meyrick, 1932

| 296 | **mixophanes** Meyrick, 1932 | Brazil (SC) |

MEA Busck, 1906
    Progona Dietz, 1905, preocc. (Berg, 1882 [Arctiidae])

| 297 | **incudella** Forbes, 1931 | Puerto Rico |
| 298 | **yunquella** Forbes, 1931 | Puerto Rico |

MONOPIS Hübner, [1825]
    Blabophanes Zeller, 1852
    Hyalospila Herrich-Schäffer, 1853
    Rhitia Walker, 1864
    Eusynopa Lower, 1903

299	**impressipenella** (Bilimek, 1867) (Ornix)	Mexico (Mor)
	cacahuamilpensis (Herrera, 1892) (Ornix)	Mexico (Mor)
300	**prometopias** (Gyen, 1913) (Amydria), **n. comb.**	Chile

MYRMECOZELA Zeller, 1852
    Psecadioides Butler, 1881
    Promasia Chrètien, 1905
    Protolophe Rebel, 1915

| 301 | **corymbota** Meyrick, 1919 | Peru |
| 302 | **renitens** Meyrick, 1922 | Brazil (Am) |

MYTHOPLASTIS Meyrick, 1919

| 303 | **chalcochra** Meyrick, 1931 | Brazil (SP) |
| 304 | **exanthes** Meyrick, 1919 | French Guiana |

NEMAPOGON Schrank, 1802
    Diaphthirusa Hübner, [1825]

305	**granella** (Linnaeus, 1758) (Tinea)	[Sweden?]
	costotristrigella (Chambers, 1873) (Tinea)	USA (Ky)
	fuscomaculella (Chambers, 1873) (Tinea)	USA (Ky)
	marmorella (Chambers, 1875) (Tinea)	Canada
	costistrigella (Dietz, 1905) (Tinea), missp.	
	fascomaculella (Dietz, 1905) (Tinea), missp.	
	nigroatomella (Dietz, 1905) (Tinea)	USA (NJ)

NIDITINEA Petersen, 1957
    Tineidia Zagulajev, 1960

306	**fuscella** (Linnaeus, 1758) (Tinea)	[Sweden?]
	spretella ([Denis & Schiffermüller], 1775) (Tinea)	[Austria]
	fuscipunctella (Haworth, 1828) (Tinea)	UK (England)
	nubilipennella (Clemens, 1859) (Tinea)	USA [Pa]
	abligatella (Walker, 1863) (Tinea)	South Africa
	ignotella (Walker, 1864) (Tinea)	Sierra Leone
	frigidella (Packard, 1867) (Oecophora)	Canada
	griseella (Chambers, 1873) (Tinea)	USA (Ky)
	eurinella (Zagulajev, 1952) (Tinea)	USSR
	distans (Gozmány, 1959) (Tinea)	Afghanistan
307	**praeumbrata** (Meyrick, 1919) (Tinea)	Guyana
	negreai Căpuse & Georgescu, 1977	Cuba

OENOE Chambers, 1874

308	**euphrantis** Meyrick, 1927	Bermuda
309	**minimella** Forbes 1930	Virgin Is. (St. John)
310	**pumiliella** (Walsingham, 1897) (Tinea), **n. comb.**	Virgin Is. (St. Croix/ St. Thomas)

| 311 | **synchorda** Meyrick, 1919 | Guyana |

OINOPHILA Stephens, 1848
    Oenophila Zeller, 1853, missp.

| 312 | **v-flavum** (Haworth, 1828) (Gracillaria) | UK (England) |

OPOGONA Zeller, 1853
    Lozostoma Stainton, 1859
    Conchyliospila Wallengren, 1861
    Cachura Walker, 1864
    Dendroneura Walsingham, 1892
    Hieroxestis Meyrick, 1892
    Exala Meyrick, 1912

313	**hemidryas** Meyrick, 1915	Guyana
314	**leptynta** Meyrick, 1915	Guyana
315	**lotoxantha** Meyrick, 1915	Guyana
316	**metanastes** Walsingham, 1914	Mexico (Tab)
317	**pelinoma** Meyrick, 1931	Peru
318	**praestans** (Walsingham, 1892) (Dendroneura)	St. Vincent
319	**pyrographa** Walsingham, 1914	Guatemala
320	**rhynchacma** Meyrick, 1920	Brazil (Am)
321	**serta** Walsingham, 1914	Mexico (Gue)
322	**simplex** (Walsingham, 1897) (Dendroneura), **n. comb.**	Dom. Rep.
323	**subcervinella** (Walker, 1863) (Tinea)	Mauritius
324	**xerota** Walsingham, 1914	Panama

OPSODOCA Meyrick, 1919

| 325 | **amentata** Meyrick, 1919 | Guyana |
| 326 | **metrodoxa** Meyrick, 1919 | Guyana |

OTOCHARES Meyrick, 1919

| 327 | **gypsopa** Meyrick, 1919 | Guyana |
| 328 | **peronacma** Meyrick, 1919 | Guyana |

PACHYDYTA Meyrick, 1922
    Pachdyta Meyrick, 1922, missp.

| 329 | **clitozona** Meyrick, 1922 | Brazil (Pa) |

PALAEPHATUS Butler, 1883

| 330 | **falsus** Butler, 1883 | Chile |

PANTHYTARCHA Meyrick, 1922

| 331 | **astrocharis** Meyrick, 1922 | Brazil (Pa) |

PEDALIOTIS Meyrick, 1930

| 332 | **dryographa** Meyrick, 1930 | Brazil (Pa) |

PERILICMETIS Meyrick, 1932

| 333 | **diplaca** Meyrick, 1932 | Brazil (SC) |

PHAEOSES Forbes, 1922

| 334 | **sabinella** Forbes, 1922 | USA (La) |

PHEREOECA Hinton & Bradley, 1956

335	**allutella** (Rebel, 1892) (Tineola)	Canary Is.
336	**dubitatrix** (Meyrick, 1932) (Tinea), **n. comb.**	Virgin Is. (St. Thomas)
	walsinghami (Busck, [1934]) (Tineola), **n. syn.**	Virgin Is. (St. Thomas)
337	**uterella** (Walsingham, 1897) (Tineola)	Brazil (Pa)
	borboropis (Meyrick, 1919), misid. [in part]	Guyana

POLYPSECTA Meyrick, 1930

| 338 | **halmeuta** Meyrick, 1930 | Brazil (Pa) |

POMPOSTOLELLA Fletcher, 1940
    Pompostola Meyrick, 1927, preocc. (Hübner, [1819]
    [Zygaenidae])

| 339 | **aeneoalbida** (Walsingham, 1897) (Ereunetis), **n. comb.** | Virgin Is. (St. Thomas) |
| 340 | **charipepla** (Meyrick, 1927) (Pompostola) | Bermuda |

PRAEACEDES Amsel, 1954
    Titaenoses Hinton & Bradley, 1956

341	**atomosella** (Walker, 1863) (Tinea)	Sierra Leone
	seminolella (Beutenmüller, 1889) (Tinea), **n. syn.**	USA (Fl)
	thecophora (Walsingham, 1908) (Tinea), **n. syn.**	Canary Is.
	despecta (Meyrick, 1919) (Tinea), **n. syn.**	Guyana
	deluccae Amsel, 1954, **n. syn.**	Malta

PROBOLOPTILA Meyrick, 1921

342	**aeolella** (Walsingham, 1897) (Ereunetis), n. comb.	Virgin Is. (St. Thomas)
343	**frontella** (Walsingham, 1897) (Ereunetis)	Virgin Is. (St. Thomas)

PROTODARCIA Forbes, 1931

344	**argyrophaea** Forbes, 1931	Puerto Rico
345	**bicolorella** Forbes, 1931	Puerto Rico
346	**plumella** (Walsingham, 1892) (Tinea)	St. Vincent
347	**tetraonella** (Walsingham, 1897) (Tinea), n. comb.	Virgin Is. (St. Croix/ St. Thomas)
348	**tischeriella** (Walsingham, 1897) (Tinea), n. comb.	Haiti/Virgin Is. (St. Thomas)

PTILOPSALTIS Meyrick, 1935

348.1	**synchorista** Meyrick, 1935	Trinidad

SCARDIA Treitschke, 1830
  Agarica Sodoffsky, 1837
  Morophaga Herrich-Schäffer, 1853
  Atabyria Snellen, 1884
  Osphretica Meyrick, 1910
  Microscardia Amsel, 1952

349	**bimendella** (Zeller, 1863) (Tinea), n. comb.	Venezuela
350	**brasiliensis** (Zagulajev, 1966) (Morophaga), n. comb.	Brazil (RS)
351	**funeratella** (Zeller, 1863) (Tinea), n. comb.	Venezuela
352	**isthmiella** Busck, 1914	Panama
353	**iulina** (Walsingham, 1914) (Phycis)	Guatemala
354	**luctuosa** (Walsingham, 1914) (Phycis)	Costa Rica/Mexico (Ver)
355	**minimella** Busck, 1914	Panama
356	**montium** (Walsingham, 1914) (Phycis)	Panama

SETIARCHA Meyrick, 1932

357	**aleuropis** Meyrick, 1932	Brazil (SC)

SETOMORPHA Zeller, 1852
  Epilegis Dietz, 1905
  Apotomia Dietz, 1905
  Semiota Dietz, 1905
  Trisyntopa Lower, 1918

358	**cycladopa** (Meyrick, 1927) (Syrmologa), n. comb.	Peru
359	**rutella** Zeller, 1852	South Africa
	insectella auct. (not Fabricius, 1794)	
	rupicella Zeller, 1852	Cuba
	operosella Zeller, 1873	USA (Tx)
	inamoenella Zeller, 1873	USA (Tx)
	ruderella Zeller, 1873	USA (Tx)
	multimaculella (Chambers, 1878) (Gelechia)	USA (Tx)
	dryas (Butler, 1881) (Chrestotes)	USA (Hi)
	corticinella Snellen, 1884	Indonesia
	bogotatella Alpheraky, 1889 (not Walker, 1864)	Canary Is.
	discipunctella Rebel, 1894	Canary Is.
	majorella Dietz, 1905	USA (Ca)
	sigmoidella Dietz, 1905	USA (Co)
	cariosella (Dietz, 1905) (Epilegis)	USA (Ca)
	fractiliniella (Dietz, 1905) (Apotomia)	USA (Ca)
	transversestrigella (Dietz, 1905) (Semiota)	USA (Ca)
	margalaestriata Keuchenius, 1917	Indonesia
	euryspoda (Lower, 1918) (Trisyntopa)	Australia
	tineoides Dammerman, 1919	Indonesia
	nitella Voute, 1935, missp.	

SYNCRATERNIS Meyrick, 1922

360	**anthestias** Meyrick, 1922	Brazil (Am)
361	**phaeospila** Meyrick, 1922	Brazil (Am)

SYRMOLOGA Meyrick, 1919

362	**chersopa** Meyrick, 1919	Guyana
363	**leucoclistra** Meyrick, 1919	Colombia
364	**spermatias** Meyrick, 1919	Guyana
365	**thriophora** Meyrick, 1919	Guyana

TAENIODICTYS Forbes, 1933

366	**sericella** Forbes, 1933	Puerto Rico

TETRAPALPUS Davis, 1972

367	**trinidadensis** Davis, 1972	Trinidad

TINEA Linnaeus, 1758 [9]
  Autoses Hübner, [1825]
  Acedes Hübner, [1825]
  Edosa Walker, 1866
  Chrysoryctis Meyrick, 1886

Dystinea Börner, 1925

368	**absolutrix** Meyrick, 1930	Peru
369	**analytica** Meyrick, 1919	Guyana
370	**auromaculata** Walsingham, 1897	Virgin Is. (St. Thomas)
371	**bivirgella** R. Felder & Rogenhofer, 1875	Colombia
	dividuella Zeller, 1877, n. syn.	Colombia
372	**boliviana** Busck, 1911	Bolivia
373	**borboropis** Meyrick, 1919	Guyana
	borboropis Meyrick, 1922, redesc.	Guyana
	barbaropis Busck, [1934], missp.	
374	**caducella** Zeller, 1877	Panama?
375	**caerula** Meyrick, 1927	Peru
376	**catalytica** Meyrick, 1919	Colombia
377	**chloroceros** Meyrick, 1919	Ecuador
378	**conchylitis** Meyrick, 1919	Peru
379	**coracopis** Meyrick, 1919	Peru
380	**cretella** Walsingham, 1897	Haiti
381	**crocodeta** Meyrick, 1927	Peru
382	**culminicola** (Staudinger, 1894) (Tineola)	Bolivia
383	**cumulatella** Zeller, 1877	Colombia
384	**dissimilis** Walsingham, 1914	Guatemala
385	**enchytopa** Meyrick, 1931	Argentina
386	**erasella** Zeller, 1863	Venezuela
387	**eriochrysa** Meyrick, 1932	Brazil (Pa)
388	**extracta** Meyrick, 1919	Guyana
389	**familiaris** Zeller, 1877	Colombia
390	**flectella** Walker, 1863	Venezuela
391	**frontestrigata** Walsingham, 1897	Virgin Is. (St. Croix)
392	**gypsomicta** Meyrick, 1931	Chile
393	**holocapna** Meyrick, 1931	Argentina
394	**isodonta** Meyrick, 1931	Chile
395	**latipennella** Zeller, 1877	Colombia
396	**minutella** (Fabricius, 1794) (Alucita)	"West Indies"
397	**montezuma** Meyrick, 1927	Bolivia
398	**murariella** Staudinger, 1859	Spain
	bipunctella (Ragonot, 1874) (Tineola)	Spain
399	**nigriceps** Zeller, 1877	Colombia
400	**nigripalpis** Walsingham, 1914	Mexico (Gue)
401	**ostiaria** Meyrick, 1927	Virgin Is. (St. Thomas)
	divisa Walsingham, 1897, preocc. (not Wollaston, 1879)	Virgin Is. (St. Thomas)
402	**oxymora** Meyrick, 1919	Peru
403	**pallescentella** Stainton, 1851	UK (England)
	nigrifoldella Gregson, 1856	UK (England)
	galeatella Mabille, 1888 [10]	"Cape Horn" [Argentina?]
	horosema Meyrick, 1931	Argentina
	stimulatrix Meyrick, 1931	Argentina
404	**pallidorsella** Zeller, 1877	Colombia
405	**perisepta** Meyrick, 1922	Ecuador
406	**phaeonephela** Meyrick, 1927	Peru
407	**platysaris** Meyrick, 1931	Argentina
408	**praestabilis** Meyrick, 1927	Peru
409	**prensoria** Meyrick, 1931	Chile
410	**scotocleptes** Meyrick, 1934	Costa Rica
411	**scrutatricella** Zeller, 1877	Colombia
412	**sequens** Meyrick, 1919	Guyana
413	**solenobiella** Walsingham, 1897	Virgin Is. (St. Croix/St. Thomas)
414	**subcuprea** Meyrick, 1932	Brazil (Am)/Peru
415	**symmeles** (Meyrick, 1919) (Lampronia)	Colombia
416	**translucens** Meyrick, 1917	Pakistan
	metonella Pierce & Metcalfe, 1934	UK (England)
	leonhardi Petersen, 1957	Yugoslavia
	margaritacea Gozmány, 1967	Zaire
	fortificata Gozmány, 1968	South Africa
417	**umbraticostella** Walsingham, 1897	St. Vincent/Virgin Is. (St. Croix/St. Thomas)
418	**xanthosomella** (Maassen, 1890) (Gelechia), n. comb.	Bolivia
419	**xenodes** Meyrick, 1909	Bolivia

TIQUADRA Walker, 1863
  Oscella Walker, 1864
  Manchana Walker, 1866
  Ventia Walker, 1866
  Acureuta Zeller, 1877

420	**aeneonivella** (Walker, 1864) (Oscella)	Venezuela
	aenconivella (Walker, 1864) (Oscella), missp.	
421	**avitella** Walker, 1866	Colombia
	ruderella (R. Felder & Rogenhofer, 1875) (Scardia)	Brazil (Am)
	aspera (Zeller, 1877) (Acureuta)	Colombia
422	**butyranthes** Meyrick, 1931	Brazil
423	**circumdata** (Zeller, 1877) (Acureuta)	Colombia
424	**crocidura** Meyrick, 1922	Brazil (SC)
425	**drapetica** Meyrick, 1919	Brazil (Pa)
426	**exercitata** Meyrick, 1922	Brazil (SC)
427	**galactura** Meyrick, 1931	Brazil (SP)
428	**inscitella** Walker, 1863	"Mexico"
429	**lentiginosa** (Zeller, 1877) (Acureuta)	Brazil (RJ)/Peru
430	**mallodeta** Meyrick, 1924	Mexico (Chia)
431	**nivosa** (R. Felder & Rogenhofer, 1875) (Scardia)	Brazil (RJ)
432	**nubilella** Amsel, 1956	Venezuela
433	**nucifraga** Meyrick, 1919	Colombia

434	**pircuniae** (Zeller, 1877) (Acureuta)	Argentina
435	**pontifica** Meyrick, 1919	French Guiana
436	**reversella** (Walker, 1866) (Ventia)	Brazil (Am)
437	**semiglobata** Meyrick, 1922	Peru
438	**stenopa** Walsingham, 1914	Guatemala
439	**syntripta** Meyrick, 1922	Brazil (Pa)
440	**vilis** Meyrick, 1922	Brazil (Am)
	coracophila Meyrick, 1932	Argentina

TRICHOPHAGA Ragonot, 1894

441	**tapetzella** (Linnaeus, 1758) (Tinea)	[Sweden?]
	tapetiella (Zeller, 1852) (Tinea), missp.	
	palaestrica (Butler, 1877) (Tinea)	New Zealand

TRIEROSTOLA Meyrick, 1932

| 442 | **remivola** Meyrick, 1932 | Brazil (Am) |

XYLESTHIA Clemens, 1859
    Xylestia Dyar, [1903], missp.

443	**horridula** (Zeller, 1877) (Ochsenheimeria)	Colombia
444	**menidias** Meyrick, 1922	Brazil (Am/Pa)/Peru
445	**pruniramiella** Clemens, 1859	USA [Pa]
	clemensella Chambers, 1873	USA [Ky]
	congeminatella Zeller, 1873	USA (Ma)
	kearfottella Dietz, 1905	USA (NJ/Pa)

XYSTROLOGA Meyrick, 1919
    Achanodes Meyrick, 1922, **n. syn.**
    Syrrhoaula Meyrick, 1932, **n. syn.**

446	**antipathetica** (Forbes, 1931) (Achanodes), **n. comb.**	Puerto Rico
447	**fulvicolor** Meyrick, 1919	Guyana
448	**grenadella** (Walsingham, 1897) (Setomorpha), **n. comb.**	Grenada
449	**invidiosa** Meyrick, 1919	Colombia
450	**lactirivis** (Meyrick, 1932) (Syrrhoaula), **n. comb.**	Brazil (Am)
451	**nigrovittata** (Walsingham, 1897) (Tinea), **n. comb.**	Virgin Is. (St. Thomas)
452	**sympathetica** (Meyrick, 1922) (Achanodes), **n. comb.**	Brazil (Am)

ZONOCHARES Meyrick, 1922

| 453 | **tetradyas** Meyrick, 1922 | Brazil (Am) |

ZYMOLOGA Meyrick, 1919

| 454 | **mylicopa** Meyrick, 1919 | Colombia |

16. PSYCHIDAE

by D. R. Davis

Penestoglossinae

AMIANTASTIS Meyrick, 1932
    Autocnaptis Meyrick, 1935

1	**manicola** Meyrick, 1932	Argentina
2	**brachycasis** Meyrick, 1932	Argentina
	sciospora (Meyrick, 1935) (Autocnaptis)	Argentina
3	**renovata** (Meyrick, 1922) (Narycia), **n. comb.**	Brazil (Am)

PLUMANA Busck, 1911
    Homilostola Meyrick, 1917

4	**aequanima** (Meyrick, 1917) (Homilostola)	Peru
5	**ascalopa** (Meyrick, 1917) (Homilostola)	French Guiana
6	**autoplecta** (Meyrick, 1917) (Homilostola)	French Guiana
7	**piperatella** Busck, 1911	French Guiana
8	**taeniata** (Meyrick, 1917) (Homilostola)	French Guiana

PERISCEPTIS Meyrick, 1931

| 9 | **horiarcha** Meyrick, 1931 | Paraguay |

PTEROGYNE Davis, 1975

| 10 | **insularis** Davis, 1975 | Dom. Rep. |

Psychinae

EPICHNOPTERIX Hübner, [1825]
    Epichnopteryx Heylaerts, 1875, missp.

| 11 | **pulla** (Esper, 1785) (Bombyx) | [Europe] |
| | fiebrigi (Köhler, 1939) (Cochliotheca) | Paraguay |

PAUCIVENA Davis, 1975

| 12 | **reticulata** Davis, 1975 | Puerto Rico |
| 13 | **hispaniolae** Davis, 1975 | Dom. Rep. |

PROCHALIA Barnes & McDunnough, 1913

| 14 | **licheniphilus** (Köhler, 1939) (Chlania [sic]), **n. comb.** | Cuba |

METAXYPSYCHE Davis, 1975

| 15 | **trinidadensis** Davis, 1975 | Trinidad |

NAEVIPENNA Davis, 1964

16	**aphaidropa** (Dyar, 1914) (Platoeceticus)	Panama
17	**cruttwellae** Davis, 1975	Trinidad
	cruttwelli Davis, 1975, missp.	

CRYPTOTHELEA Duncan, 1841
    Platoeceticus Packard, 1869
    Cryptotheles Costa Lima, 1945, missp.

18	**surinamensis** (Möschler, 1878) (Psyche)	Surinam
19	**watsoni** (Jones, 1923) (Psyche)	Haiti
20	**macleayi** (Guilding, 1827) (Oiketicus)	"West Indies" [St. Vincent]
	macleaii (Westwood, 1854) (Oiketicus), missp.	
	macleayii (Meyrick & Lower, 1907) (Eurycyttara [sic]), missp.	
21	**hoffmanni** (Köhler, 1939) (Platoeceticus)	Cuba
22	**congregata** (Jones, 1945) (Platoeceticus)	Mexico (Mor)
23	**symmicta** (Dyar, 1914) (Platoeceticus)	Panama
24	**gloverii** (Packard, 1869) (Platoeceticus)	USA (Fl)
	jonesi (Barnes & Benjamin, 1922) (Maratha)	USA (Tx)
	pizote (Schaus, 1927) (Chalia)	Guatemala

DENDROPSYCHE Jones, 1926

| 25 | **venezuelae** Davis, 1975 | Venezuela |
| 26 | **burrowsi** Jones, 1926 | Guyana |

LUMACRA Davis, 1964

27	**brasiliensis** (Heylaerts, 1884) (Eumeta)	Brazil
	marona (Schaus, 1905) (Platoeceticus)	French Guiana
	costaricensis (Schaus, 1911) (Platoeceticus)	Costa Rica
28	**kuenckelii** (Heylaerts, 1901) (Chalia)	Argentina
	klinckelii (Heylaerts, 1901) (Chalia), incorr. spell.	
	küenkeli (Köhler, 1924) (Chalia), missp.	
	künckeli (Köhler, 1939) (Chalia), missp.	
29	**haitiensis** Davis, 1964	Haiti
30	**quadridentata** Davis, 1964	French Guiana
31	**hyalinacra** Davis, 1964	El Salvador
32	**leucobasilaris** Davis, 1975	Venezuela

CURTORAMA Davis, 1964

33	**cassiae** (Weyenbergh, 1884) (Platoeceticus)	Argentina
	rebeli (Köhler, 1924) (Chalia)	Argentina
	rugosus (Köhler, 1931) (Platoeceticus)	Argentina

ASTALA Davis, 1964

34	**confederata** (Grote & C. Robinson, 1868) (Psyche)	USA (Tx)
	lepidopteris (Dyar, 1926) (Pachythelia)	Mexico (Col)
35	**hoffmanni** (Vázquez, 1941) (Eurukuttarus)	Mexico (Pue)
36	**polingi** (Barnes & Benjamin, 1924) (Eurukuttarus)	USA (Az)
37	**tristis** (Schaus, 1901) (Chalia)	Mexico (Ver)
38	**zacualpania** (Dyar, 1916) (Chalia)	Mexico (Mor?)
39	**vigasi** (Schaus, 1901) (Chalia)	Mexico (Ver)

Oiketicinae

THANATOPSYCHE Butler, 1882

| 40 | **chilensis** (Philippi, 1860) (Psyche) | Chile |
| 41 | **canescens** Butler, 1882 | Chile |

ANIMULA Herrich-Schäffer, 1856
    Subgenus **Animula** Herrich-Schäffer, 1856

42	**limpia** Dognin, 1894	Ecuador
43	**microptera** (Schaus, 1905) (Thyridopteryx)	French Guiana
44	**dichroa** Herrich-Schäffer, 1856	Venezuela

Subgenus **Artipenna** Davis, 1964

| 45 | **seitzi** (Gaede, 1936) (Thyridopteryx) | Brazil (RJ) |

BIOPSYCHE Dyar, 1905

| 46 | **thoracica** (Grote, 1865) (Hymenopsyche) | Cuba |
| 47 | **apicalis** (Hampson, 1904) (Thanatopsyche) | Bahamas |

OIKETICUS Guilding, 1827
    Oeketicus Lefebre, 1842, missp.
    Oeceticus Herrich-Schäffer, 1858, missp.
    Oiketikus Seitz, 1919, missp.
    Oicocestis Köhler, 1924, missp.
    Oiceticus Lahille, 1926, missp.

Subgenus **Paraoiketicus** Davis, 1964

48	**bergii** (Weyenbergh, 1884) (Psyche)	Argentina
	oviformis Köhler, 1939	Argentina
49	**borsanii** Köhler, 1953	Argentina
50	**geyeri** Berg, 1877	"Patagonia" [Argentina]
	tabacillus Weyenbergh, 1884	Argentina
	jonesi Schaus, 1896	Brazil (SP)
	thoracica (Schaus, 1905) (Thanatopsyche)	Guyana
51	**zihuatanejensis** Vázquez, 1951	Mexico (Gue)

Subgenus **Oiketicus** Guilding, 1827

52	**toumeyi** Jones, 1922	USA (Az)
	mortonjonesi Vázquez, 1949	Mexico (Mex)
53	**assimilis** Vázquez, 1942	Mexico (Mic?/Mor?)
54	**specter** Schaus, 1905	Venezuela
55	**townsendi** Townsend, 1894	
	* a) **townsendi** Townsend, 1894	USA (NM)
	bonniwelli Barnes & Benjamin, 1924	USA (Tx)
	* b) **dendrokomos** Jones, 1926	USA (Tx)
	c) **mexicanus** Gaede, 1936	Mexico
	multidentatus Vázquez, 1942	Mexico (Ver)
*56	**abbotii** Grote, 1880	USA (Tx)
	abboti Holland, 1905, missp.	
57	**kirbyi** Guilding, 1827	"West Indies" [St. Vincent]
	kirbii Walker, 1855, missp.	
	fulgurator Herrich-Schäffer, 1856	Brazil
	fulgerator Grote, 1865 (Oeceticus [sic]), missp.	
	poeyi Lucas, 1857	Cuba
	gigantea (Zeller, 1871) (Psyche)	Brazil (Pe)
	orizavae Schaus, 1901	Mexico (Ver)
	sinaloanus Vázquez, 1942	Mexico (Sin)
	ochoterenai Vázquez, 1942	Mexico (Sin)
	fasciculatus Vázquez, 1942	Mexico (Ver)
58	**platensis** Berg, 1883	Argentina
	kerbyi Lahille, 1926 (Oiceticus [sic]), missp.	
	(not Guilding, 1827)	

THYRIDOPTERYX Stephens, 1834
Hymenopsyche Grote, 1865

*59	**meadi** H. Edwards, 1881	USA (Ca)
	meadii Kirby, 1892, missp.	
60	**ephemeraeformis** (Haworth, 1803) (Sphinx)	UK [England]
	coniferarum (Packard, 1864) (Oeceticus [sic])	USA
	pallidovenata Grossbeck, 1917	USA (Fl)
	vernalis Jones, 1923	USA (De)

Unplaced Species

61	"Psychoglene" **basinigra** R. Felder, 1874	Brazil
62	"Psyche" **burmeisteri** Weyenbergh, 1884	Argentina
63	"Platoeceticus" **chaquensis** Köhler, 1939	Argentina
64	"Chalia" **daguerrei** Köhler, 1939	Argentina
65	"Chalia" **dispar** Köhler, 1939	Argentina
66	"Oiketicus" **elegans** Köhler, 1931	Argentina
67	"Oiketicus" **ginocchionus** Köhler, 1952	Argentina
68	"Zamopsyche" **haywardi** Köhler, 1939	Argentina
69	"Oiketicus" **horni** Köhler, 1938	Argentina
70	"Oiketicus" **lizeri** Köhler, 1939	Argentina
71	"Platoeceticus" **tandilensis** Köhler, 1931	Argentina
72	"Oiketicus" **westwoodii** Berg, 1882	Argentina
73	"Clania" **yamorkinei** Köhler, 1953	Argentina

17. ARRHENOPHANIDAE

by D. R. Davis

ARRHENOPHANES Walsingham, 1913

1	**perspicilla** (Stoll, 1790) (Bombyx)	Surinam
	inca Meyrick, 1913	Peru
2	**volcanica** Walsingham, 1913	Panama

CNISSOSTAGES Zeller, 1863

3	**mastictor** Bradley, 1951	Costa Rica
4	**oleagina** Zeller, 1863	Venezuela

ECPATHOPHANES Bradley, 1951

5	**anachoreta** Bradley, 1951	Colombia
6	**chiquita** (Busck, 1914) (Arrhenophanes)	Panama

HARMACLONA Busck, 1914 [11]
Ptychoxena Meyrick, 1916

7	**cossidella** Busck, 1914	Panama

18. AMPHITHERIDAE

by J. B. Heppner

DASYCAREA Zeller, 1877 [12]

1	**viridisquamata** Zeller, 1877	Colombia

19. LYONETIIDAE

by D. R. Davis & S. E. Miller

Bedelliinae

EUPRORA Busck, 1906

1	**argentiliniella** Busck, 1906	USA (Tx)

PHILONOME Chambers, 1874
Phillonome Chambers, 1875, missp.
Eurynome Chambers, 1875, preocc. (Leach, 1814
   [Crustacea])
Busckia Dyar, [1903]

2	**cuprescens** Walsingham, 1914	Mexico (Gue)
3	**spectata** Meyrick, 1920	Brazil (Am)

BEDELLIA Stainton, 1849

4	**somnulentella** (Zeller, 1847) (Lyonetia)	Italy

Lyonetiinae

LYONETIA Hübner, [1825]
Argyromiges Curtis, 1829
Argyromis Stephens, 1829
Lyoneta Matsumura, 1931, missp.

5	**myura** Meyrick, 1931	Peru

Cemiostominae

LEUCOPTERA Hübner, [1825]
Cemiostoma Zeller, 1848

6	**entemopa** Walsingham, 1914	Mexico (Gue)
7	**salicis** Walsingham, 1914	Mexico (DF)

PERILEUCOPTERA Silvestri, 1943

8	**coffeella** (Guérin-Méneville, 1842) (Elachista)	Guadeloupe?
	noctuella (Madinier, 1870), unavail. (Art.11 [3])	Martinique

COMPSOSCHEMA Walsingham, 1897

9	**bimarginellum** Walsingham, 1897	Virgin Is. (St. Thomas)

HAPALOTHYMA Meyrick, 1920

10	**ioplocama** Meyrick, 1920	Guyana
11	**xanthochorda** Meyrick, 1920	Guyana

THOMICTIS Meyrick, 1920

12	**ephorista** Meyrick, 1920	Guyana

Bucculatriginae

BUCCULATRIX Zeller, 1848
Ceroclastis Zeller, 1848
Bacculatrix Flint, Noble, & Shaw, 1978, missp.

13	**flexuosa** Walsingham, 1897	Virgin Is. (St. Thomas)
14	**gossypiella** Morrill, 1927	Mexico (Son)
15	**rhombophora** Meyrick, 1926	Bermuda
16	**stictopus** Walsingham, 1914	Mexico (Tab)
17	**subnitens** Walsingham, 1914	Mexico (Tab)
18	**thurberiella** Busck, 1914	USA (Az)
19	**unipuncta** Walsingham, 1897	Virgin Is. (St. Thomas)

20. GRACILLARIIDAE

by D. R. Davis & S. E. Miller

Gracillariinae

CALOPTILIA Hübner, [1825]
Poeciloptilia Hübner, [1825]
Ornix Treitschke, 1833
Coriscium Zeller, 1839
Antiolopha Meyrick, 1894

1	**aeneocapitella** (Walsingham, 1891) (Gracilaria [sic]), n. comb.	St. Vincent
2	**aeolastis** (Meyrick, 1920) (Gracilaria [sic]), n. comb.	Brazil (Am)
3	**burserella** (Busck, 1900) (Gracilaria [sic])	USA (Fl)
4	**callichora** (Meyrick, 1915) (Gracilaria [sic]), n. comb.	Guyana
5	**camaronae** (Zeller, 1877) (Gracilaria [sic]), n. comb.	Colombia

6   chloroptilia (Meyrick, 1915) (Gracilaria [sic]), **n. comb.**   Guyana
7   eolampis (Meyrick, 1915) (Gracilaria [sic]), **n. comb.**   Guyana
8   hexameris (Meyrick, 1921) (Gracilaria [sic]), **n. comb.**   Brazil (Am)
9   immuricata (Meyrick, 1915) (Gracilaria [sic]), **n. comb.**   Peru
10   oriarcha (Meyrick, 1915) (Gracilaria [sic]), **n. comb.**   Peru
11   pastranai (Bourquin, 1962) (Gracilaria [sic]), **n. comb.**   Argentina
12   perseae (Busck, 1920) (Gracilaria [sic])   USA (Fl)
13   phiaropis (Meyrick, 1921) (Gracilaria [sic]), **n. comb.**   Peru
14   pneumatica (Meyrick, 1920) (Gracilaria [sic]), **n. comb.**   Brazil (Pa)
15   semiclausa (Meyrick, 1921) (Gracilaria [sic]), **n. comb.**   Brazil (Am)
16   similatella (Zeller, 1877) (Gracilaria [sic]), **n. comb.**   Colombia
17   viridula (Zeller, 1877) (Gracilaria [sic]), **n. comb.**   Colombia

EUCOSMOPHORA Walsingham, 1897

18   cupreella Walsingham, 1897   Jamaica
19   dives Walsingham, 1897   Grenada
20   ornata Walsingham, 1897   Grenada

NEUROSTROTA Ely, 1918
    Neurostrata Ely, 1918, missp.

21   gunniella (Busck, 1906) (Acrocercops)   USA (Tx)
22   pithecolobiella Busck, [1934] (Neurostrata [sic])   Cuba

PARECTOPA Clemens, 1860
    Euspilopteryx Zeller, 1847, emend. (not Stephens, 1835)
    Euspilapteryx Spuler, 1910, missp. (not Stephens, 1835)

23   dactylota Meyrick, 1915   Ecuador
24   exorycha Meyrick, 1928   Brazil (RS)
25   heptametra Meyrick, 1915   Colombia
26   lithocolletina (Zeller, 1877) (Gracilaria [sic])   Colombia
27   lithomacha Meyrick, 1915   Ecuador
28   nesitis (Walsingham, 1897) (Gracilaria [sic])   Virgin Is. (St. Thomas)
29   pselaphotis Meyrick, 1915   Ecuador
30   pulverella (Walsingham, 1897) (Gracilaria   Dom. Rep./Virgin Is.
        [sic])
31   quadristrigella (Zeller, 1877) (Gracilaria [sic])   Colombia
32   refulgens Meyrick, 1915   Ecuador
33   rotigera Meyrick, 1931   Chile
34   trichophysa Meyrick, 1915   Peru
35   undosa (Walsingham, 1897) (Gracilaria [sic])   Haiti/Virgin Is.
36   viminea Meyrick, 1915   Peru

NEUROBATHRA Ely, 1918

37   curcassi Busck, [1934]   Cuba

PENICA Walsingham, 1914

38   peritheta Walsingham, 1914   Mexico (Gue)

PARORNIX Spuler, 1910

39   errantella (Walsingham, 1897) (Ornix)   Virgin Is. (St. Thomas)
40   micrura Walsingham, 1914   Mexico (Gue)

CHILOCAMPYLA Busck, 1900

41   psidiella Busck, [1934]   Cuba

DIALECTICA Walsingham, 1897
    Eutrichocnemis Spuler, 1910

42   permixtella Walsingham, 1897   Dom. Rep./Grenada
43   rendalli Walsingham, 1897   Jamaica
44   sanctaecrucis Walsingham, 1897   Virgin Is.

ACROCERCOPS Wallengren, 1881

45   achnodes Meyrick, 1915   Ecuador
46   albomarginata (Walsingham, 1897) (Coris-   Virgin Is. (St. Thomas)
        cium)
47   anthogramma Meyrick, 1921   Brazil (Am)
48   apicepunctella (Walsingham, 1891) (Gracilaria [sic])   St. Vincent
49   argocosma Meyrick, 1915   Ecuador
50   asaphogramma Meyrick, 1920   Brazil (Pa)
51   atalantis Meyrick, 1924   Costa Rica
52   attenuata (Walsingham, 1897) (Coriscium)   Virgin Is. (St. Thomas)
53   breyeri Bourquin, 1962   Argentina
54   caementosa Meyrick, 1915   Peru
55   camptochrysa Meyrick, 1921   Brazil (Am)
56   chalinopa Meyrick, 1920   Brazil (Pa)
57   charitopis Meyrick, 1915   Guyana
58   chloronympha Meyrick, 1921   Brazil (Am)
59   chrysocosma Meyrick, 1915   Guyana
60   cirrhantha Meyrick, 1915   Guyana
61   cissiella Busck, [1934]   Cuba
62   clitoriella Busck, [1934]   Cuba
63   clytosema Meyrick, 1920   Brazil (Am)
64   contorta Meyrick, 1920   Brazil (Pa)
65   cordiella Busck, [1934]   Cuba
66   crotalistis Meyrick, 1915   Peru

67   cyclogramma Meyrick, 1921   Peru
68   cymella Forbes, 1931   Puerto Rico
69   demotes Walsingham, 1914   Mexico (Tab)
70   desmochares Meyrick, 1921   Brazil (Am)
71   eclampsis Durrant, 1914   Panama
72   encentris Meyrick, 1915   Guyana
73   eurychalca Meyrick, 1920   Brazil (Pa)
74   fasciculata Meyrick, 1915   Guyana
75   gemmans Walsingham, 1914   Mexico (Gue)
76   hapsidota Meyrick, 1915   Guyana
77   hastigera Meyrick, 1915   Ecuador
78   helicomitra Meyrick, 1924   Brazil (Ba)
79   hippuris Meyrick, 1915   Peru
80   inconspicua Forbes, 1930   Puerto Rico
81   insulella (Walsingham, 1892) (Zarathra)   St. Vincent
82   ipomoeae Busck, [1934]   Cuba
83   leucographa Clarke, 1953   Argentina
84   leuconota (Zeller, 1877) (Gracilaria [sic])   Colombia
85   luctuosa Meyrick, 1915   Guyana
86   maranthaceae Busck, [1934]   Cuba
87   marmaritis Walsingham, 1914   Mexico (Gue)
88   melanactis Meyrick, 1915   Guyana
89   melanocosma Meyrick, 1920   Brazil (Pa)
90   melantherella Busck, [1934]   Cuba
91   microphis Meyrick, 1921   Brazil (Am)
92   nolickeniella (Zeller, 1877) (Gracilaria [sic])   Colombia
93   obversa Meyrick, 1915   Guyana
94   perturbata Meyrick, 1921   Peru/Brazil (Am)
95   piligera Meyrick, 1915   Colombia
96   pontifica Forbes, 1931   Puerto Rico
97   pylonias Meyrick, 1921   Peru
98   ramigera Meyrick, 1920   Brazil (Pa)
99   rhynchograpta Meyrick, 1920   Brazil (Pa)
100   serrigera Meyrick, 1915   Ecuador/Peru
101   sortis Meyrick, 1915   Ecuador
102   stalagmitis Meyrick, 1915   Guyana
103   taeniarcha Meyrick, 1932   Brazil (Am)
104   trimetala Meyrick, 1915   Guyana
105   undifraga Meyrick, 1931   Haiti
106   urbanella (Zeller, 1877) (Gracilaria [sic])   Colombia
107   xeniella (Zeller, 1877) (Gracilaria [sic])   Colombia
108   xystrota Meyrick, 1915   Guyana
109   zebrulella Forbes, 1931   Puerto Rico

CUPHODES Meyrick, 1897
    Phrixosceles Meyrick, 1908

110   paragrapta (Meyrick, 1915) (Phrixosceles), **n. comb.**   Guyana

SPANIOPTILA Walsingham, 1897

111   cordicaria Meyrick, 1920   Brazil (Pa)
112   eucnemis Walsingham, 1914   Mexico (Gue)
113   nemeseta Meyrick, 1920   Brazil (Am)
114   spinosum Walsingham, 1897   Virgin Is. (St. Thomas)

LEUCANTHIZA Clemens, 1859

115   forbesi Bourquin, 1962   Argentina

MARMARA Clemens, 1863
    Aesyle Chambers, 1875

116   affirmata (Meyrick, 1918) (Parectopa)   Peru
117   ischnotoma (Meyrick, 1915) (Parectopa)   Guyana
118   isortha (Meyrick, 1915) (Parectopa)   Guyana
119   phaneropis (Meyrick, 1915) (Parectopa)   Ecuador
120   stemonodes (Meyrick, 1915) (Parectopa)   Ecuador

Lithocolletinae

CREMASTOBOMBYCIA Braun, 1908

121   lantanella Busck, 1910 [14]   USA (Hi)

PHYLLONORYCTER Hübner, [1822]
    Lithocolletis Hübner, [1825]
    Eucestis Hübner, [1825]
    Lithocolletes Matsumura, 1907, missp.
    Phyllorycter Walsingham, 1914, missp.

122   albimacula (Walsingham, 1897) (Litho-   Virgin Is. (St. Thomas)
        colletis), **n. comb.**
123   antitoxa (Meyrick, 1915) (Lithocolletis), **n. comb.**   Peru
124   argentifrontella (Walsingham, 1897) (Lith-   Virgin Is. (St. Thomas)
        ocolletis)
125   chalcobaphes Walsingham, 1914   Mexico (Gue)
126   clerotoma (Meyrick, 1915) (Lithocolletis), **n. comb.**   Ecuador
126.1   durangensis Deschka, 1982   Mexico (Dur)
127   epispila (Meyrick, 1915) (Lithocolletis), **n. comb.**   Ecuador
128   iriphanes (Meyrick, 1915) (Lithocolletis), **n. comb.**   Peru
129   oxygrapta (Meyrick, 1915) (Lithocolletis), **n. comb.**   Peru
130   pictus Walsingham, 1914   Mexico (Gue)
131   solani (E. M. Hering, 1958) (Lithocolletis), **n. comb.**   Argentina
132   stigmaphyllae Busck, [1934]   Cuba

133	**tenuicaudella** (Walsingham, 1897) (Litho-colletis), **n. comb.**	Virgin Is. (St. Croix)

CAMERARIA Chapman, 1902

*134	**agrifoliella** (Braun, 1908) (Lithocolletis)	USA (Ca)

PORPHYROSELA Braun, 1908

135	**desmodiella** (Clemens, 1859) (Lithocolletis)	USA (Pa)
	gregariella (Murtfeldt, 1881) (Lithocolletis)	USA (Mo)
136	**minuta** Clarke, 1953	Argentina

Phyllocnistinae

PHYLLOCNISTIS Zeller, 1848
 Phylloenistis Chambers, 1875, missp.
 Phylloetis Chambers, 1876, missp.
 Phyllocnitis Busck, 1900, missp.

137	**abatiae** E. M. Hering, 1958	Argentina
138	**auriinea** Zeller, 1877	Colombia
139	**baccharidis** E. M. Hering, 1958	Argentina
140	**bourquini** Pastrana, 1960	Argentina
141	**dorcas** Meyrick, 1915	Guyana
142	**meliacella** Becker, 1974	Costa Rica
143	**rotans** Meyrick, 1915	Ecuador
[144	**rubella** (Blanchard, 1852) (Elachista) [15]	Chile]
145	**sciophanta** Meyrick, 1915	Peru
146	**sexangula** Meyrick, 1915	Peru
147	**wygodzinskyi** E. M. Hering, 1958	Argentina

Gelechioidea

21. OECOPHORIDAE

by V. O. Becker [16]

Depressariinae

Tribe Depressariini

EXAERETIA Stainton, 1849[17]
 Depressariodes Turati, 1924
 Martyrhilda Clarke, 1941

1	**ammitis** (Meyrick, 1931) (Depressaria)	Argentina
2	**ascetica** (Meyrick, 1926) (Depressaria)	Colombia
3	**baleni** (Zeller, 1877) (Depressaria)	Colombia
4	**lusciosa** (Meyrick, 1915) (Depressaria)	Peru
5	**mesosceptra** (Meyrick, 1915) (Depressaria)	Peru
6	**relegata** (Meyrick, 1922) (Depressaria)	Chile (Juan Fernandez Is.)
7	**rubristricta** (Walsingham, 1912) (Agonopteryx [sic])	Guatemala
8	**significa** (Meyrick, 1915) (Depressaria)	Ecuador

TALITHA Clarke, 1978[18]

9	**anomala** Clarke, 1978	Chile

Tribe Amphisbatini

AFDERA Clarke, 1978

10	**orphnaea** (Meyrick, 1931) (Cryptolechia)	Chile

ANCIPITA Busck, 1914

11	**atteria** Busck, 1914	Panama

ATHRINACIA Walsingham, 1911

12	**cosmophragma** Meyrick, 1922	Brazil (Pa)
13	**leucographa** Walsingham, 1911	Mexico (Gue)
14	**psephophragma** Meyrick, 1929	Brazil (Am)
15	**trifasciata** Walsingham, 1911	Mexico (Gue)
16	**xanthographa** Walsingham, 1911	Mexico (Gue)

AUXOTRICHA Meyrick, 1931

17	**ochrogypsa** Meyrick, 1931	Peru

CHARIPHYLLA Meyrick, 1921

18	**closterias** Meyrick, 1921	Peru

COMOTECHNA Meyrick, 1920

19	**corculata** Meyrick, 1921	Brazil (Pa)
20	**dentifera** Meyrick, 1921	Peru
21	**ludicra** Meyrick, 1920	Guyana
22	**parmifera** Meyrick, 1921	Brazil (Pa)
23	**scutulata** Meyrick, 1921	Brazil (Am)
24	**semiberbis** Meyrick, 1921	Peru

COMPSISTIS Meyrick, 1888

25	**caerulipalpis** Meyrick, 1921	Peru
26	**homochorda** Meyrick, 1921	Peru
27	**labyrinthias** Meyrick, 1921	Peru
28	**macrochorda** Meyrick, 1921	Peru
29	**malacoscia** Meyrick, 1921	Peru

COPTOTELIA Zeller, 1863
 Hyphypena Warren, 1889

30	**allardi** Clarke, 1951	Peru
31	**bipunctalis** (Warren, 1889) (Hyphypena)	Brazil (Am)
	byrsocyma (Meyrick, 1921) (Hypercallia)	Brazil (Pa)
32	**calidaria** (Meyrick, 1921) (Hypercallia)	Brazil (Am)
33	**chaldaica** Meyrick, 1913	Argentina
34	**colpodes** (Walsingham, 1912) (Hyphypena)	Panama
35	**complicata** Clarke, 1951	Brazil (SC)
36	**cyathopa** Meyrick, 1913	Colombia
37	**cyathopoides** Clarke, 1951	Ecuador
38	**elena** Clarke, 1951	Brazil (Am)
39	**fenestrella** Zeller, 1863	Venezuela
40	**gioia** Clarke, 1951	Bolivia
41	**margaritacea** (Meyrick, 1924) (Hypercallia)	Bolivia
42	**nigriplaga** Dognin, 1904	Colombia
	prominula Meyrick, 1913	Colombia
43	**pecten** Clarke, 1951	Guatemala
44	**perseaphaga** Clarke, 1951	Costa Rica
45	**terminalis** Clarke, 1951	Mexico

COSTOMA Busck, 1914
 Phalarotarsa Meyrick, 1924

46	**basirosella** Busck, 1914	Panama
47	**cirrhophaea** (Meyrick, 1924) (Phalarotarsa)	Bolivia
48	**flavicosta** (R. Felder & Rogenhofer, 1875) (Crypto-lechia)	Brazil (Am)

CRYPTOLECHIA Zeller, 1852

49	**asemanta** Dognin, 1905	Ecuador
50	**citrodeta** Meyrick, 1921	Brazil (Pa)
51	**diplosticha** Meyrick, 1926	Colombia
52	**epidesma** Walsingham, 1912	Mexico (Tab)
53	**vallifera** Meyrick, 1914	Guyana
53	**eucharistis** Meyrick, 1931	Argentina
54	**glischrodes** Meyrick, 1931	Argentina
55	**holopyrrha** Meyrick, 1912	Colombia
56	**hydara** Walsingham, 1912	Guatemala
57	**ichnitis** Meyrick, 1918	French Guiana
58	**microglyptis** Meyrick, 1936	Colombia
59	**pateropa** Meyrick, 1931	Brazil (SP)
60	**pentathlopa** Meyrick, 1933	Brazil (SC)
61	**percnocoma** Meyrick, 1930	Brazil (RJ)
62	**praevecta** Meyrick, 1929	Colombia
63	**remotella** (Staudinger, 1899) (Depressaria)	Argentina
64	**sciodeta** Meyrick, 1930	Brazil (RJ)
65	**semibrunnea** Dognin, 1905	Colombia
66	**taphrocopa** Meyrick, 1926	Colombia
67	**transfossa** Meyrick, 1926	Peru
68	**veniflua** Meyrick, 1914	Colombia

DOINA Clarke, 1978

69	**annulata** Clarke, 1978	Chile
70	**asperula** Clarke, 1978	Chile
71	**edmondsii** (Butler, 1883) (Depressaria)	Chile
72	**eremnogramma** Clarke, 1978	Chile
73	**flinti** Clarke, 1978	Chile
74	**glebula** Clarke, 1978	Chile
75	**inconspicua** Clarke, 1978	Chile
76	**increta** (Butler, 1883) (Orthotelia)	Chile
77	**lagneia** Clarke, 1978	Chile
78	**paralagneia** Clarke, 1978	Chile
79	**phaobregna** Clarke, 1978	Chile
80	**scariphista** (Meyrick, 1931) (Cryptolechia)	Chile
81	**subicula** Clarke, 1978	Chile
82	**trachycantha** Clarke, 1978	Chile
83	**truncata** Clarke, 1978	Chile

DOSHIA Clarke, 1978

84	**miltopeza** Clarke, 1978	Chile

ECTAGA Walsingham, 1912

85	**canescens** Walsingham, 1912	Mexico (Gue)
86	**lenta** Clarke, 1956	Argentina
87	**lictor** Walsingham, 1912	Guatemala
88	**promeces** Walsingham, 1912	Guatemala

ERITHYMA Meyrick, 1914

89	**cyanoplecta** Meyrick, 1914	Guyana

90	**trabeella** (R. Felder & Rogenhofer, 1875) (Oecophora) Brazil (Am)	
	polychroma Meyrick, 1914	Guyana
	trabeata Meyrick, 1922, missp.	

EUPRAGIA Walsingham, 1911

91	**oxinopa** Meyrick, 1929	Colombia
92	**solida** Walsingham, 1911	Mexico (Tab)

FILINOTA Busck, 1911
    Lupercalia Busck, 1912
    Mnesichara Walsingham, 1912 [45]

93	**brunniceps** (R. Felder & Rogenhofer, 1875) (Lecitho-cera)	Colombia
94	**cassiteranthes** Meyrick, 1932	Bolivia
95	**dyctiota** (Walsingham, 1912) (Mnesichara) [45]	Guatemala
96	**gratiosa** (R. Felder & Rogenhofer, 1875) (Lecithocera)	Colombia
	peruviella Busck, 1911	Peru
97	**hermosella** Busck, 1911	French Guiana
98	**ignita** (Busck, 1912) (Lupercalia)	Panama
99	**ithymetra** Meyrick, 1926 [45]	Colombia
100	**lamprocosma** Meyrick, 1916	French Guiana
101	**regifica** Meyrick, 1921	Brazil (Am)
102	**rhodograpta** Meyrick, 1915	Guyana
103	**sphenoplecta** Meyrick, 1921	Brazil (Am)
104	**vociferans** Meyrick, 1930	Brazil (RJ)

GNATHOTONA Meyrick, 1931

105	**thermopsamma** Meyrick, 1931	Paraguay

GONADA Busck, 1911

106	**cabima** Busck, 1912	Panama
107	**falculinella** Busck, 1911	French Guiana
108	**flavidorsis** Meyrick, 1930	Brazil (RJ)
109	**phosphorodes** Meyrick, 1922	French Guiana
110	**pyronota** Meyrick, 1924	Peru
111	**rubens** Meyrick, 1916	French Guiana

GONIONOTA Zeller, 1877

112	**acrocosma** (Meyrick, 1912) (Coptotelia)	Colombia
113	**aethographa** Clarke, 1971	Costa Rica
114	**aethoptera** Clarke, 1971	Venezuela
115	**amauroptera** Clarke, 1971	Argentina
116	**amphicrena** (Meyrick, 1912) (Coptotelia)	Colombia
117	**anelicta** (Meyrick, 1926) (Hypercallia)	Bolivia
118	**anisodes** (Meyrick, 1916) (Hypercallia)	French Guiana
119	**argopleura** Clarke, 1971	Brazil (SC)
120	**autocrena** (Meyrick, 1930) (Hypercallia)	Brazil (RJ)
121	**borquiniella** (Köhler, 1940) (Hypercallia)	Argentina
122	**bourquini** Clarke, 1964	Brazil (SC)
123	**captans** (Meyrick, 1931) (Hypercallia)	Brazil (SP)
124	**citronota** (Meyrick, 1932) (Hypercallia)	Brazil (SC)
125	**cologramma** Clarke, 1971	Venezuela
126	**comastis** Meyrick, 1909	Peru
127	**confinella** (R. Felder & Rogenhofer, 1875) (Hypercallia)	Colombia
128	**constellata** (Meyrick, 1912) (Coptotelia)	Colombia
129	**contrasta** Clarke, 1964	Peru
130	**cristata** Walsingham, 1912	Panama
131	**cyanaspis** (Meyrick, 1909) (Doleromima)	Peru
132	**determinata** Clarke, 1964	Guyana
133	**dissita** Clarke, 1964	Trinidad
134	**dryodesma** (Meyrick, 1916) (Hypercallia)	French Guiana
	dryocrypta (Meyrick, 1931) (Hypercallia)	Guatemala
135	**eremia** Clarke, 1971	French Guiana
136	**erotopis** (Meyrick, 1926) (Hypercallia)	Bolivia
137	**erythroleuca** (Meyrick, 1928) (Hypercallia)	Peru
138	**eurydryas** (Meyrick, 1926) (Hypercallia)	Colombia
139	**euthyrsa** (Meyrick, 1930) (Hypercallia)	Ecuador
140	**excavata** Clarke, 1964	Mexico (Ver)
141	**extima** Clarke, 1964	Costa Rica
142	**festicola** (Meyrick, 1924) (Hypercallia)	Peru
143	**fimbriata** Clarke, 1964	Panama
144	**gaiophanes** Clarke, 1971	Brazil (SC)
145	**habristis** (Meyrick, 1914) (Hypercallia)	Guyana
146	**hydrogramma** (Meyrick, 1912) (Coptotelia)	Colombia
147	**hypoleuca** Clarke, 1971	Venezuela
148	**hyptiotes** Clarke, 1964	Mexico (Ver)
149	**incalescens** (Meyrick, 1914) (Hypercallia)	Colombia
150	**incisa** Meyrick, 1909	Bolivia
151	**incontigua** Clarke, 1964	Venezuela
152	**insignata** Clarke, 1971	Ecuador
153	**insulana** Clarke, 1968	Dominica
154	**intonans** (Meyrick, 1933) (Hypercallia)	Argentina
155	**ioleuca** (Meyrick, 1912) (Coptotelia)	Argentina
156	**isastra** (Meyrick, 1926) (Hypercallia)	Colombia
157	**isodryas** (Meyrick, 1921) (Hypercallia)	Brazil (Pa)
158	**isophylla** Meyrick, 1909	Peru
159	**lecithitis** (Meyrick, 1912) (Coptotelia)	Argentina
160	**leucoporpa** (Meyrick, 1926) (Hypercallia)	Colombia
161	**lichenista** (Meyrick, 1926) (Hypercallia)	Colombia

162	**luteola** (R. Felder & Rogenhofer, 1875) (Carcina)	Brazil (Am)
163	**melobaphes** Walsingham, 1912	Panama
164	**menura** Clarke, 1971	Panama
165	**militaris** (Meyrick, 1914) (Hypercallia)	Colombia
166	**mimulina** (Butler, 1883) (Agriocoma)	Chile
	araucana (Bartlett-Calvert, 1893) (Agriocoma)	Chile
167	**mitis** (Meyrick, 1914) (Hypercallia)	Peru
168	**notodontella** Zeller, 1877	Colombia
169	**oligarcha** (Meyrick, 1913) (Coptotelia)	Peru
170	**oriphanta** (Meyrick, 1928) (Hypercallia)	Colombia
171	**oxybela** Clarke, 1971	Peru
172	**paravexillata** Clarke, 1971	Venezuela
173	**periphereia** Clarke, 1964	Ecuador
174	**persistis** (Meyrick, 1914) (Hypercallia)	Peru
175	**phocodes** Meyrick, 1909	Peru
176	**phthiochroma** Clarke, 1971	Bolivia
177	**pialea** (Meyrick, 1921) (Hypercallia)	Brazil (Pa)
178	**poecilia** Clarke, 1971	Venezuela
179	**praeclivis** (Meyrick, 1921) (Hypercallia)	Peru
180	**prolectans** (Meyrick, 1926) (Hypercallia)	Colombia
181	**pyrocausta** (Meyrick, 1931) (Hypercallia)	Colombia
182	**pyrrhotrota** (Meyrick, 1932) (Hypercallia)	Bolivia
183	**rhacina** Walsingham, 1912	Guatemala
184	**rosacea** (Forbes, 1931) (Hypercallia)	Haiti
185	**satrapis** (Meyrick, 1914) (Hypercallia)	Colombia
186	**saulopis** Meyrick, 1909	Peru
187	**selene** Clarke, 1971	Brazil (SC)
188	**sphenogramma** Clarke, 1971	Venezuela
189	**teganitis** Meyrick, 1909	Peru
190	**tenebralis** (Hampson, 1906) (Salobrena)	Brazil (RJ)
	charagma Clarke, 1971	Brazil (SC)
191	**transversa** Clarke, 1971	Brazil (Pr)
192	**uberrima** (Meyrick, 1914) (Hypercallia)	Peru
193	**vexillata** (Meyrick, 1913) (Coptotelia)	Peru
194	**vivida** (Meyrick, 1924) (Hypercallia)	Bolivia

HABROPHYLAX Meyrick, 1931

195	**chalcochtha** Meyrick, 1931	Brazil (SP)

HAMADERA Busck, 1914

196	**aurea** Busck, 1914	Panama

HASTAMEA Fletcher, 1929, repl. name
    Hasta Busck, 1911, preocc. (Kirkaldy, 1906 [Hemiptera])

197	**argentidorsella** (Busck, 1911) (Hasta)	Brazil (Pr)

HIMOTICA Meyrick, 1912

198	**thyrsitis** Meyrick, 1912	Brazil (SP)

HYPERCALLIA Stephens, 1834
    Agriocoma Zeller, 1877
    Brachyplatea Zeller, 1877
    Eumimographe Dognin, 1905 [19]

199	**alexandra** (Meyrick, 1909) (Gonionota)	Peru
200	**argyropa** Meyrick, 1914	Peru
201	**arista** Walsingham, 1912	Mexico (Ver)
202	**bruneri** Busck, [1934]	Cuba
203	**catenella** (Zeller, 1877) (Agriocoma)	Peru
204	**chionastra** Meyrick, 1926	Colombia
205	**chionopis** Meyrick, 1916	French Guiana
206	**citroclista** Meyrick, 1930	Brazil (RJ)
207	**cnephaea** (Walsingham, 1912) (Gonionota)	Panama
208	**crocatella** Zeller, 1877	Colombia
209	**cupreata** (Dognin, 1905) (Eumimographe)	Ecuador
210	**diplotrocha** Meyrick, 1937	Argentina
211	**gnorisma** (Walsingham, 1912) (Coptotelia)	Mexico (Ver)
212	**halobapta** Meyrick, 1930	Brazil (RJ)
213	**heliodepta** Meyrick, 1932	Mexico
214	**heliomima** Meyrick, 1930	Colombia
215	**heterochroma** Clarke, 1971	Venezuela
216	**incensella** (Zeller, 1877) (Brachyplatea)	Peru
217	**longimaculata** (Dognin, 1905) (Eumimographe)	Ecuador
218	**loxochorda** Meyrick, 1926	Colombia
219	**lydia** (Druce, 1901) (Atteria), **n. comb.** [20]	Ecuador
220	**miltopa** (Meyrick, 1912) (Coptotelia)	Colombia
	tunicata (Busck, 1914) (Cryptolechia)	Panama
221	**miniata** (Dognin, 1905) (Cryptolechia)	Ecuador
222	**niphocycla** Meyrick, 1926	Colombia
223	**obliquistriga** Dognin, 1905	Ecuador
224	**orthochaeta** (Meyrick, 1913) (Coptotelia)	Peru
225	**phlebotes** (Walsingham, 1912) (Gonionota)	Guatemala
226	**psittacopa** (Meyrick, 1912) (Coptotelia)	Colombia
227	**rhodosarca** (Walsingham, 1912) (Cryptolechia)	Guatemala
	rhodosarea Meyrick, 1922, missp.	
228	**syntoma** (Walsingham, 1912) (Hyphypena)	Panama

IDIOCRATES Meyrick, 1909

229	**balanitis** Meyrick, 1909	Bolivia

LUCYNA Clarke, 1978

| 230 | **fenestella** (Zeller, 1874) (Cryptolechia) | Chile |
| | **thyridopa** (Meyrick, 1912) (Coptotelia), repl. name | Chile |

MACHIMIA Clemens, 1860

231	**aethostola** Meyrick, 1931	Brazil (SP)
232	**anthracospora** Meyrick, 1934	Brazil (SC)
233	**caduca** (Walsingham, 1912) (Cryptolechia)	Guatemala
234	**chorrera** (Busck, 1914) (Cryptolechia)	Panama
235	**cruda** Meyrick, 1926	Colombia
236	**desertorum** (Berg, 1875) (Cryptolechia)	Argentina
237	**diagrapha** Meyrick, 1931	Paraguay
238	**dolopis** (Walsingham, 1912) (Cryptolechia)	Mexico (Gue)
239	**eothina** Meyrick, 1920	French Guiana
240	**ignicolor** (Busck, 1914) (Cryptolechia)	Panama
241	**illuminella** (Busck, 1914) (Cryptolechia)	Panama
241.1	**intaminata** Meyrick, 1922	Brazil
242	**morata** Meyrick, 1912	Argentina
243	**neuroscia** Meyrick, 1930	Brazil (Pa)
244	**notella** (Busck, 1914) (Cryptolechia)	Panama
245	**oxybela** Meyrick, 1931	Brazil
246	**peperita** (Walsingham, 1912) (Cryptolechia)	Guatemala
247	**perianthes** Meyrick, 1922	French Guiana
248	**pyrocalyx** Meyrick, 1922	Brazil (SC)
249	**pyrograpta** Meyrick, 1932	Bolivia
250	**rogifera** Meyrick, 1914	Guyana
251	**sejunctella** (Walker, 1864) (Gelechia)	Brazil (Am)
252	**trigama** (Meyrick, 1928) (Cryptolechia)	USA (Tx)
253	**trunca** Meyrick, 1930	Brazil (Pa)

MAESARA Clarke, 1968

| 254 | **gallegoi** Clarke, 1968 | Colombia |

MELANEULIA Butler, 1883

| 255 | **hecate** Butler, 1883 | Chile |

MUNA Clarke, 1978

| 256 | **zostera** Clarke, 1978 | Chile |

NEDENIA Clarke, 1978

| 257 | **rhodochra** Clarke, 1978 | Chile |

NEMATOCHARES Meyrick, 1931

| 258 | **citraulax** Meyrick, 1931 | Brazil |

OSMARINA Clarke, 1978

| 259 | **argilla** Clarke, 1978 | Chile |

PALINORSA Meyrick, 1924

260	**acritomorpha** Clarke, 1964	Peru
261	**literatella** (Busck, 1911) (Pleurota)	French Guiana
262	**raptans** (Meyrick, 1920) (Orsotricha)	Peru
263	**zonaria** Clarke, 1964	Bolivia

PERZELIA Clarke, 1978

| 264 | **arda** Clarke, 1978 | Chile |

PHILTRONOMA Meyrick, 1914

| 265 | **roseicorpus** (Dognin, 1910) (Oecophora) | French Guiana |

PHOLCOBATES Meyrick, 1931

| 266 | **flagelliformis** Meyrick, 1931 | Brazil (ES) |

PHYTOMIMIA Walsingham, 1912

267	**chlorophylla** Walsingham, 1912	Guatemala
	redundans Walsingham, 1912	Guatemala
	silvicolor Meyrick, 1932	Panama
268	**cynegetis** Meyrick, 1932	Peru
269	**pyrrhophthalma** Meyrick, 1932	Costa Rica

PISINIDEA Butler, 1883

| 270 | **viridis** Butler, 1883 | Chile |

PROFILINOTA Clarke, 1973

| 271 | **phillita** Clarke, 1973 | Venezuela |

PSEUDOCENTRIS Meyrick, 1921

| 272 | **testudinea** Meyrick, 1921 | Peru |

PSILOCORSIS Clemens, 1860
    Paepia Walker, 1864
    Hagno Chambers, 1872

273	**argyropasta** Walsingham, 1912	Mexico (Gue)
274	**carpocapsella** (Walker, 1864) (Paepia)	Brazil (Am)
275	**exagitata** (Meyrick, 1926) (Cryptolechia)	Colombia
276	**indalma** Walsingham, 1912	Guatemala
277	**melanophthalma** (Meyrick, 1929) (Cryptolechia)	Colombia
278	**minerva** (Meyrick, 1928) (Cryptolechia)	Colombia
279	**propriella** Zeller, 1877	Colombia
280	**purpurascens** Walsingham, 1912	Guatemala

PSITTACASTIS Meyrick, 1909
    Necedes Walsingham, 1912

281	**argentata** Meyrick, 1921	Brazil (Am)
282	**championella** (Walsingham, 1912) (Necedes)	Guatemala
283	**cocae** (Busck, 1931) (Eucleodora)	Peru
284	**cosmodoxa** Meyrick, 1921	Peru
285	**eumolybda** Meyrick, 1926	Peru
286	**eurychrysa** Meyrick, 1909	Bolivia
287	**gaulica** Meyrick, 1909	Bolivia
288	**incisa** (Walsingham, 1912) (Necedes)	Mexico (Tab)
289	**molybdaspis** Meyrick, 1926	Peru
290	**pictrix** Meyrick, 1921	Colombia
291	**propriella** (Walker, 1864) (Gelechia)	Brazil (Am)
292	**pyrsophanes** Meyrick, 1936	Peru
293	**stigmaphylli** (Walsingham, 1912) (Necedes)	Jamaica
	brevipalpis (Walsingham, 1912) (Necedes)	Jamaica
294	**superatella** (Walker, 1864) (Gelechia)	Brazil (Am)
	ambigua (R. Felder & Rogenhofer, 1875)	Brazil (Am)
	(Oecophora)	
295	**trierica** Meyrick, 1909	Bolivia

RHINDOMA Busck, 1914

| 296 | **rosapicella** Busck, 1914 | Panama |

SCOLIOGRAPHA Meyrick, 1916

| 297 | **argospila** Meyrick, 1916 | French Guiana |

TARUDA Walker, 1864
    Ecliptoloma Zeller, 1877

298	**apicella** Walker, 1864	Brazil (Am)
299	**cuneatella** Walker, 1864	Brazil (Am)
	obydella (R. Felder & Rogenhofer, 1875)	Brazil (Am)
	(Rhinosia)	
300	**haemoplecta** Meyrick, 1931	Brazil (ES)
301	**hemiommata** (Zeller, 1877) (Ecliptoloma)	Brazil
302	**leucochna** Meyrick, 1921	Peru
303	**oblitella** (R. Felder & Rogenhofer, 1875) (Lecitho-	Brazil (Am)
	cera)	

TRYCHERODES Meyrick, 1914
    Teratomorpha Walsingham, 1912, preocc. (Turner,
    1896 [Oecophoridae])

304	**albifrons** (Walsingham, 1912) (Teratomorpha)	Mexico (Tab)
305	**chilibrella** (Busck, 1914) (Teratomorpha)	Panama
306	**producta** (Walsingham, 1912) (Teratomorpha)	Guatemala

ZEMIOCRITA Meyrick, 1933

| 307 | **spermatopis** Meyrick, 1933 | Argentina |

Ethmiinae

ERYSIPTILA Meyrick, 1914 [21]
    Cyrictodes Meyrick, 1926

| 308 | **clevelandi** (Busck, 1914) (Borkhausenia) | Panama |
| | phormophora (Meyrick, 1926) (Cyrictodes) | Costa Rica |

ETHMIA Hübner, [1819]
    Psecadia Hübner, [1825]
    Anesychia Hübner, [1825]
    Disthymnia Hübner, [1825]
    Melanoleuca Stephens, 1829
    Aedia Duponchel, 1836
    Chalybe Duponchel, 1836
    Azinis Walker, 1863
    Tamarrha Walker, 1864
    Ceratophysetis Meyrick, 1887
    Theoxenia Walsingham, 1887
    Babaiaxa Busck, 1902
    Wiltshireia Amsel, 1949

309	**abraxasella** (Walker, 1864) (Psecadia)	
	a) **abraxasella** (Walker, 1864) (Psecadia)	Dom. Rep.
	aureoapicella (Möschler, 1890) (Psecadia)	Puerto Rico
	abraxella Meyrick, 1914, missp.	

	b) **clarissa** Busck, 1914	Cuba
310	**albicostella** (Beutenmüller, 1889) (Psecadia)	USA (Co)
311	**angustalatella** Powell, 1973	Mexico (NL)
312	**arctostaphylella** (Walsingham, 1880) (Psecadia)	USA (Ca)
	obscurella (Beutenmüller, 1888) (Psecadia)	USA (Ca)
	mediella Busck, 1913	USA (Ca)
313	**baja** Powell, 1973	Mexico (BCS)
314	**baliostola** Walsingham, 1912	Costa Rica
	baliostoma Busck, 1914, missp.	
315	**bittenella** (Busck, 1906) (Tamarrha)	USA (Tx)
316	**calumniella** Powell, 1973	Brazil (Pa)
317	**catapeltica** Meyrick, 1924	Costa Rica
318	**cellicoma** Meyrick, 1931	Paraguay
319	**chalcodora** Meyrick, 1912	Argentina
320	**chalcogramma** Powell, 1973	Bolivia
321	**chemsaki** Powell, 1959	Costa Rica
322	**clarkei** Powell, 1973	Mexico (QR)
323	**clava** Powell, 1973	Mexico (Ver)
324	**confusella** (Walker, 1863) (Hyponomeuta)	Dom. Rep.
	strigosella (Walker, 1864) (Cryptolechia)	Dom. Rep.
	ingricella (Möschler, 1890) (Hyponomeuta)	Puerto Rico
	strigosa (Cockerell, 1891) (Psecadia), missp.	
325	**confusellastra** Powell, 1973	Mexico (Yuc)
326	**conglobata** Meyrick, 1912	Colombia
*327	**coquillettella** Busck, 1907	USA (Ca)
328	**cordia** Powell, 1973	Mexico (Yuc)
329	**coronata** Walsingham, 1912	Mexico (Gue)
	abdominella Busck, 1912	Mexico (Gue)
330	**cubensis** Busck, [1934]	Cuba
331	**cupreonivella** (Walsingham, 1880) (Psecadia)	Brazil (ES)
332	**cyanea** Walsingham, 1912	Mexico (Ver)
333	**cypraeella** (Zeller, 1863) (Psecadia)	Venezuela
334	**cypraspis** Meyrick, 1930	Brazil (Pa)
335	**davisella** Powell, 1973	Mexico (Tam)
336	**delliella** (Fernald, 1891) (Psecadia)	USA (Tx)
337	**discostrigella** (Chambers, 1877) (Anesychia)	
	a) **discostrigella** (Chambers, 1877) (Anesychia)	USA (Co)
	b) **subcaerulea** (Walsingham, 1880) (Psecadia)	USA (Ca)
338	**duckworthi** Powell, 1973	Panama
339	**elutella** Busck, 1914	Panama
340	**epilygella** Powell, 1973	Brazil (SC)
341	**exornata** (Zeller, 1877) (Psecadia)	Peru
	exornatella Busck, 1906, missp.	
342	**farrella** Powell, 1973	Jamaica
343	**festiva** Busck, 1914	Panama
	xantholitha Meyrick, 1929	Colombia
344	**flavicaudata** Walsingham, 1912	Mexico (Ver)
345	**fritillella** Powell, 1973	Brazil (SC)
346	**gelidella** (Walker, 1864) (Tamarrha)	Jamaica
347	**gigantea** Busck, 1914	Mexico (Ver)
348	**hagenella** (Chambers, 1878) (Anesychia)	
	a) **hagenella** (Chambers, 1878) (Anesychia)	USA (Tx)
	b) **josephinella** Dyar, 1902	USA (NM)
349	**hammella** Busck, 1910	Costa Rica
350	**heptastica** Walsingham, 1912	Mexico (Gue)
351	**hieroglyphica** Powell, 1973	Bolivia
352	**hiramella** Busck, 1914	Cuba
353	**hodgesella** Powell, 1973	USA (Az)
354	**howdeni** Powell, 1973	Mexico (Sin)
355	**humilis** Powell, 1973	Jamaica
356	**iridella** Powell, 1973	Mexico (Pue)
357	**janzeni** Powell, 1973	Mexico (Oax)
358	**joviella** Walsingham, 1897	Grenada
359	**julia** Powell, 1973	Puerto Rico
360	**kirbyi** (Möschler, 1890) (Psecadia)	Puerto Rico
361	**lichyi** Powell, 1973	Venezuela
362	**linda** Busck, 1914	Venezuela
363	**linsdalei** Powell, 1973	Mexico (Oax)
364	**mansita** Busck, 1914	Mexico (Pue)
365	**marmorea** (Walsingham, 1888) (Psecadia)	USA (Az)
366	**mnesicosma** Meyrick, 1924	Costa Rica
367	**mulleri** Busck, 1910	Mexico (Pue)
	mülleri Busck, 1912, missp. (incorr. spell.)	
368	**nigritaenia** Powell, 1973	Mexico (Yuc)
369	**nivosella** (Walker, 1864) (Tamarrha)	Jamaica
	adustella (Zeller, 1877) (Psecadia)	Puerto Rico
	niveosella (Busck, 1906) (Tamarrha), missp.	
370	**notatella** (Walker, 1863) (Psecadia)	Dom. Rep.
	xanthorrhoa (Zeller, 1877) (Psecadia)	Puerto Rico
371	**notomurinella** Powell, 1973	Argentina
372	**omega** Powell, 1973	Brazil (RS)
373	**oterosella** Busck, [1934]	Cuba
374	**pala** Powell, 1973	Mexico (Sin)
375	**papiella** Powell, 1973	Mexico (Sin)
376	**paucella** (Walker, 1863) (Hyponomeuta)	Dom. Rep.
377	**penthica** Walsingham, 1912	Mexico (Oax)
378	**perpulchra** Walsingham, 1912	Mexico (Ver)
379	**phoenicura** Meyrick, 1932	Mexico (BCS)
380	**phylacis** Walsingham, 1912	
	a) **phylacis** Walsingham, 1912	Mexico [Sin]
	b) **ornata** Busck, [1934]	Cuba
381	**phylacops** Powell, 1973	Mexico (Yuc)
382	**piperella** Powell, 1973	Jamaica
383	**plaumanni** Powell, 1973	Brazil (SC)
384	**playa** Powell, 1973	Mexico (Sin)
385	**prattiella** Busck, 1915	USA (Tx)
386	**proximella** Busck, 1912	Mexico (Pue)
387	**punctessa** Powell, 1973	Mexico (NL)
388	**sandra** Powell, 1973	El Salvador
389	**scutula** Powell, 1973	Mexico (Sin)
390	**scythropa** Walsingham, 1912	Costa Rica
391	**semilugens** (Zeller, 1872) (Psecadia)	USA (Tx)
	multipunctella (Chambers, 1874) (Anesychia)	USA (Tx)
	semiopaca (Grote, 1881) (Psecadia)	USA (Co)
	plumbeella (Beutenmüller, 1889) (Psecadia)	USA (Tx)
392	**semiombra** Dyar, 1902	
	a) **semiombra** Dyar, 1902	USA (Tx)
	b) **nebulombra** Powell, 1973	Mexico (Yuc)
393	**semitenebrella** Dyar, 1902	USA (Az)
394	**similatella** Busck, 1920	Guatemala
395	**sphenisca** Powell, 1973	Mexico (Dur)
396	**striatella** Busck, 1913	Mexico (Pue)
397	**submissa** Busck, 1914	Cuba
398	**subnigritaenia** Powell, 1973	Mexico (DF)
399	**subsimilis** Walsingham, 1897	Jamaica
400	**terpnota** Walsingham, 1912	Costa Rica
401	**transversella** Busck, 1914	Costa Rica
402	**ungulatella** Busck, 1914	Panama
403	**volcanella** Powell, 1973	Mexico (Oax)
404	**wellingi** Powell, 1973	Mexico (Yuc)
405	**zebrata** Powell, 1959	Mexico (Pue)

MACROCIRCA Meyrick, 1931[22]

406	**strabo** Meyrick, 1931	Argentina

Peleopodinae

DURRANTIA Busck, 1908
Dolidiria Busck, 1912

407	**acompsa** Walsingham, 1912	Panama
	monotona (Amsel, 1956) (Stenoma)	Venezuela
408	**amabilis** Walsingham, 1912	Guatemala
409	**arcanella** (Busck, 1912) (Dolidiria)	Panama
410	**flaccescens** Meyrick, 1925	Peru
411	**pugnax** Walsingham, 1912	Guatemala
412	**resurgens** Walsingham, 1912	Guatemala

PELEOPODA Zeller, 1877
Theatria Walsingham, 1912

413	**convoluta** Duckworth, 1970	Venezuela
414	**lobitarsis** Zeller, 1877	Panama
	lobitarsus Hodges, 1974, missp.	
415	**navigatrix** (Meyrick, 1912) (Xylorycta)	Colombia
416	**semocrossa** Meyrick, 1930	Bolivia
417	**spudasma** (Walsingham, 1912) (Theatria)	Panama

PSEUDEROTIS Clarke, 1956

418	**cannescens** Clarke, 1956	Argentina
419	**thamnolopha** (Meyrick, 1932) (Asapharca)	Costa Rica

SCHISTONOEA Forbes, 1931

420	**fulvidella** (Walsingham, 1897) (Brachmia)	Virgin Is. (St. Thomas)

Stenomatinae

ANADASMUS Walsingham, 1897

421	**accurata** (Meyrick, 1916) (Stenoma)	French Guiana
	perjura (Meyrick, 1925) (Stenoma)	Brazil (Am)
422	**anceps** (Butler, 1877) (Cryptolechia)	Brazil (Am)
	praeceps (Meyrick, 1915) (Stenoma)	French Guiana
423	**arenosa** (Meyrick, 1916) (Stenoma)	French Guiana
424	**byrsinites** (Meyrick, 1912) (Stenoma)	Colombia
425	**caliginea** (Meyrick, 1930) (Stenoma)	Brazil (SP)
	lianthes (Meyrick, 1932) (Stenoma)	Brazil (SC)
426	**capnocrossa** (Meyrick, 1925) (Stenoma)	Brazil (Am)
427	**chlorothrota** (Meyrick, 1932) (Stenoma)	Bolivia
428	**endochra** (Meyrick, 1925) (Stenoma)	Brazil (Am)
429	**germinans** (Meyrick, 1925) (Stenoma)	Colombia
430	**incitatrix** (Meyrick, 1925) (Stenoma)	Argentina
431	**ischioptila** (Meyrick, 1925) (Stenoma)	Colombia
432	**leontodes** (Meyrick, 1915) (Gonioterma)	Surinam
433	**lithogypsa** (Meyrick, 1932) (Stenoma)	Brazil (Pa)
434	**nonagriella** (Walker, 1864) (Cryptolechia)	Brazil (Am)
435	**obmutescens** (Meyrick, 1916) (Stenoma)	French Guiana
	pleutotricha (Meyrick, 1925) (Stenoma)	Brazil (Am)
436	**paurocentra** (Meyrick, 1912) (Stenoma)	Colombia
437	**pelinitis** (Meyrick, 1912) (Stenoma)	Colombia
438	**pelodes** (Walsingham, 1913) (Stenoma)	Panama
	scortea (Meyrick, 1915) (Stenoma)	Brazil (Am)
439	**plebicola** (Meyrick, 1918) (Stenoma)	French Guiana
440	**quadratella** (Walker, 1864) (Cryptolechia)	Brazil (Pa)
	hebes (Dognin, 1905) (Mesoptycha)	Colombia

	gerda (Busck, 1914) (Gonioterma)	Panama
441	sororia (Zeller, 1877) (Cryptolechia)	Colombia
	catapsecta (Meyrick, 1915) (Stenoma)	Guyana
442	vacans (Meyrick, 1916) (Stenoma)	French Guiana
443	venosella (Walker, 1864) (Cryptolechia)	Brazil (Am)
	pleximorpha (Meyrick, 1930) (Stenoma)	Bolivia
	dictyogramma (Meyrick, 1932) (Stenoma)	Brazil (Am)

ANAPATRIS Meyrick, 1932

444	chersopsamma Meyrick, 1932	Panama

ANTAEOTRICHA Zeller, 1854

Mesoptycha Zeller, 1854
Brachiloma Clemens, 1863
Harpalyce Chambers, 1874
Ide Chambers, 1880
Antoeotricha Walsingham, 1881, missp.
Aedemoses Walsingham, 1912
Athleta Walsingham, 1912
Prasolithites Meyrick, 1912
Aphanoxena Meyrick, 1915
Psephomeres Meyrick, 1916
Eumiturga Meyrick, 1925

445	acrograpta (Meyrick, 1915) (Aphanoxena)	Guyana
446	acronephela Meyrick, 1915	Guyana
447	actista (Meyrick, 1913) (Stenoma)	Venezuela
448	addon (Busck, 1911) (Stenoma)	Guyana
	cicadella (Sepp, [1852]-55) (Phalaena), preocc. (not Sepp, [1832-40])	Surinam
449	adductella (Walker, 1864) (Cryptolechia)	Brazil (Am)
450	adjunctella (Walker, 1864) (Cryptolechia)	Brazil (Am)
	additella (Walker, 1864) (Cryptolechia)	Brazil (Am)
	absconditella (Walker, 1864) (Cryptolechia)	Brazil (Am)
451	admixta (Walsingham, 1913) (Stenoma)	Mexico (Gue)
452	adornata (Meyrick, 1915) (Stenoma)	Peru
453	aequabilis (Meyrick, 1916) (Stenoma)	French Guiana
454	aerinotata (Butler, 1877) (Cryptolechia)	Brazil (Am)
	sperata (Busck, 1911) (Stenoma)	French Guiana
455	affinis R. Felder & Rogenhofer, 1875	Brazil (Am)
	tanysta Meyrick, 1915	Guyana
	anticharis (Meyrick, 1916) (Stenoma)	French Guiana
456	aggravata (Meyrick, 1916) (Stenoma)	French Guiana
457	aglypta Meyrick, 1925	Brazil (Am)
458	albicilla (Zeller, 1854) (Cryptolechia)	Venezuela
	albicella Busck, [1934], missp.	
459	albifrons Zeller, 1877	Brazil
460	albilimbella (R. Felder & Rogenhofer, 1875) (Gelechia)	Brazil (Am)
	nuntia (Meyrick, 1925) (Stenoma)	Peru
461	albitincta (Meyrick, 1930) (Stenoma)	Brazil (ES)
	pauroconis (Meyrick, 1932) (Stenoma)	Brazil (SC)
462	albovenosa Zeller, 1877	Peru
463	amicula Zeller, 1877	Panama
464	ammodes (Walsingham, 1913) (Stenoma)	Mexico (Tab)
465	amphilyta Meyrick, 1916	French Guiana
466	amphizyga Meyrick, 1930	Brazil (Pa)
467	anaclintris Meyrick, 1916	French Guiana
468	arachnia (Meyrick, 1915) (Stenoma)	Guyana
469	aratella (Walker, 1864) (Cryptolechia)	Brazil (Am)
470	argocorys (Meyrick, 1931) (Stenoma)	Brazil (ES)
471	arystis Meyrick, 1915	Guyana
472	assecta Zeller, 1877	Peru
473	astynoma (Meyrick, 1915) (Aphanoxena)	Guyana
474	atmospora (Meyrick, 1925) (Stenoma)	Colombia
475	axena Meyrick, 1916	French Guiana
476	balanocentra (Meyrick, 1915) (Aphanoxena)	Guyana
477	ballista (Meyrick, 1916) (Stenoma)	French Guiana
478	basalis Zeller, 1854	Brazil
	harpobathra Meyrick, 1916	Argentina
479	basiferella (Walker, 1864) (Cryptolechia)	Brazil (Am)
480	basilaris (Busck, 1914) (Stenoma)	Panama
481	basirubrella (Walker, 1864) (Cryptolechia)	Brazil (Am)
482	bathrotoma (Meyrick, 1925) (Stenoma)	Brazil (Am)
483	biarcuata Meyrick, 1926	Colombia
	stenobathra Meyrick, 1932	Panama
	vogli Amsel, 1956	Venezuela
484	bicolor (Zeller, 1839) (Stenoma)	Brazil
	dissimilis Kearfott, 1911	Brazil (SP)
	annixa Meyrick, 1918	Brazil (SP)
485	bilinguis (Meyrick, 1918) (Stenoma)	French Guiana
486	binubila Zeller, 1854	Brazil (Pa)
	aporodes Meyrick, 1915	Surinam
487	bipupillata Meyrick, 1930	Brazil (Am)/Paraguay
488	bracatingae (Köhler, 1943) (Stenoma)	Argentina
489	brachysaris Meyrick, 1916	French Guiana
490	brochota Meyrick, 1915	Peru
491	caenochytis (Meyrick, 1915) (Stenoma)	Guyana
492	camarina Meyrick, 1915	Guyana
493	campylodes Meyrick, 1916	French Guiana
494	cantharitis (Meyrick, 1916) (Aphanoxena)	French Guiana
495	caprimulga (Walsingham, 1912) (Stenoma)	Mexico (Ver)
496	capsiformis (Meyrick, 1930) (Stenoma)	Brazil (Ba)
497	capsulata Meyrick, 1918	French Guiana
498	carabodes (Meyrick, 1915) (Stenoma)	Guyana
499	carabophanes Meyrick, 1932	Colombia
500	carbasea (Meyrick, 1915) (Stenoma)	Brazil (RJ)
501	caryograpta (Meyrick, 1930) (Stenoma)	Brazil (SP)
502	cathagnista Meyrick, 1925	Brazil (Am)
503	catharactis Meyrick, 1930	Brazil
504	cedroxyla Meyrick, 1930	Brazil (SP)
505	celidotis Meyrick, 1925	Peru
506	ceratistes (Walsingham, 1912) (Stenoma)	Mexico (Gue)
507	chalastis (Meyrick, 1915) (Stenoma)	Guyana
508	chalinophanes (Meyrick, 1931) (Stenoma)	Bolivia
509	chilosema (Meyrick, 1918) (Stenoma)	French Guiana
510	christocoma Meyrick, 1915	Peru
511	cicadella (Sepp, [1832-40]) (Phalaena)	Surinam
512	cirrhoxantha (Meyrick, 1915) (Stenoma)	French Guiana
513	cleopatra Meyrick, 1925	Brazil
514	cnemosaris (Meyrick, 1925) (Stenoma)	Brazil (Pa)
515	colposaris (Meyrick, 1925) (Stenoma)	Brazil (Am)
516	comosa (Walsingham, 1912) (Stenoma)	Mexico (Ver)
517	compsographa Meyrick, 1916	French Guiana
518	compsoneura (Meyrick, 1925) (Stenoma)	French Guiana
519	confixella (Walker, 1864) (Cryptolechia)	Brazil (Am)
520	congelata Meyrick, 1926	Peru
521	coniopa (Meyrick, 1925) (Stenoma)	Brazil (Pa)
522	constituta (Meyrick, 1925) (Stenoma)	French Guiana
523	constricta (Meyrick, 1926) (Stenoma)	Colombia
524	conturbatella (Walker, 1864) (Cryptolechia)	Brazil (Am)
	illucidella (Walker, 1864) (Cryptolechia)	Brazil (Am)
525	copromima Meyrick, 1930	French Guiana
	citrophaea (Meyrick, 1931) (Stenoma)	Brazil (RJ)
526	coriodes Meyrick, 1915	Guyana
527	corvigera Meyrick, 1915	Guyana
528	cosmoterma Meyrick, 1930	Brazil (RS)
529	costatella (Walker, 1864) (Cryptolechia)	Brazil (Am)
530	cremastis (Meyrick, 1925) (Stenoma)	Peru
531	cryeropis Meyrick, 1925	Mexico (Gue)
532	crysiphaea (Meyrick, 1925) (Stenoma)	Brazil (Pa)
533	cyclobasis Meyrick, 1930	Brazil (SP)
534	cycnolopha (Meyrick, 1925) (Stenoma)	Peru
535	cycnomorpha Meyrick, 1925	Brazil (Pa)
536	cymogramma (Meyrick, 1925) (Stenoma)	Peru
537	cyprodeta Meyrick, 1930	Brazil (RS)
538	deltopis Meyrick, 1915	Guyana
539	demas (Busck, 1911) (Stenoma)	French Guiana
540	demotica (Walsingham, 1912) (Stenoma)	Mexico (Gue)
541	deridens Meyrick, 1925	Bolivia
542	desecta (Meyrick, 1918) (Stenoma)	French Guiana
543	destillata (Zeller, 1877) (Cryptolechia)	Panama
544	diacta (Meyrick, 1916) (Stenoma)	French Guiana
545	diffracta Meyrick, 1916	French Guiana
546	diplarcha Meyrick, 1916	Guyana
	arachniotis Meyrick, 1930	Brazil (Pa)
547	diplophaea Meyrick, 1916	French Guiana
548	diplosaris (Meyrick, 1915) (Stenoma)	Guyana
549	dirempta (Zeller, 1855) (Cryptolechia)	Brazil
550	discalis (Busck, 1914) (Stenoma)	Panama
551	discolor (Walsingham, 1912) (Stenoma)	Guatemala
552	disjecta (Zeller, 1854) (Cryptolechia)	Brazil
553	dissona (Meyrick, 1925) (Stenoma)	Brazil (Am)
554	doleopis (Meyrick, 1915) (Stenoma)	Guyana
555	dromica (Meyrick, 1925) (Stenoma)	Brazil (Am)
556	elaeodes (Walsingham, 1913) (Stenoma)	Mexico (Ver)
557	elatior (R. Felder & Rogenhofer, 1875) (Cryptolechia)	Brazil (Am)
558	encyclia Meyrick, 1915	Colombia
559	enodata Meyrick, 1916	French Guiana
560	epicrossa (Meyrick, 1932) (Stenoma)	Peru
561	epignampta Meyrick, 1916	Peru
562	episimbla (Meyrick, 1915) (Aphanoxena)	Guyana
563	ergates (Walsingham, 1913) (Stenoma)	Mexico (Tab)
564	erotica (Meyrick, 1916) (Stenoma)	French Guiana
565	eucoma Meyrick, 1925	Brazil (Am)
566	euthrinca Meyrick, 1915	Colombia
567	exasperata (Meyrick, 1916) (Stenoma)	French Guiana
568	excisa Meyrick, 1916	French Guiana
569	extenta (Busck, 1920) (Stenoma)	Guatemala
	ptilocrates Meyrick, 1932	Panama
570	exusta Meyrick, 1916	French Guiana
571	falsidica (Meyrick, 1915) (Stenoma)	Surinam
572	fascicularis Zeller, 1854	Brazil
	leprosa (R. Felder & Rogenhofer, 1875) (Cryptolechia)	Brazil (Am)
	gunni (Busck, 1911) (Stenoma)	French Guiana
	leucogramma Meyrick, 1915	Guyana
573	fasciata (Busck, 1911) (Stenoma)	French Guiana
574	filiferella (Walker, 1864) (Cryptolechia)	Brazil (Am)
	nivititurella (Walker, 1864) (Gelechia)	Brazil (Am)
	menestella (Walsingham, 1913) (Stenoma)	Panama
575	flocculosa (Meyrick, 1925) (Eumiturga)	Brazil (Am)
576	forreri (Walsingham, 1913) (Stenoma)	Mexico (Dur)
577	fractilinea (Walsingham, 1912) (Stenoma)	Mexico (Tab)
578	fractinubes (Walsingham, 1912) (Stenoma)	Panama
579	fraterna (R. Felder & Rogenhofer, 1875) (Cryptolechia)	Brazil (Am)

	perfusa Meyrick, 1916	French Guiana
580	**frontalis** (Zeller, 1855) (Cryptolechia)	"Mexico"
581	**fulta** Meyrick, 1926	Colombia
582	**fumifica** (Walsingham, 1912) (Stenoma)	Mexico (Ver)
	submersa (Meyrick, 1915) (Stenoma)	Guyana
*583	**furcata** (Walsingham, 1889) (Stenoma)	USA (Az)
*584	**fuscorectangulata** Duckworth, 1964	USA (Az)
585	**generatrix** Meyrick, 1926	Brazil (RS)
586	**glaphyrodes** (Meyrick, 1913) (Stenoma)	Brazil (Am)
587	**glaucescens** (Meyrick, 1916) (Stenoma)	French Guiana
588	**glycerostoma** Meyrick, 1915	Colombia
589	**graphopterella** (Walker, 1864) (Cryptolechia)	Brazil (Am)
	crocuta (R. Felder & Rogenhofer, 1875) (Cryptolechia)	Brazil (Am)
590	**gravescens** Meyrick, 1926	Colombia
591	**griseanomima** Busck, 1934, repl. name	Surinam
	griseana (Sepp, [1852]-55) (Phalaena), preocc. (not Fabricius, 1794)	Surinam
592	**gubernatrix** Meyrick, 1925	Peru
593	**gymnolopha** Meyrick, 1925	Brazil (Am)
594	**gypsoterma** (Meyrick, 1915) (Stenoma)	Guyana
595	**habilis** (Meyrick, 1915) (Stenoma)	Guyana
596	**haesitans** (Walsingham, 1912) (Aedemoses)	Mexico (Dur)
	hessitans (Heinrich, 1921) (Aedemoses), missp.	
	hesitans Busck, 1934, missp.	
597	**haplocentra** Meyrick, 1925	Brazil (Pa)
598	**hapsicora** Meyrick, 1915	Brazil (SP)
599	**helicias** Meyrick, 1916	French Guiana
600	**hemibathra** Meyrick, 1932	"Mexico"
601	**hemiscia** (Walsingham, 1912) (Stenoma)	Guatemala
602	**herilis** R. Felder & Rogenhofer, 1875	Brazil
	basimacula Möschler, 1882	Surinam
	xanthoptila Meyrick, 1912	Surinam
603	**heterosaris** (Meyrick, 1915) (Stenoma)	Guyana
604	**himaea** Meyrick, 1916	French Guiana
605	**homologa** (Meyrick, 1915) (Aphanoxena)	Guyana
606	**horizontias** (Meyrick, 1925) (Stenoma)	Brazil (Am)
607	**humerella** (Walker, 1864) (Cryptolechia)	Brazil (Am)
	intermedia (R. Felder & Rogenhofer, 1875) (Cryptolechia)	"Ceylon" [S. Am.]
	luscina (Zeller, 1877) (Stenoma)	Panama
	meridiana (Meyrick, 1915) (Stenoma)	Guyana
608	**hyalophanta** (Meyrick, 1932) (Stenoma)	Peru
609	**hydrophora** Meyrick, 1925	Peru
610	**ianthina** (Walsingham, 1913) (Stenoma)	Panama
611	**illepida** Meyrick, 1916	French Guiana
	martini Amsel, 1956	Venezuela
612	**imminens** (Meyrick, 1915) (Stenoma)	Surinam
613	**immota** Meyrick, 1916	French Guiana
614	**impactella** (Walker, 1864) (Cryptolechia)	Brazil (Am)
615	**impedita** (Meyrick, 1915) (Stenoma)	Peru
616	**incisurella** (Walker, 1864) (Cryptolechia)	Brazil (Am)
617	**incompleta** Meyrick, 1932	"Mexico"
618	**incongrua** Meyrick, 1932	Peru
619	**incrassata** Meyrick, 1916	French Guiana
620	**indicatella** (Walker, 1864) (Cryptolechia)	Brazil (Am)
621	**infecta** (Meyrick, 1930) (Stenoma)	Brazil (Pa)
622	**infrenata** (Meyrick, 1918) (Stenoma)	French Guiana
623	**innexa** (Meyrick, 1925) (Stenoma)	Peru
624	**inquinula** Zeller, 1854	Brazil (Pa)
625	**insidiata** (Meyrick, 1916) (Stenoma)	French Guiana
626	**insimulata** Meyrick, 1916	Colombia
627	**intersecta** (Meyrick, 1916) (Stenoma)	French Guiana
628	**iopetra** (Meyrick, 1932) (Stenoma)	Guatemala
629	**ioptila** (Meyrick, 1915) (Stenoma)	Guyana
630	**iras** Meyrick, 1926	Peru
*631	**irene** (Barnes & Busck, 1920) (Stenoma)	USA (Tx)
632	**irenias** (Meyrick, 1916) (Stenoma)	French Guiana
633	**isochyta** (Meyrick, 1915) (Stenoma)	Guyana
634	**isomeris** (Meyrick, 1912) (Stenoma)	Brazil (RJ)
635	**isoplintha** (Meyrick, 1925) (Stenoma)	Brazil (Am)
636	**isoporphyra** (Meyrick, 1932) (Asapharcha)	Costa Rica
637	**isosticta** (Meyrick, 1932) (Stenoma)	"Mexico"
638	**isotona** Meyrick, 1932	Panama
639	**ithytona** Meyrick, 1929	Colombia
640	**juvenalis** (Meyrick, 1930) (Stenoma)	Brazil (Pa)
641	**lacera** (Zeller, 1877) (Auxocrossa)	S. Am. (?)
642	**lampyridella** (Busck, 1914) (Stenoma)	Panama
643	**lathiptila** (Meyrick, 1915) (Stenoma)	Guyana
644	**laudata** Meyrick, 1916	French Guiana
645	**laxa** (Meyrick, 1915) (Stenoma)	Venezuela
646	**lebetias** (Meyrick, 1915) (Stenoma)	French Guiana
647	**lecithaula** Meyrick, 1915	Guyana
648	**lepidocarpa** (Meyrick, 1930) (Stenoma)	Brazil (Pa)
649	**leptogramma** (Meyrick, 1916) (Psephomeres)	French Guiana
*650	**leucillana** (Zeller, 1854) (Cryptolechia)	USA (Ga)
	algidella (Walker, 1864) (Cryptolechia)	Canada
651	**leucocryptis** (Meyrick, 1932) (Stenoma)	Colombia
652	**lignicolor** Zeller, 1877	Peru
	emollita Meyrick, 1926	Colombia
*653	**lindseyi** (Barnes & Busck, 1920) (Stenoma)	USA (Az)
654	**lophoptycha** (Meyrick, 1925) (Stenoma)	Brazil (Am)
655	**lophosaris** (Meyrick, 1925) (Stenoma)	Brazil (Pa)
656	**loxogrammos** (Zeller, 1854) (Cryptolechia)	Brazil

657	**lucrosa** (Meyrick, 1925) (Stenoma)	Brazil (Am)
658	**lunimaculata** (Dognin, 1913) (Stenoma)	Colombia
659	**lysimeris** Meyrick, 1915	Peru
660	**machetes** (Walsingham, 1912) (Stenoma)	Mexico (Gue)
661	**macronota** (Meyrick, 1912) (Stenoma)	Colombia
662	**malachita** Meyrick, 1915	Guyana
663	**manceps** Meyrick, 1925	Peru
664	**marmorea** R. Felder & Rogenhofer, 1875	Brazil (Am)
	mentigera Meyrick, 1926	Bolivia
	quatiens Meyrick, 1930	Brazil (Pa)
	thalamobathra Meyrick, 1932	Bolivia
665	**melanarma** Meyrick, 1916	French Guiana
666	**melanopis** Meyrick, 1909	Peru
667	**melinopa** (Meyrick, 1925) (Stenoma)	Brazil (Pa)
668	**mendax** (Zeller, 1855) (Cryptolechia)	Brazil (MG)
	crypsithias (Meyrick, 1930) (Stenoma)	Brazil (MG)
669	**mesosaris** (Meyrick, 1925) (Stenoma)	French Guiana
670	**mesostrota** Meyrick, 1912	Venezuela
671	**microtypa** (Meyrick, 1915) (Stenoma)	Guyana
672	**milictis** Meyrick, 1925	Brazil (Am)
673	**mitratella** (Busck, 1914) (Stenoma)	Panama
674	**modulata** (Meyrick, 1915) (Stenoma)	Guyana
675	**monocolona** Meyrick, 1932	Bolivia
	monoclona Busck, 1934, missp.	
676	**monosaris** (Meyrick, 1915) (Stenoma)	Guyana
677	**mundella** (Walker, 1864) (Cryptolechia)	Brazil (Am)
	contortella (Walker, 1864) (Paepia)	Brazil (Am)
678	**murinella** (Walker, 1864) (Cryptolechia)	Brazil (Am)
679	**mustela** (Walsingham, 1912) (Stenoma)	Panama
680	**navicularis** (Meyrick, 1930) (Stenoma)	Brazil (SP)
681	**neocrossa** (Meyrick, 1925) (Stenoma)	Peru
682	**nephelocyma** (Meyrick, 1930) (Stenoma)	Brazil (SC)
683	**nerteropa** Meyrick, 1915	Peru
684	**neurographa** Meyrick, 1923	Brazil (RJ)
685	**nictitans** (Zeller, 1854) (Mesoptycha)	Brazil (Pa)
	heteropa (Meyrick, 1913) (Stenoma)	Guyana
686	**nimbata** Meyrick, 1925	Peru
687	**nitescens** Meyrick, 1925	Brazil (Pa)
688	**nitidorella** (Walker, 1864) (Cryptolechia)	Brazil (Am)
689	**nitrota** Meyrick, 1916	French Guiana
690	**notogramma** (Meyrick, 1930) (Stenoma)	Brazil (Pa)
691	**notosaris** (Meyrick, 1925) (Stenoma)	Brazil (Pa)
692	**notosemia** (Zeller, 1877) (Stenoma)	Colombia
693	**nuclearis** Meyrick, 1913	Peru
694	**obtusa** (Meyrick, 1916)(Stenoma)	French Guiana
695	**ocellifer** (Walsingham, 1912) (Stenoma)	Mexico (Dur)
696	**ogmolopha** (Meyrick, 1930) (Stenoma)	Brazil (Am)
697	**ogmosaris** (Meyrick, 1915) (Stenoma)	Guyana
698	**ophrysta** Meyrick, 1912	Surinam
699	**orgadopa** (Meyrick, 1925) (Stenoma)	Brazil (Pa)
700	**orthophaea** Meyrick, 1930	Brazil (SP)
701	**orthotona** Meyrick, 1916	French Guiana
702	**orthriopa** Meyrick, 1925	Brazil (Pa)
703	**ostodes** (Walsingham, 1913) (Stenoma)	Guatemala
704	**ovulifera** (Meyrick, 1925) (Stenoma)	Peru
705	**oxycentra** Meyrick, 1916	French Guiana
706	**oxydecta** (Meyrick, 1915) (Stenoma)	Guyana
707	**oxyschista** (Meyrick, 1925) (Stenoma)	Brazil (Am)
708	**pactota** Meyrick, 1915	Guyana
709	**palaestrias** Meyrick, 1916	French Guiana
710	**pallicosta** (R. Felder & Rogenhofer, 1875) (Cryptolechia)	Brazil (Am)
711	**paracrypta** Meyrick, 1915	Guyana
712	**paracta** (Meyrick, 1915) (Stenoma)	Peru
713	**parastis** Gyen, 1913	Chile
714	**particularis** (Zeller, 1877) (Cryptolechia)	Panama
715	**pellocoma** (Meyrick, 1915) (Aphanoxena)	Guyana
716	**percnocarpa** (Meyrick, 1925) (Stenoma)	Brazil (Am)
717	**percnogona** Meyrick, 1925	Peru
718	**periphrictis** (Meyrick, 1915) (Stenoma)	Guyana
719	**phaeoneura** (Meyrick, 1913) (Stenoma)	Guyana
720	**phaeoplintha** (Meyrick, 1915) (Stenoma)	Guyana
721	**phaeosaris** Meyrick, 1915	Guyana
722	**phaselodes** (Meyrick, 1931) (Stenoma)	Brazil (ES)
	phaseolodes (Busck, 1934) (Stenoma), missp.	
723	**phaula** (Walsingham, 1912) (Stenoma)	Guatemala
724	**phollicodes** (Meyrick, 1916) (Stenoma)	French Guiana
725	**phryactis** Meyrick, 1925	Peru
726	**planicoma** (Meyrick, 1925) (Stenoma)	Brazil (Pa)
727	**platydesma** Meyrick, 1915	Guyana
728	**plerotis** Meyrick, 1925	Peru
729	**plesistia** (Meyrick, 1930) (Stenoma)	Brazil (RJ)
	ptilallactis (Meyrick, 1930) (Stenoma)	Brazil (ES)
730	**plumosa** (Busck, 1914) (Stenoma)	Panama
731	**polyglypta** (Meyrick, 1915) (Stenoma)	Guyana
732	**praecisa** Meyrick, 1912	Brazil (RJ)
733	**praerupta** Meyrick, 1915	Guyana
734	**pratifera** (Meyrick, 1925) (Stenoma)	Costa Rica
735	**prosora** (Walsingham, 1912) (Stenoma)	Panama
736	**protosaris** Meyrick, 1915	Guyana
737	**pseudochyta** Meyrick, 1915	Guyana
738	**ptycta** (Walsingham, 1912) (Athleta)	Guatemala
	cenotes (Walsingham, 1912) (Athleta)	Guatemala
	dryotechna (Meyrick, 1915) (Stenoma)	Guyana

739	pumilis (Busck, 1914) (Catarata)	Panama
740	purulenta Zeller, 1877	Brazil
741	pyrgota (Meyrick, 1930) (Stenoma)	Brazil
742	pyrobathra (Meyrick, 1931) (Stenoma)	Brazil (Ba)
743	pythonaea Meyrick, 1916	French Guiana
744	quiescens (Meyrick, 1916) (Stenoma)	French Guiana
745	radicalis (Zeller, 1877) (Stenoma)	Panama
746	radicicola Meyrick, 1932	Peru
747	reciprocella (Walker, 1864) (Cryptolechia)	Brazil (Pa)
748	reductella (Walker, 1864) (Cryptolechia)	Brazil (Am)
	phoebe (Busck, 1911) (Stenoma)	French Guiana
749	refractrix Meyrick, 1930	Brazil
750	renselariana (Stoll, 1781) (Phalaena)	Surinam
	bahiensis (Perty, [1833]) (Pyralis)	Brazil (Ba)
751	reprehensa Meyrick, 1926	Brazil (RS)
	acrobapta Meyrick, 1933	Argentina
752	resiliens Meyrick, 1925	Brazil (Pa)
753	rhipidaula (Meyrick, 1915) (Stenoma)	Guyana
754	ribbei Zeller, 1877	Panama
755	rostriformis (Meyrick, 1916) (Stenoma)	French Guiana
756	sana Meyrick, 1926	Colombia
757	sarcinata Meyrick, 1918	French Guiana
758	sardania Meyrick, 1925	Brazil (Pa)
759	scapularis (Meyrick, 1918) (Stenoma)	French Guiana
760	sciospila (Meyrick, 1930) (Stenoma)	Brazil (ES)
761	segmentata (Meyrick, 1915) (Stenoma)	Guyana
762	sellifera Meyrick, 1925	Brazil (Am)
763	semicinerea Zeller, 1877	Panama
	hemitephras Meyrick, 1930	Brazil (Pa)
764	semiovata Meyrick, 1926	Colombia
765	semisignella (Walker, 1864) (Cryptolechia)	Brazil (Am)
	consociella (Walker, 1864) (Cryptolechia), preocc. (not Walker, 1864)	Brazil (Am)
	batesella (Walker, 1866) (Cryptolechia), repl. name	Brazil (Am)
	consonella (Busck, 1934) (Stenoma), repl. name	Brazil (Am)
	semisiquella (Busck, 1934) (Stenoma), missp.	
766	serangodes Meyrick, 1915	Panama
767	serarcha Meyrick, 1930	Brazil (MG)
768	similis (Busck, 1911) (Stenoma)	French Guiana
769	smileuta Meyrick, 1915	Guyana
770	sortifera Meyrick, 1930	Bolivia
771	sparganota Meyrick, 1915	Guyana
772	spermolitha (Meyrick, 1915) (Stenoma)	Guyana
773	spurca (Zeller, 1855) (Cryptolechia)	Venezuela
	humeriferella (Walker, 1864) (Cryptolechia)	Brazil (Am)
	erschoffii (Zeller, 1877) (Stenoma)	Colombia
774	spurcatella (Walker, 1864) (Cryptolechia)	Brazil?
	chloromis (Meyrick, 1915) (Stenoma)	Guyana
775	staurota Meyrick, 1916	French Guiana
776	sterrhomitra (Meyrick, 1925) (Stenoma)	Brazil (Am)
777	stigmatias (Walsingham, 1913) (Stenoma)	Guatemala
778	stringens Meyrick, 1925	Brazil (Am)
779	stygeropa (Meyrick, 1925) (Stenoma)	Brazil (Am)
780	subdulcis (Meyrick, 1925) (Stenoma)	Brazil (Pa)
	remorsa (Meyrick, 1925) (Stenoma)	Bolivia
781	substricta Meyrick, 1918	French Guiana
782	suffumigata Walsingham, 1897	Grenada
783	superciliosa Meyrick, 1918	French Guiana
784	synercta Meyrick, 1925	Brazil (Pa)
785	tectoria (Meyrick, 1915) (Stenoma)	Guyana
786	teleosema Meyrick, 1925	Brazil (Pa)
787	tempestiva (Meyrick, 1916) (Stenoma)	French Guiana
788	tephrodesma (Meyrick, 1916) (Stenoma)	French Guiana
789	tetrapetra (Meyrick, 1925) (Stenoma)	Brazil (Am)
790	thammii Zeller, 1877	Peru
791	thapsinopa Meyrick, 1916	French Guiana
	clivosa Meyrick, 1918	French Guiana
792	theoretica Meyrick, 1932	Panama
793	thesmophora Meyrick, 1915	Guyana
794	thylacosaris (Meyrick, 1915) (Stenoma)	Guyana
795	thysanodes (Meyrick, 1915) (Stenoma)	Guyana
	cyanopis (Meyrick, 1915) (Stenoma)	Guyana
796	tibialis Zeller, 1877	Brazil
797	tinactis (Meyrick, 1915) (Stenoma)	Guyana
798	tornogramma Meyrick, 1925	Brazil (Pa)
799	tractrix Meyrick, 1925	Brazil (Pa)
800	tremulella (Walker, 1864) (Cryptolechia)	Brazil (Am)
	chelobathra Meyrick, 1916	French Guiana
801	tribomias (Meyrick, 1915) (Stenoma)	Guyana
802	tricapsis (Meyrick, 1930) (Stenoma)	Brazil (SP)
803	trichonota Meyrick, 1926	Brazil (RS)
804	triplectra (Meyrick, 1915) (Stenoma)	Guyana
805	triplintha (Meyrick, 1916) (Aphanoxena)	French Guiana
806	tripustulella (Walker, 1864) (Cryptolechia)	Brazil (Am)
807	trisecta (Walsingham, 1912) (Athleta)	Mexico (Tab)
808	trisinuata Meyrick, 1930	Brazil
	raricilia Meyrick, 1930	Brazil (MG)
809	tritogramma Meyrick, 1925	Brazil (Am)
810	trivallata Meyrick, 1934	Costa Rica
811	trochoscia Meyrick, 1915	Guyana
812	tumens (Meyrick, 1916) (Stenoma)	French Guiana
813	umbratella Walker, 1864	Brazil (Am)

	pallulella (Busck, 1914) (Stenoma)	Panama
	lacertosa Meyrick, 1915	Guyana
814	umbriferella (Walker, 1864) (Cryptolechia)	Brazil (Am)
815	unisecta (Meyrick, 1930) (Stenoma)	Brazil (ES)
816	vacata Meyrick, 1925	Grenada
817	vannifera (Meyrick, 1915) (Stenoma)	Peru
	asphalopis (Meyrick, 1925) (Stenoma)	French Guiana
818	venatum (Busck, 1911) (Stenoma)	French Guiana
819	venezuelensis Amsel, 1956	Venezuela
820	virens (Meyrick, 1912) (Prasolithites)	Colombia
	viridis (Busck, 1914) (Stenoma)	Panama
821	walchiana (Stoll, 1782) (Phalaena)	Surinam
	dorsella (Fabricius, 1787) (Tinea)	French Guiana
	griseana (Fabricius, 1794) (Pyralis)	"West Indies"
	dorsella (Fabricius, 1794) (Tinea), redesc.	French Guiana
	lativittella (Walker, 1864) (Cryptolechia)	Brazil
	suppressella (Walker, 1864) (Cryptolechia)	Brazil (Am)
	glaciata Meyrick, 1909	Bolivia
	carphitis Meyrick, 1912	Brazil (SC)
	dynastis Meyrick, 1915	Peru
	ampherista Meyrick, 1925	Surinam
	forsteri Amsel, 1956	Venezuela
822	xanthopetala (Meyrick, 1931) (Stenoma)	Brazil
823	xuthosaris Meyrick, 1925	Brazil (Am)
824	xylocosma Meyrick, 1916	French Guiana
825	xylurga (Meyrick, 1913) (Stenoma)	Peru
826	zanclogramma (Meyrick, 1915) (Stenoma)	Guyana
827	zelleri Walsingham & Durrant, 1896	Panama
	fumipennis (Busck, 1914) (Stenoma)	Panama
828	zelotes (Walsingham, 1912) (Stenoma)	Mexico (Gue)

BAEONOMA Meyrick, 1916

829	euphanes Meyrick, 1916	French Guiana
830	favillata (Meyrick, 1915) (Stenoma)	Peru
831	helotypa Meyrick, 1916	French Guiana
832	holarga Meyrick, 1916	French Guiana
833	infamis Meyrick, 1925	Brazil (Am)
834	leucodelta (Meyrick, 1914) (Machimia)	Guyana
835	leucophaeella (Walker, 1864) (Cryptolechia)	Brazil (Pa)
836	mastodes Meyrick, 1916	French Guiana
837	orthozona Meyrick, 1916	French Guiana
838	suavis Meyrick, 1916	French Guiana

CATARATA Walsingham, 1912

839	lepisma Walsingham, 1912	Panama
840	obnubila Busck, 1914	Panama
841	stenota Walsingham, 1912	Guatemala

CERCONOTA Meyrick, 1915
Pomphocrita Meyrick, 1930
Cerconata Busck, 1934, missp.

842	acajuti Becker, 1971	Brazil (Pr)
843	achatina (Zeller, 1855) (Cryptolechia)	Colombia
	lembifera (Meyrick, 1915) (Stenoma)	Guyana
	punicea (Meyrick, 1916) (Stenoma)	French Guiana
844	agraria (Meyrick, 1925) (Stenoma)	Bolivia
845	anonella (Sepp, [1852]-55) (Phalaena)	Surinam
	hamon (Busck, 1911) (Stenoma)	French Guiana
	strophalodes (Meyrick, 1915) (Stenoma)	Peru
846	aphanes (Walsingham, 1912) (Stenoma)	Panama
847	armiferella (Walker, 1864) (Cryptolechia)	Brazil (Am)
848	atricassis (Meyrick, 1916) (Stenoma)	French Guiana
849	bathyphaea (Meyrick, 1932) (Stenoma)	Panama
850	brachyplaca (Meyrick, 1926) (Ptilogenes)	Brazil (Am)
851	capnosphaera (Meyrick, 1916) (Stenoma)	French Guiana
852	carbonifer (Busck, 1914) (Stenoma)	Panama
853	censoria (Meyrick, 1915) (Stenoma)	Guyana
854	certiorata (Meyrick, 1932) (Stenoma)	Brazil (SC)
855	congressella (Walker, 1864) (Cryptolechia)	Brazil (Am)
	cycloptila (Meyrick, 1915) (Stenoma)	Guyana
	tyroxesta (Meyrick, 1925) (Stenoma)	Brazil (Am)
	omphacopa (Meyrick, 1931) (Stenoma)	Bolivia
856	consobrina (Meyrick, 1915) (Stenoma)	Guyana
857	dimorpha Duckworth, 1962	Ecuador
858	dryoscia (Meyrick, 1932) (Stenoma)	"Mexico"
859	ebenocista (Meyrick, 1928) (Ptilogenes)	French Guiana
860	emma (Busck, 1911) (Gonioterma)	French Guiana
	physotricha (Meyrick, 1915) (Stenoma)	Venezuela
861	eriacma (Meyrick, 1915) (Stenoma)	Guyana
862	fermentata (Meyrick, 1916) (Stenoma)	French Guiana
863	figularis (Meyrick, 1918) (Stenoma)	French Guiana
864	flexibilis (Meyrick, 1916) (Stenoma)	French Guiana
865	fulminata (Meyrick, 1916) (Stenoma)	French Guiana
866	fusigera (Meyrick, 1915) (Stenoma)	French Guiana
867	hexascia (Meyrick, 1925) (Stenoma)	Brazil (Am)
868	horometra (Meyrick, 1925) (Stenoma)	Peru
869	hydrelaeas (Meyrick, 1931) (Stenoma)	French Guiana
870	impressella (Walker, 1864) (Cryptolechia)	Brazil (Pa)
	prasoleuca (Meyrick, 1916) (Agriophara)	French Guiana
871	inturbatella (Walker, 1864) (Cryptolechia)	Brazil (Am)
	xanthobyrsa (Meyrick, 1915) (Stenoma)	Guyana

872	**ischnoscia** (Meyrick, 1932) (Stenoma)	Brazil (SC)
873	**languescens** (Meyrick, 1915) (Stenoma)	Guyana
874	**lutulenta** (Zeller, 1877) (Cryptolechia)	Brazil
875	**lysalges** (Walsingham, 1913) (Gonioterma)	Panama
876	**machinatrix** (Meyrick, 1925) (Stenoma)	Colombia
877	**melema** (Walsingham, 1913) (Gonioterma)	Panama
	cora (Busck, 1914) (Gonioterma)	Panama
878	**minna** (Busck, 1914) (Gonioterma)	Panama
	orthridia (Meyrick, 1916) (Stenoma)	French Guiana
879	**miseta** (Walsingham, 1913) (Stenoma)	Costa Rica
880	**myrodora** (Meyrick, 1925) (Stenoma)	Brazil (Am)
881	**nimbosa** (Zeller, 1877) (Cryptolechia)	Peru
	vanis (Busck, 1911) (Stenoma)	French Guiana
	bythochroa (Meyrick, 1915) (Gonioterma)	Guyana
882	**nitens** (Butler, 1877) (Cryptolechia)	Brazil (Am)
883	**noverca** (Meyrick, 1916) (Agriophara)	French Guiana
884	**nymphas** (Meyrick, 1916) (Stenoma)	French Guiana
885	**obsordescens** (Meyrick, 1930) (Pomphocrita)	Brazil (Pa)
886	**oceanitis** (Meyrick, 1916) (Stenoma)	French Guiana
887	**palliata** (Walsingham, 1913) (Stenoma)	Guatemala
888	**phaeophanes** (Meyrick, 1912) (Stenoma)	Colombia
889	**ptilosema** Meyrick, 1918	French Guiana
890	**recurrens** (Meyrick, 1925) (Stenoma)	Bolivia
891	**recurvella** (Walker, 1864) (Cryptolechia)	Brazil (Am)
	pseudacma (Meyrick, 1918) (Stenoma)	French Guiana
892	**robiginosa** (Meyrick, 1925) (Stenoma)	Brazil (Am)
893	**rosacea** (Butler, 1877) (Cryptolechia)	Brazil (Am)
	erotarcha (Meyrick, 1915) (Stenoma)	French Guiana
894	**sciaphilina** (Zeller, 1877) (Stenoma)	Panama
	torophragma (Meyrick, 1915) (Stenoma)	Guyana
895	**scolopacina** (Walsingham, 1913) (Stenoma)	Panama
896	**seducta** (Meyrick, 1918) (Stenoma)	French Guiana
897	**siraphora** (Meyrick, 1915) (Stenoma)	French Guiana
	leucosaris (Meyrick, 1925) (Stenoma)	Peru
898	**sphragidopis** (Meyrick, 1915) (Stenoma)	Guyana
899	**stylonota** (Meyrick, 1915) (Stenoma)	Guyana
900	**tabida** (Butler, 1877) (Cryptolechia)	Brazil (Am)
	salutaris (Butler, 1877) (Cryptolechia)	Brazil (Am)
	maroni (Busck, 1911) (Stenoma)	French Guiana
	astacopis (Meyrick, 1930) (Stenoma)	Brazil (Pa)
901	**tholodes** (Meyrick, 1915) (Stenoma)	Guyana
902	**tinctipennis** (Butler, 1877) (Cryptolechia)	Brazil (Am)
903	**tricharacta** (Meyrick, 1925) (Stenoma)	Brazil (Am)
904	**trichoneura** (Meyrick, 1913) (Stenoma)	Venezuela
905	**tridesma** Meyrick, 1915	Guyana
906	**trizeucta** (Meyrick, 1930) (Ptilogenes)	Brazil (Pa)
907	**trochistis** (Meyrick, 1916) (Stenoma)	French Guiana
908	**trymalopa** (Meyrick, 1925) (Stenoma)	Colombia
909	**tumulata** (Meyrick, 1916) (Stenoma)	French Guiana

CHLAMYDASTIS Meyrick, 1916
Ptilogenes Meyrick, 1917

910	**acronitis** (Busck, 1911) (Stenoma)	French Guiana
911	**anamochla** (Meyrick, 1929) (Ptilogenes)	Colombia
912	**ancalota** (Meyrick, 1916) (Agriophara)	French Guiana
913	**apoclina** (Meyrick, 1929) (Ptilogenes)	Colombia
914	**arenaria** (Walsingham, 1913) (Stenoma)	Brazil (ES)
	vividella (Busck, 1914) (Stenoma)	Panama
915	**argocymba** (Meyrick, 1926) (Ptilogenes)	Brazil (Pa)
916	**batrachopis** (Meyrick, 1913) (Agriophara)	Peru
917	**bifida** (Meyrick, 1916) (Agriophara)	French Guiana
918	**byssophanes** (Meyrick, 1926) (Ptilogenes)	Brazil (Am)
919	**caecata** (Meyrick, 1916) (Agriophara)	French Guiana
920	**chionoptila** (Meyrick, 1926) (Ptilogenes)	Brazil (Am)
921	**chionosphena** (Meyrick, 1931) (Ptilogenes)	French Guiana
922	**chlorosticta** (Meyrick, 1913) (Agriophara)	Peru
923	**complexa** (Meyrick, 1916) (Agriophara)	French Guiana
924	**crateroptila** (Meyrick, 1918) (Stenoma)	French Guiana
925	**curviliniella** (Busck, 1914) (Catarata)	Panama
926	**cystiodes** (Meyrick, 1916) (Agriophara)	French Guiana
927	**deflexa** (Meyrick, 1916) (Agriophara)	French Guiana
928	**deflua** (Meyrick, 1918) (Ptilogenes)	French Guiana
929	**discors** (Meyrick, 1913) (Agriophara)	Peru
930	**disticha** (Meyrick, 1916) (Agriophara)	French Guiana
931	**dominicae** Duckworth, 1969	Dominica
932	**dryosphaera** (Meyrick, 1926) (Ptilogenes)	Brazil (Am)
933	**elaeostola** (Meyrick, 1930) (Ptilogenes)	Brazil (Pa)
934	**epophrysta** (Meyrick, 1909) (Stenoma)	Peru
935	**forcipata** (Meyrick, 1913) (Agriophara)	Colombia
936	**fragmentella** (Dognin, 1913) (Stenoma)	French Guiana
	ponderata (Meyrick, 1916) (Agriophara)	French Guiana
937	**funicularis** (Meyrick, 1926) (Ptilogenes)	French Guiana
938	**galeomorpha** (Meyrick, 1931) (Ptilogenes)	Brazil (Pe)
939	**gemina** (Zeller, 1855) (Cryptolechia)	Colombia
940	**hemichlora** (Meyrick, 1916) (Agriophara)	French Guiana
941	**ichthyodes** (Meyrick, 1926) (Ptilogenes)	Peru
942	**illita** (Meyrick, 1926) (Ptilogenes)	Peru
943	**inscitum** (Busck, 1911) (Stenoma)	French Guiana
944	**inspectrix** (Meyrick, 1916) (Agriophara)	French Guiana
945	**lactis** (Busck, 1911) (Stenoma)	French Guiana
946	**leptobelisca** (Meyrick, 1929) (Ptilogenes)	Colombia
947	**leucoplasta** (Meyrick, 1926) (Ptilogenes)	Brazil (Am)
948	**leucoptila** (Meyrick, 1918) (Ptilogenes)	French Guiana

	laetifica (Busck, 1920) (Stenoma)	Costa Rica
949	**lichenias** (Meyrick, 1916) (Agriophara)	French Guiana
950	**lithograpta** (Meyrick, 1913) (Agriophara)	Peru
951	**melanometra** (Meyrick, 1926) (Ptilogenes)	Colombia
952	**melanonca** (Meyrick, 1915) (Agriophara)	Guyana
953	**mendoron** (Busck, 1911) (Stenoma)	French Guiana
954	**metacymba** (Meyrick, 1916) (Agriophara)	French Guiana
955	**metacystis** (Meyrick, 1918) (Ptilogenes)	French Guiana
956	**metamochla** (Meyrick, 1931) (Ptilogenes)	Brazil
957	**mochlopa** (Meyrick, 1915) (Agriophara)	Guyana
958	**molinella** (Stoll, 1781) (Phalaena)	Surinam
	apicalis (Busck, 1911) (Stenoma)	French Guiana
959	**monastra** (Meyrick, 1909) (Stenoma)	Peru
960	**morbida** (Zeller, 1877) (Cryptolechia)	Peru
961	**mysticopis** (Meyrick, 1926) (Ptilogenes)	Peru
962	**nestes** (Busck, 1911) (Stenoma)	French Guiana
963	**obnupta** (Meyrick, 1916) (Agriophara)	French Guiana
964	**ommatopa** (Meyrick, 1926) (Ptilogenes)	Bolivia
965	**ophiopa** (Meyrick, 1916) (Agriophara)	French Guiana
966	**orion** (Busck, 1920) (Stenoma)	Guatemala
	rufispinis (Meyrick, 1932) (Ptilogenes)	Colombia
967	**oxyplaca** (Meyrick, 1929) (Ptilogenes)	Colombia
968	**paradromis** (Meyrick, 1915) (Agriophara)	Colombia
969	**perducta** (Meyrick, 1916) (Agriophara)	French Guiana
970	**phytoptera** (Busck, 1914) (Stenoma)	Panama
971	**platyspora** (Meyrick, 1932) (Ptilogenes)	Brazil (SP)
	amblystoma (Meyrick, 1936) (Ptilogenes)	Brazil (RS)
972	**plocogramma** (Meyrick, 1915) (Agriophara)	Guyana
973	**poliopa** (Meyrick, 1916) (Agriophara)	Colombia
974	**praenubila** (Meyrick, 1926) (Ptilogenes)	Brazil (Am)
975	**prudentula** (Meyrick, 1926) (Ptilogenes)	Peru
976	**ptilopa** (Meyrick, 1913) (Agriophara)	Colombia
977	**rhomaeopa** (Meyrick, 1931) (Ptilogenes)	Brazil (ES)
978	**scutellata** (Meyrick, 1916) (Agriophara)	French Guiana
979	**smodicopa** (Meyrick, 1915) (Agriophara)	Brazil (Am)
980	**spectrophthalma** (Meyrick, 1932) (Stenoma)	Bolivia
981	**squamosa** (Walsingham, 1892) (Diastoma)	St. Vincent
982	**stagnicolor** (Meyrick, 1926) (Ptilogenes)	Brazil (Am)
983	**steloglypta** (Meyrick, 1931) (Ptilogenes)	French Guiana
984	**strabonia** (Meyrick, 1930) (Ptilogenes)	Brazil (Pa)
985	**synedra** (Meyrick, 1916) (Agriophara)	Paraguay
986	**trastices** (Busck, 1911) (Stenoma)	French Guiana
	aphrogenes (Meyrick, 1915) (Agriophara)	Guyana
987	**tritypa** (Meyrick, 1909) (Stenoma)	Peru
988	**truncatula** (Meyrick, 1913) (Agriophara)	Venezuela
989	**tryphon** (Busck, 1920) (Stenoma)	Guatemala
990	**ungulifera** (Meyrick, 1929) (Ptilogenes)	Colombia
991	**xylinaspis** (Meyrick, 1915) (Agriophara)	Peru

DIASTOMA Möschler, 1882

992	**nubilella** Möschler, 1882	Surinam

ENERGIA Walsingham, 1912

993	**inopina** Walsingham, 1912	Panama
994	**subversa** Walsingham, 1912	Mexico (Ver)

FALCULINA Zeller, 1877

995	**antitypa** Meyrick, 1917	French Guiana
996	**bella** Duckworth, 1966	Brazil (Am)
997	**caustopis** Meyrick, 1932	Brazil (Am)
998	**kasyi** Duckworth, 1966	Surinam
999	**lepidota** Meyrick, 1916	French Guiana
1000	**ochricostata** Zeller, 1877	C. Am. (?)

GONIOTERMA Walsingham, 1897

1001	**advocata** (Meyrick, 1916) (Stenoma)	French Guiana
1002	**aesiocopia** (Walsingham, 1913) (Stenoma)	Mexico (Ver)
	aphrogramma (Meyrick, 1929) (Stenoma)	Panama
1003	**alsiosum** Walsingham, 1913	Panama
1004	**anna** Busck, 1911	Surinam
1005	**argicerauna** (Meyrick, 1925) (Stenoma)	Colombia
1006	**bolistis** (Meyrick, 1925) (Stenoma)	Brazil (Am)
1007	**bryophanes** (Meyrick, 1915) (Stenoma)	French Guiana
1008	**burmanniana** (Stoll, 1781) (Phalaena)	Surinam
	tortricella (Walker, 1864) (Cryptolechia)	Brazil (Am)
1009	**chlorina** (Kearfott, 1911) (Stenoma)	Brazil (SP)
1010	**choleroptila** (Meyrick, 1915) (Stenoma)	Guyana
1011	**chromolitha** (Meyrick, 1925) (Stenoma)	Bolivia
1012	**compressa** (Walsingham, 1913) (Stenoma)	Mexico (Tab)
	cacoeciella (Amsel, 1956) (Stenoma)	Venezuela
1013	**conchita** Busck, 1920	Guatemala
	desidiosa (Meyrick, 1925) (Stenoma)	Colombia
1014	**crocoptila** (Meyrick, 1915) (Stenoma)	Guyana
1015	**descitum** Walsingham, 1913	Panama
1016	**diatriba** (Walsingham, 1913) (Stenoma)	Guatemala
1017	**dimetropis** (Meyrick, 1932) (Stenoma)	Mexico (Gue)
1018	**expansa** (Meyrick, 1915) (Stenoma)	Brazil (RJ)
1019	**exquisita** Duckworth, 1964	Brazil (SP)
1020	**fastigata** (Meyrick, 1915) (Stenoma)	Guyana
1021	**gubernata** (Meyrick, 1915) (Stenoma)	French Guiana

1022	**ignobilis** (Zeller, 1854) (Cryptolechia)	Brazil
	pauperatella (Walker, 1864) (Cryptolechia)	Brazil (Am)
	anita Busck, 1920	Guatemala
1023	**indecora** (Zeller, 1854) (Cryptolechia)	Brazil
1024	**inga** Busck, 1911	French Guiana
1025	**latipennis** (Zeller, 1877) (Cryptolechia)	Colombia
	algosa (Meyrick, 1916) (Stenoma)	French Guiana
1026	**linteata** (Meyrick, 1916) (Stenoma)	French Guiana
1027	**notifera** (Meyrick, 1915) (Stenoma)	Paraguay
1028	**pacatum** Walsingham, 1913	Guatemala
1029	**periscelta** (Meyrick, 1915) (Stenoma)	Peru
1030	**phortax** Meyrick, 1915	Guyana
	ochrosaris (Meyrick, 1925) (Stenoma)	Brazil (Am)
1031	**pleonastes** (Meyrick, 1915) (Stenoma)	French Guiana
1032	**projecta** (Meyrick, 1915) (Stenoma)	French Guiana
1033	**seppiana** (Stoll, 1781) (Phalaena)	Surinam
	platycolpa (Meyrick, 1915) (Stenoma)	French Guiana
	sceptrifera (Meyrick, 1916) (Stenoma)	French Guiana

HYALOPSEUSTIS Meyrick, 1925

1034	**vitrea** Meyrick, 1925	Peru

LETHATA Duckworth, 1964

1035	**aletha** Duckworth, 1967	Peru
1036	**amazona** Duckworth, 1967	Brazil (Am)
1037	**angusta** Duckworth, 1967	Brazil (Pr)
1038	**anophthalma** (Meyrick, 1931) (Stenoma)	Paraguay
	badiella (Amsel, 1956) (Stenoma)	Venezuela
	maculata Duckworth, 1964	Brazil (SC)
1039	**aromatica** (Meyrick, 1915) (Stenoma)	Brazil (SP)
1040	**asthenopa** (Meyrick, 1916) (Stenoma)	French Guiana
1041	**bovinella** (Busck, 1914) (Stenoma)	Panama
	curiata (Meyrick, 1929) (Stenoma)	Panama
	indistincta (Amsel, 1956) (Stenoma)	Venezuela
1042	**buscki** Duckworth, 1964	Belize
1043	**dispersa** Duckworth, 1967	Brazil (MT)
1044	**fernandezyepezi** Duckworth, 1967	Venezuela
1045	**fusca** Duckworth, 1964	Brazil (Pa)
1046	**glaucopa** (Meyrick, 1912) (Stenoma)	Colombia
1047	**gypsolitha** (Meyrick, 1931) (Stenoma)	Paraguay
1048	**herbacea** (Meyrick, 1931) (Stenoma)	Brazil (SP)
1049	**illustra** Duckworth, 1967	Peru
1050	**irresoluta** Duckworth, 1967	Peru
1051	**lanosa** Duckworth, 1967	Panama
1052	**leucothea** (Busck, 1914) (Stenoma)	Panama
1053	**monopa** Duckworth, 1967	Brazil (RJ)
1054	**mucida** Duckworth, 1967	Colombia
1055	**myopina** (Zeller, 1877) (Cryptolechia)	Brazil
1056	**myrochroa** (Meyrick, 1915) (Stenoma)	Venezuela
1057	**obscura** Duckworth, 1967	Peru
1058	**oculosa** Duckworth, 1967	Brazil (SP)
1059	**optima** Duckworth, 1967	Peru
1060	**psidii** (Sepp, [1852]-55) (Phalaena)	Surinam
	invigilans (Meyrick, 1915) (Stenoma)	French Guiana
1061	**pyrenodes** (Meyrick, 1915) (Stenoma)	Argentina
1062	**ruba** Duckworth, 1964	Brazil (SC)
1063	**satyropa** (Meyrick, 1915) (Stenoma)	French Guiana
1064	**sciophthalma** (Meyrick, 1931) (Stenoma)	Brazil
1065	**striolata** (Meyrick, 1932) (Stenoma)	Brazil (SC)
1066	**trochalosticta** (Walsingham, 1913) (Stenoma)	Panama
	trochilosticta (Busck, 1934) (Stenoma), missp.	

LOXOTOMA Zeller, 1854

1067	**elegans** Zeller, 1854	Colombia
	rhodanthes Meyrick, 1915	Guyana
1068	**seminigrescens** Meyrick, 1932	Brazil (RJ)

MENESTA Clemens, 1860

1069	**astronoma** (Meyrick, 1909) (Stenoma)	Bolivia
1070	**cinereocervina** (Walsingham, 1892) (Gelechia)	St. Vincent
1071	**succinctella** (Walker, 1864) (Gelechia)	Brazil (Am)

MOTHONICA Walsingham, 1912

1072	**cubana** Duckworth, 1969	Cuba
1073	**fluminata** (Meyrick, 1912) (Stenoma)	Colombia
1074	**ocellea** Forbes, 1930	Guatemala
1075	**periapta** Walsingham, 1912	Costa Rica

MYSAROMIMA Meyrick, 1926

1076	**liquescens** Meyrick, 1926	Colombia

ORPHNOLECHIA Meyrick, 1909

1077	**acridula** (Meyrick, 1918) (Stenoma)	French Guiana
1078	**anaphanta** (Meyrick, 1915) (Stenoma)	Brazil (Pa)
1079	**crypsiphragma** Meyrick, 1909	Bolivia
1080	**neastra** (Meyrick, 1915) (Stenoma)	Guyana

PARASCAEAS Meyrick, 1936

1081	**uranophanes** (Meyrick, 1931) (Stenoma)	Colombia
	cyanolampra Meyrick, 1936	Panama

PARASPASTIS Meyrick, 1915

1082	**circographa** Meyrick, 1915	Guyana

PETALOTHYRSA Meyrick, 1921

1083	**microphthalma** Meyrick, 1921	Brazil (Am)

PETASANTHES Meyrick, 1925

1084	**leucactis** Meyrick, 1925	Ecuador

PHELOTROPA Meyrick, 1915

1085	**conversa** Meyrick, 1923	French Guiana
1986	**oenodes** Meyrick, 1915	Guyana

PROMENESTA Busck, 1914

1087	**autampyx** Meyrick, 1925	Peru
	citroscia Meyrick, 1931	Brazil (SP)
1088	**callichlora** Meyrick, 1915	Guyana
1089	**capnocoma** (Meyrick, 1931) (Stenoma)	Brazil (ES)
1090	**chrysampyx** Meyrick, 1915	Guyana
1091	**haplodoxa** Meyrick, 1925	Brazil (Pa)
1092	**isotrocha** Meyrick, 1918	Argentina
1093	**leucomias** Meyrick, 1925	Brazil (Pa)
1094	**lithochroma** Busck, 1914	Panama
1095	**marginella** Busck, 1914	Panama
1096	**solella** (Walker, 1864) (Gelechia)	Brazil (Am)
1097	**triacmopa** (Meyrick, 1931) (Stenoma)	Paraguay

RECTIOSTOMA Becker, 1982, repl. name [23]
     Setiostoma Zeller, 1875, preocc. (R. Felder &
     Rogenhofer, 1875 [Glyphipterigidae])

1098	**argyrobasis** (Duckworth, 1971) (Setiostoma)	Venezuela
1099	**callidora** (Meyrick, 1909) (Setiostoma)	Bolivia
1100	**chrysibasis** (Duckworth, 1971) (Setiostoma)	Brazil (SC)
1101	**cirrhobasis** (Duckworth, 1971) (Setiostoma)	El Salvador
1102	**cnecobasis** (Duckworth, 1971) (Setiostoma)	Bolivia
1103	**earobasis** (Duckworth, 1971) (Setiostoma)	Bolivia
1104	**eusema** (Walsingham, 1914) (Setiostoma)	Guatemala
1105	**fernaldella** (Riley, 1889) (Setiostoma)	USA (Ca)
1106	**flinti** (Duckworth, 1971) (Setiostoma)	Mexico (SLP)
1107	**haemitheia** (R. Felder & Rogenhofer, 1875) (Setiostoma)	Colombia
	chlorobasis (Zeller, 1875) (Setiostoma)	[Peru]
	haemotheia (Meyrick, 1913) (Setiostoma), missp.	
	dietzi (Duckworth, 1971) (Setiostoma)	Colombia
1108	**leuconympha** (Meyrick, 1921) (Setiostoma)	Brazil (Am)
1109	**ochrobasis** (Duckworth, 1971) (Setiostoma)	Bolivia
1110	**silvibasis** (Duckworth, 1971) (Setiostoma)	Venezuela
1111	**thiobasis** (Duckworth, 1971) (Setiostoma)	Brazil (SC)
1112	**xanthobasis** (Zeller, 1875) (Setiostoma)	USA (Tx)
1113	**xuthobasis** (Duckworth, 1971) (Setiostoma)	Colombia

RHODANASSA Meyrick, 1915

1114	**io** (Busck, 1911) (Stenoma)	French Guiana
	callimnestra Meyrick, 1915	French Guiana

STENOMA Zeller, 1839 [24]
     Auxocrossa Zeller, 1854
     Auxocrassa Walsingham, 1882, missp.

1115	**acontiella** (Walker, 1864) (Cryptolechia)	Brazil (Am)
1116	**acratodes** Meyrick, 1916	French Guiana
1117	**adminiculata** Meyrick, 1915	French Guiana
1118	**adoratrix** Meyrick, 1925	Bolivia
1119	**adulans** Meyrick, 1925	Peru
	malacoxesta Meyrick, 1930	Brazil (Pa)
1120	**adustella** (Walker, 1864) (Cryptolechia)	Venezuela
1121	**adytodes** Meyrick, 1925	Peru
1122	**affirmatella** Busck, 1914	Panama
1123	**aggregata** Meyrick, 1916	French Guiana
1124	**albida** (Walker, 1864) (Cryptolechia)	Brazil (Am)
1125	**alligans** (Butler, 1877) (Cryptolechia)	Brazil (Am)
	obelodes Meyrick, 1915	Guyana
1126	**alluvialis** Meyrick, 1925	Peru
1127	**ambiens** Meyrick, 1923	French Guiana
1128	**amphiptera** Meyrick, 1913	Peru
1129	**anaxesta** Meyrick, 1915	Guyana
1130	**ancillaris** Meyrick, 1916	French Guiana
1131	**anconitis** Meyrick, 1915	Guyana
	aconitis Clarke, 1955, missp.	
1132	**ancylacma** Meyrick, 1925	Peru
1133	**anetodes** Meyrick, 1915	Guyana
1134	**annosa** (Butler, 1877) (Cryptolechia)	Brazil (Am)
	cirrhogramma Meyrick, 1930	Brazil (Ba)

	sublunaris Meyrick, 1930	Brazil (Pa)
	agathelpis Meyrick, 1932	Brazil (SC)
1135	antitacta Meyrick, 1925	Peru
1136	aphrophanes Meyrick, 1918	Brazil (SP)
1137	aplytopis Meyrick, 1930	Brazil (Pa)
1138	apsorrhoa Meyrick, 1915	Guyana
1139	aptila Meyrick, 1915	Guyana
1140	argillacea (Zeller, 1877) (Cryptolechia)	Peru
1141	argospora Meyrick, 1915	Guyana
1142	armata (Zeller, 1877) (Cryptolechia)	Brazil (RJ)
1143	ascodes Meyrick, 1915	Guyana
1144	assignata Meyrick, 1918	French Guiana
1145	aterpes Walsingham, 1913	Mexico (Ver)
1146	augescens Meyrick, 1925	Peru
1147	auricoma Meyrick, 1930	Brazil (ES)
1148	aztecana Walsingham, 1913	Mexico (Gue)
1149	balanoptis Meyrick, 1932	Peru
1150	baliandra Meyrick, 1915	Guyana
	theobromae (Busck, 1920) (Zetesima)	Surinam
1151	bathrocentra Meyrick, 1915	Guyana
1152	bathrogramma (Meyrick, 1912) (Orphnolechia)	Venezuela
1153	bathyntis Meyrick, 1931	Brazil (ES)
1154	benigna Meyrick, 1916	French Guiana
1155	biannulata Meyrick, 1930	Brazil (Pa)
1156	bicensa Meyrick, 1915	Brazil (SP)
1157	biseriata (Zeller, 1877) (Cryptolechia)	Brazil
1158	bisignata Meyrick, 1916	French Guiana
1159	blandula Meyrick, 1915	Venezuela
1160	bryocosma Meyrick, 1916	French Guiana
1161	bryoxyla Meyrick, 1915	Peru
1162	byssina (Zeller, 1855) (Cryptolechia)	Brazil (Pa)
	tetragonella (Walker, 1864) (Cryptolechia)	Brazil (Am)
	isabella (R. Felder & Rogenhofer, 1875) (Cryptolechia)	Brazil (Am)
1163	bythitis Meyrick, 1915	Guyana
1164	caesarea Meyrick, 1915	Guyana
1165	caesia Meyrick, 1915	Guyana
1166	callicoma Meyrick, 1916	French Guiana
1167	camarodes Meyrick, 1915	French Guiana
1168	camptospila Meyrick, 1925	Brazil (Am)
1169	cana (R. Felder & Rogenhofer, 1875) (Cryptolechia)	Brazil (Am)
	octacentra Meyrick, 1915	Peru
	antilyra Meyrick, 1925	French Guiana
1170	canonias Meyrick, 1913	French Guiana
1171	capnobola Meyrick, 1913	Guyana
1172	caryodesma Meyrick, 1925	Peru
1173	cassigera Meyrick, 1915	French Guiana
1174	castellana Meyrick, 1916	French Guiana
1175	catenifer Walsingham, 1912	Guatemala
1176	catharmosta Meyrick, 1915	Guyana
1177	cathosiota Meyrick, 1925	Brazil (Am)
1178	chalepa Walsingham, 1913	Panama
1179	chalybaeella (Walker, 1864) (Cryptolechia)	Brazil (Am)
1180	charitarcha Meyrick, 1915	Guyana
1181	chionogramma (Meyrick, 1909) (Orphnolechia)	Bolivia
	rectificata Meyrick, 1925	Peru
1182	chloroloba Meyrick, 1915	Peru
1183	chloroplaca (Meyrick, 1915) (Gonioterma)	Guyana
1184	chloroxantha Meyrick, 1925	Brazil (RS)
1185	cholerocrossa Meyrick, 1930	Brazil (ES)
1186	chromatopa Meyrick, 1930	Brazil
1187	chromotechna Meyrick, 1925	Brazil (Am)
1188	citroxantha Meyrick, 1916	French Guiana
1189	claripennis Busck, 1914	Panama
1190	clysmographa Meyrick, 1925	Peru
1191	codicata Meyrick, 1916	French Guiana
1192	colligata Meyrick, 1915	Guyana
1193	collybista (Meyrick, 1915) (Gonioterma)	Peru
1194	columbaris Meyrick, 1909	Peru
1195	comma Busck, 1911	French Guiana
	melanocrypta Meyrick, 1915	Guyana
1196	commutata (Meyrick, 1926) (Eumiturga)	Brazil (Am)
1197	completella (Walker, 1864) (Cryptolechia)	Brazil (Am)
1198	compsocharis Meyrick, 1925	Bolivia
1199	compsocoma Meyrick, 1930	Brazil (Pa)
1200	condemnatrix Meyrick, 1930	Colombia
1201	congrua Meyrick, 1925	Peru
1202	coniophaea Meyrick, 1930	Brazil (Pa)
1203	consociella (Walker, 1864) (Cryptolechia)	Brazil (Am)
	petrina Walsingham, 1912	Guatemala
1204	conveniens Meyrick, 1925	Brazil (Am)
1205	convexicostata (Zeller, 1877) (Cryptolechia)	Brazil (RJ)
	liniella Busck, 1910	Costa Rica
	cantatrix Meyrick, 1925	Bolivia
1206	corvula (Meyrick, 1912) (Antaeotricha)	Colombia
1207	crambina Busck, 1920	Mexico (Gue)
1208	crepitans Meyrick, 1918	French Guiana
	crepitana Busck, 1934, missp.	
1209	crocosticta Meyrick, 1925	Peru
1210	crypsangela Meyrick, 1932	Peru
1211	crypsastra Meyrick, 1915	Guyana
1212	crypsetaera Meyrick, 1925	Brazil (Am)
1213	curtipennis (Butler, 1877) (Cryptolechia)	Brazil (Am)
1214	cyanarcha Meyrick, 1915	Guyana

1215	cycnographa Meyrick, 1930	Brazil (Pa)
1216	cymbalista Meyrick, 1918	French Guiana
1217	cyphoxantha Meyrick, 1931	Brazil (ES)
1218	dasyneura Meyrick, 1923	French Guiana
1219	decora (Zeller, 1854) (Cryptolechia)	Brazil
1220	delphinodes Meyrick, 1925	Peru
1221	deltomis Meyrick, 1925	Brazil (Am)
1222	deuteropa Meyrick, 1931	Paraguay
1223	diametrica Meyrick, 1926	Colombia
1224	dilinopa Meyrick, 1925	Brazil (Am)
1225	diorista (Meyrick, 1929) (Ptilogenes)	Colombia
1226	discrepans Meyrick, 1925	Peru
	discrepana Busck, 1934, missp.	
1227	dispilella (Walker, 1866) (Cryptolechia)	[Brazil]
1228	dorcadopa Meyrick, 1916	French Guiana
1229	dryaula Meyrick, 1925	Brazil (Pa)
1230	dryoconis Meyrick, 1930	Bolivia
1231	dryocosma Meyrick, 1918	French Guiana
1232	elaeurga Meyrick, 1926	Bolivia
1233	embythia Meyrick, 1916	French Guiana
1234	eminens Meyrick, 1918	French Guiana
1235	emphatica Meyrick, 1916	French Guiana
1236	empyrota Meyrick, 1915, repl. name	Argentina
	tortricella (Staudinger, 1899) (Cryptolechia), preocc. (not Walker, 1864)	Argentina
1237	enumerata Meyrick, 1932	Brazil (RS)
1238	epicnesta Meyrick, 1915	Guyana
1239	epicta Walsingham, 1912	Mexico (Tab)
1240	epipacta Meyrick, 1915	Guyana
1241	eumenodora Meyrick, 1937	Argentina
1242	eusticta Meyrick, 1916	French Guiana
1243	eva Meyrick, 1915	Guyana
1244	evanescens (Butler, 1877) (Cryptolechia)	Brazil (Am)
1245	exarata (Zeller, 1854) (Cryptolechia)	Brazil (Pa)
1246	exempta Meyrick, 1925	Brazil
1247	exhalata Meyrick, 1915	French Guiana
1248	explicita Meyrick, 1930	Brazil (Pa)
1249	externella (Walker, 1864) (Cryptolechia)	Brazil (Am)
	megaspilella (Walker, 1864) (Cryptolechia)	Brazil (Am)
	aggerata Meyrick, 1915	Guyana
1250	fallax (Butler, 1877) (Cryptolechia)	Brazil (Am)
1251	farraria Meyrick, 1915	Brazil (SP)
1252	fassliana (Dognin, 1913) (Hilarographa)	Colombia
1253	fenestra Busck, 1914	Panama
	lithoxesta Meyrick, 1915	Guyana
1254	ferculata Meyrick, 1923	French Guiana
1255	ferrocanella (Walker, 1864) (Cryptolechia)	Brazil (Am)
	marcida (Butler, 1877) (Cryptolechia)	Brazil (Am)
	ferricanella Walsingham, 1913, emend.	
1256	finitrix Meyrick, 1925	Colombia
1257	frondifer Busck, 1914	Panama
1258	fulcrata Meyrick, 1915	French Guiana
1259	funerana (Sepp, [1832-40]) (Phalaena)	Surinam
1260	fusistrigella (Walker, 1864) (Cryptolechia)	Brazil (Am)
	faecosa (R. Felder & Rogenhofer, 1875) (Cryptolechia)	Brazil [Am?]
1261	futura Meyrick, 1913	Peru
1262	gemellata Meyrick, 1916	French Guiana
1263	grandaeva (Zeller, 1854) (Cryptolechia)	Brazil (Pa)
	chrysogastra Meyrick, 1915	French Guiana
1264	graphica Busck, 1920	Costa Rica
1265	gymnastis Meyrick, 1915	Guyana
1266	halmas Meyrick, 1925	Peru
1267	haploxyla Meyrick, 1915	Guyana
1268	harpoceros Meyrick, 1930	Brazil (Pa)
1269	hectorea (Meyrick, 1915) (Gonioterma)	Peru
1270	hemilampra Meyrick, 1915	French Guiana
1271	hemiphanta Meyrick, 1925	Brazil (Am)
1272	herifuga Meyrick, 1932	Peru
1273	hesmarcha (Meyrick, 1930) (Ptilogenes)	Brazil (Pa)
1274	heteroxantha Meyrick, 1931	Paraguay
1275	himerodes Meyrick, 1916	French Guiana
1276	holcadica Meyrick, 1916	French Guiana
1277	holophaea (Meyrick, 1916) (Baeonoma)	French Guiana
1278	homala Walsingham, 1912	Mexico
1279	hopfferi (Zeller, 1854) (Auxocrossa)	Brazil (Pa)
	phyllocosma Meyrick, 1916	French Guiana
1280	hoplitica Meyrick, 1925	Brazil (Am)
1281	horocharis Meyrick, 1930	Brazil (Pa)
1282	horocyma Meyrick, 1925	Brazil (Am)
1283	hospitalis Meyrick, 1915	Brazil (RJ)
1284	hyacinthitis Meyrick, 1930	Guyana
1285	hyalocryptis Meyrick, 1930	Brazil (Pa)
1286	hydraena Meyrick, 1916	French Guiana
1287	hypocirrha Meyrick, 1930	Brazil (Am)
1288	iatma Meyrick, 1915	French Guiana
1289	icteropis Meyrick, 1925	Brazil (Pa)
1290	immersa Walsingham, 1913	Mexico (Gue)
1291	immunda (Zeller, 1854) (Cryptolechia)	Brazil (RJ)
	tectella (Walker, 1864) (Cryptolechia)	Brazil (Am)
	perophora Meyrick, 1915	Peru
1292	impressella (Busck, 1914) (Gonioterma)	Panama
	cecropia Meyrick, 1916, repl. name	Panama
1293	impurata Meyrick, 1915	Guyana

1294	**inardescens** Meyrick, 1925	Brazil (Am)
1295	**inflata** (Butler, 1877) (Cryptolechia)	Brazil (Am)
	stella (Busck, 1911) (Gonioterma)	French Guiana
	delenita (Butler, 1877) (Cryptolechia)	French Guiana
1296	**infusa** Meyrick, 1916	French Guiana
1297	**injucunda** Meyrick, 1925	Peru
1298	**invulgata** Meyrick, 1915	Venezuela
1299	**iocoma** Meyrick, 1915	French Guiana
1300	**iopercna** Meyrick, 1932	Peru
1301	**iostalacta** Meyrick, 1925	Peru
1302	**irascens** Meyrick, 1930	Brazil (Pa)
1303	**jucunda** Meyrick, 1915	Peru
1304	**klemaniana** (Stoll, 1782) (Phalaena)	Surinam
1305	**lapidea** Meyrick, 1916	French Guiana
1306	**lapilella** (Busck, 1914) (Catarata)	Panama
	involucralis Meyrick, 1931	Brazil (ES)
1307	**latitans** (Dognin, 1905) (Cryptolechia)	Brazil
1308	**lavata** Walsingham, 1913	Mexico (Tab)
1309	**leptogma** Meyrick, 1925	Colombia
1310	**leucana** (Sepp, [1832–40]) (Phalaena)	Surinam
	tenera (Zeller, 1854) (Cryptolechia)	Brazil (Pa)
	virginalis (Butler, 1877) (Cryptolechia)	Brazil (Am)
	neanica Walsingham, 1913	Panama
1311	**leucaniella** (Walker, 1864) (Cryptolechia)	Venezuela
1312	**libertina** Meyrick, 1916	French Guiana
1313	**litura** Zeller, 1839	Brazil
1314	**lucidiorella** (Walker, 1864) (Cryptolechia)	Brazil (Pa)
	cretifera (R. Felder & Rogenhofer, 1875) (Cryptolechia)	Colombia
	javarica (Butler, 1877) (Cryptolechia)	Brazil (Am)
	paramochla Meyrick, 1918	French Guiana
1315	**luctifica** Zeller, 1877	Panama
1316	**macraulax** Meyrick, 1930	Brazil (RJ)
1317	**macroptycha** Meyrick, 1930	Panama
	spermidias Meyrick, 1932	Panama
1318	**melanesia** Meyrick, 1912	Colombia
1319	**melanixa** Meyrick, 1912	Colombia
	acrosticta Walsingham, 1913	Guatemala
1320	**meligrapta** Meyrick, 1925	Brazil
1321	**melixesta** Meyrick, 1925	Colombia
1322	**meridogramma** Meyrick, 1930	Brazil (Pa)
1323	**methystica** Meyrick, 1930	Brazil (Pa)
1324	**metroleuca** Meyrick, 1930	Brazil (Pa)
1325	**meyeriana** (Stoll, 1781) (Phalaena)	Surinam
1326	**milichodes** Meyrick, 1915	Colombia
1327	**minor** Busck, 1914	Panama
1328	**mniodora** Meyrick, 1925	Colombia
1329	**mendula** Meyrick, 1916	French Guiana
1330	**muscula** Zeller, 1877	Panama
	sciocnesta Meyrick, 1925	Brazil (Am)
1331	**myrrhinopa** Meyrick, 1932	Brazil (Am)
1332	**nebrita** Walsingham, 1913	Panama
1333	**negotiosa** Meyrick, 1925	Brazil (Am)
1334	**neopercna** Meyrick, 1930	Brazil (Pa)
1335	**neoptila** Meyrick, 1925	Brazil (Pa)
1336	**neurocentra** Meyrick, 1925	Peru
1337	**neurotona** (Meyrick, 1915) (Athleta)	Guyana
1338	**nigricans** (Busck, 1914) (Athleta)	Panama
1339	**niphacma** Meyrick, 1916	French Guiana
1340	**niphochlaena** (Meyrick, 1926) (Ptilogenes)	Peru
1341	**nycteropa** Meyrick, 1915	Guyana
1342	**oblita** (Butler, 1877) (Cryptolechia)	Brazil (Am)
	patula Meyrick, 1916	French Guiana
1343	**obovata** Meyrick, 1931	Brazil (RS)
1344	**ochlodes** Walsingham, 1912	Panama
1345	**ochricollis** Zeller, 1877	Panama
	marginata Busck, 1914	Panama
	atmodes Meyrick, 1915	Peru
	thymiota Meyrick, 1915	Guyana
1346	**ochropa** Walsingham, 1913	Panama
	ocellata (Busck, 1914) (Catarata)	Panama
1347	**ochrothicta** Meyrick, 1925	Brazil (Pa)
	ochrothicata Busck, 1934, missp.	
1348	**orneopis** Meyrick, 1925	Brazil (Am)
	arridens Meyrick, 1931	French Guiana
1349	**orthocapna** Meyrick, 1912	Guyana
1350	**orthographa** Meyrick, 1925	Brazil (Am)
1351	**ortholampra** Meyrick, 1930	Brazil (SP)
1352	**ovatella** (Walker, 1864) (Cryptolechia)	Brazil (Am)
	genetta (R. Felder & Rogenhofer, 1875) (Cryptolechia)	Brazil (Am)
1353	**oxyscia** Meyrick, 1923	French Guiana
1354	**pantogenes** Meyrick, 1930	Brazil (SC)
1355	**paracapna** Meyrick, 1915	Guyana
1356	**paraplecta** Meyrick, 1925	Brazil (Am)
1357	**pardalodes** Meyrick, 1918	French Guiana
1358	**paropta** Meyrick, 1916	French Guiana
1359	**patens** Meyrick, 1913	Peru
1360	**peccans** (Butler, 1877) (Cryptolechia)	Brazil (Pa)
	binodis Meyrick, 1915	French Guiana
1361	**periaula** Meyrick, 1916	French Guiana
1362	**peridesma** Meyrick, 1925	Brazil (Pa)
1363	**perirrhoa** Meyrick, 1930	French Guiana
1364	**perjecta** Meyrick, 1931	Brazil
1365	**peronia** Busck, 1913	Guyana
	ebria Meyrick, 1915	Venezuela
1366	**persita** Meyrick, 1915	Peru
1367	**pertinax** Meyrick, 1915	Peru
1368	**phaeomystis** Meyrick, 1925	Peru
1369	**phalacropa** Meyrick, 1932	Panama
1370	**phylloxantha** Meyrick, 1933	Argentina
1371	**picrantis** Meyrick, 1930	Brazil (Pa)
1372	**picta** (Zeller, 1854) (Cryptolechia)	Brazil (SP)
1373	**plagosa** (Zeller, 1877) (Antaeotricha)	Brazil
1374	**platyphylla** Meyrick, 1916	French Guiana
1375	**platyterma** Meyrick, 1915	Guyana
1376	**plurima** Walsingham, 1912	Guatemala
1377	**porphyrastis** Meyrick, 1915	Guyana
1378	**praecauta** Meyrick, 1916	French Guiana
1379	**procritica** Meyrick, 1925	Brazil (Am)
1380	**promotella** Zeller, 1877	Panama
	associata Meyrick, 1925	Peru
1381	**psalmographa** Meyrick, 1931	Brazil (ES)
1382	**psilomorpha** Meyrick, 1915	Peru
1383	**ptychobathra** Meyrick, 1930	Brazil (Pa)
1384	**ptychocentra** Meyrick, 1916	French Guiana
1385	**ptychophthalma** Meyrick, 1930	Bolivia
1386	**pustulatella** (Walker, 1864) (Cryptolechia)	Brazil (Am)
1387	**pyramidea** Walsingham, 1913	Guatemala
1388	**pyrrhias** Meyrick, 1915	Guyana
1389	**pyrrhonota** Meyrick, 1915	Guyana
1390	**receptella** (Walker, 1864) (Cryptolechia)	Brazil (Am)
1391	**recondita** Meyrick, 1915	Guyana
1392	**redintegrata** Meyrick, 1925	Bolivia
1393	**regesta** Meyrick, 1926	Peru
1394	**relata** Meyrick, 1925	Brazil (Am)
1395	**residuella** (Zeller, 1877) (Cryptolechia)	S. Am.
1396	**rhodocolpa** Meyrick, 1916	French Guiana
	orthroptila Meyrick, 1936	Peru
1397	**rhothiodes** Meyrick, 1915	Brazil (SP)
1398	**rosa** (Busck, 1911) (Gonioterma)	French Guiana
1399	**sagax** Busck, 1914	Panama
1400	**salome** Busck, 1911	Brazil (Pr)
	clavifera Meyrick, 1912	Brazil (SP)
1401	**salubris** Meyrick, 1925	Brazil (Am)
1402	**satelles** Meyrick, 1925	Brazil (Am)
1403	**sciogama** Meyrick, 1930	Brazil (Ba)
1404	**scitiorella** (Walker, 1864) (Cryptolechia)	Brazil (Am)
	laeviuscula Zeller, 1877	Colombia
	felix Busck, 1914	Panama
	argotoma Meyrick, 1915	Guyana
1405	**scoriodes** (Meyrick, 1915) (Orphnolechia)	Guyana
	avida Meyrick, 1915	Guyana
1406	**secundata** Meyrick, 1925	Peru
1407	**sematopa** Meyrick, 1915	Guyana
1408	**sequestra** Meyrick, 1918	French Guiana
1409	**sericata** (Butler, 1877) (Cryptolechia)	Brazil (Am)
	eminula Meyrick, 1915	French Guiana
1410	**sesquitertia** (Zeller, 1854) (Cryptolechia)	Brazil
	graphiphorella (Walker, 1864) (Cryptolechia)	[Brazil]
1411	**sexmaculata** (Dognin, 1904) (Cryptolechia)	Ecuador
1412	**simplex** Busck, 1914	Panama
1413	**sommerella** (Zeller, 1877) (Cryptolechia)	Brazil
	nymphotima Meyrick, 1931	Brazil (Ba)
	xylograpta Meyrick, 1931	Brazil (SP)
1414	**spodinopis** Meyrick, 1931	Brazil
1415	**stabilis** (Butler, 1877) (Cryptolechia)	Brazil (Am)
	chionodora Meyrick, 1915	Guyana
	rita (Busck, 1920) (Gonioterma)	Guyana
1416	**staudingerana** (Maassen, 1890) (Tortrix)	Colombia
	contophora Meyrick, 1915	French Guiana
	heterosema Meyrick, 1930	Brazil (Pa)
1417	**stephanodes** Meyrick, 1931	French Guiana
1418	**straminella** (Walker, 1864) (Cryptolechia)	Brazil (Am)
1419	**strenuella** (Walker, 1864) (Cryptolechia)	Brazil (Am)
1420	**striatella** Busck, 1914	Panama
1421	**strigivenata** (Butler, 1877) (Cryptolechia)	Brazil (Am)
	urbana (Butler, 1877) (Cryptolechia)	Brazil (Am)
	entephras Meyrick, 1915	Guyana
	porinodes Meyrick, 1915	Peru
	clarkei Amsel, 1956	Venezuela
1422	**stupefacta** Meyrick, 1916	French Guiana
1423	**subita** Meyrick, 1925	Bolivia
1424	**sublimbata** (Zeller, 1877) (Antaeotricha)	Panama
1425	**subnotatella** (Walker, 1864) (Cryptolechia)	Brazil (Am)
1426	**surinamella** (Möschler, 1882) (Cryptolechia)	Surinam
	expilata Meyrick, 1915	French Guiana
1427	**sustentata** Meyrick, 1926	Colombia
1428	**symmicta** Walsingham, 1913	Panama
1429	**symphonica** Meyrick, 1916	French Guiana
1430	**symposias** (Meyrick, 1915) (Gonioterma)	Colombia
	nepheloleuca Meyrick, 1932	Costa Rica
1431	**syngraphopis** Meyrick, 1930	Brazil (Pa)
1432	**tetrabola** Meyrick, 1913	Peru
1433	**thaleropa** Meyrick, 1916	French Guiana
1434	**thespia** Meyrick, 1915	French Guiana
1435	**thologramma** Meyrick, 1932	Peru
1436	**thoristes** Busck, 1911	French Guiana

1437	**thylacandra** Meyrick, 1915	French Guiana
1438	**tolmeta** Walsingham, 1912	Panama
1439	**trichocolpa** Meyrick, 1915	French Guiana
	centrodina Meyrick, 1916	French Guiana
1440	**trichorda** Meyrick, 1912	Colombia
1441	**trilineata** (Butler, 1877) (Cryptolechia)	Brazil (Am)
1442	**tripustulata** (Zeller, 1854) (Cryptolechia)	Venezuela
1443	**trirecta** Meyrick, 1931	Brazil
1444	**tristrigata** (Zeller, 1854) (Cryptolechia)	Brazil
	aphanodesma Meyrick, 1915	French Guiana
1445	**tyrocrossa** Meyrick, 1925	Brazil (Am)
1446	**ulosema** Meyrick, 1930	Brazil (RJ)
1447	**umbrinervis** Meyrick, 1930	Brazil (Pa)
1448	**uncticoma** Meyrick, 1916	French Guiana
1449	**unguentata** Meyrick, 1930	Brazil (Pa)
1450	**uruguayensis** (Berg, 1885) (Cryptolechia)	Uruguay
1451	**vaccula** Walsingham, 1913	Mexico (Dur)
1452	**vaga** (Butler, 1877) (Cryptolechia)	Brazil (Am)
	licmaea Meyrick, 1915	Guyana
1453	**vapida** (Butler, 1877) (Cryptolechia)	Brazil (Am)
	acribota Meyrick, 1930	Brazil (Pa)
1454	**vasifera** Meyrick, 1925	Colombia
	unisignis Meyrick, 1932	Bolivia
1455	**ventilatrix** Meyrick, 1916	French Guiana
1456	**vexata** Meyrick, 1915	Guyana
1457	**vinifera** Meyrick, 1916	French Guiana
1458	**viridiceps** (R. Felder & Rogenhofer, 1875) (Anatolmis)	Colombia
	erythropennis (Dognin, 1914) (Dasycera)	Colombia
1459	**vita** (Busck, 1911) (Gonioterma)	French Guiana
1460	**vitreola** Meyrick, 1925	Brazil (Pa)
1461	**vivax** Busck, 1914	Panama
1462	**volitans** Meyrick, 1925	Peru
1463	**xanthophaeella** (Walker, 1864) (Paepia)	Brazil (Am)
1464	**xylinopa** Meyrick, 1925	Bolivia
1465	**ybyrajuba** Becker, 1971	Brazil (MT)
1466	**zephyritis** Meyrick, 1925	Brazil (Pa)
1467	**zobeida** Meyrick, 1931	Mexico (Tab)
	orthopa Meyrick, 1932	Panama

THIOSCELIS Meyrick, 1909

1468	**directrix** Meyrick, 1909	Peru
1469	**fuscata** Duckworth, 1967	Peru
1470	**geranomorpha** Meyrick, 1932	Brazil (RJ)
1471	**lipara** Duckworth, 1967	Brazil (Am)
1472	**whalleyi** Duckworth, 1967	Brazil (MG)

TIMOCRATICA Meyrick, 1912
    Lychnocrates Meyrick, 1926

1473	**agramma** Becker, 1982	Brazil (ES)
1474	**albella** (Zeller, 1839) (Depressaria)	Surinam
1475	**albitogata** Becker, 1982	Brazil (Pr)
1476	**amseli** Duckworth, 1962, repl. name	Venezuela
	albella Amsel, 1956, preocc. (not Zeller, 1839)	Venezuela
1477	**anelaea** (Meyrick, 1932) (Stenoma)	Brazil (Pa)
1478	**argonais** (Meyrick, 1925) (Stenoma)	Brazil (Am)
	argonias Clarke, 1955, missp.	
1479	**bicornuta** Becker, 1982	Brazil (RJ)
1480	**butyrota** (Meyrick, 1929) (Stenoma)	Colombia
	syndicastis (Meyrick, 1929) (Stenoma)	Colombia
1481	**constrictivalva** Becker, 1982	Ecuador
1482	**effluxa** (Meyrick, 1930) (Lychnocrates)	Bolivia
1483	**fraternella** (Busck, 1910) (Stenoma)	Costa Rica
1484	**fuscipalpalis** Becker, 1982	Venezuela
1485	**grandis** (Perty, [1833]) (Yponomeuta)	Brazil (Pi)
1486	**guarani** Becker, 1982	Argentina
1487	**isarga** (Meyrick, 1925) (Stenoma)	Bolivia
1488	**leucocapna** (Meyrick, 1926) (Lychnocrates)	Colombia
1489	**leucorectis** (Meyrick, 1925) (Stenoma)	Brazil (MG)
1490	**longicilia** Becker, 1982	Colombia
1491	**loxotoma** (Busck, 1909) (Stenoma)	Mexico
1492	**macroleuca** (Meyrick, 1932) (Stenoma)	Bolivia
1493	**major** (Busck, 1911) (Stenoma)	Peru
1494	**maturescens** (Meyrick, 1925) (Stenoma)	French Guiana
1495	**megaleuca** (Meyrick, 1912) (Stenoma)	Colombia
1496	**melanocosta** Becker, 1982	Brazil (DF)
1497	**melanostriga** Becker, 1982	Brazil (SC)
1498	**meridionalis** Becker, 1982	Brazil (Pr)
1499	**monotonia** (Strand, 1911) (Cryptolechia)	Ecuador
	isographa Meyrick, 1912	Venezuela
	claudescens Meyrick, 1925	Peru
	crassa Meyrick, 1925	Brazil (Pa)
1500	**nivea** Becker, 1982	Brazil (DF)
1501	**palpalis** (Zeller, 1877) (Cryptolechia)	Brazil
	auxoleuca (Meyrick, 1925) (Stenoma)	Brazil (MG)
	haywardi Busck, 1939	Argentina
1502	**parvifusca** Becker, 1982	Costa Rica
1503	**parvileuca** Becker, 1982	Brazil (SP)
1504	**philomela** (Meyrick, 1925) (Stenoma)	Peru
1505	**pompeiana** Meyrick, 1925	Peru
1506	**spinignatha** Becker, 1982	Peru
1507	**subovalis** (Meyrick, 1932) (Stenoma)	Brazil (Am)
	stomatocosma (Meyrick, 1932) (Stenoma)	Brazil (Am)

1508	**titanoleuca** Becker, 1982	Peru
1509	**venifurcata** Becker, 1982	Brazil (DF)
1510	**xanthosoma** (Dognin, 1913) (Stenoma)	French Guiana
	a) **xanthosoma** (Dognin, 1913) (Stenoma)	French Guiana
	sacra (Meyrick, 1918) (Stenoma)	French Guiana
	b) **leucocephala** Becker, 1982	Venezuela
1511	**xanthotarsa** Becker, 1982	Panama

ZETESIMA Walsingham, 1912

1512	**albipes** (R. Felder & Rogenhofer, 1875) (Simaethis)	Brazil (Am)
1513	**lasia** Walsingham, 1912	Panama
	patellifera (Meyrick, 1931) (Stenoma)	Brazil (RS)
1514	**portentosa** Busck, 1914	Panama
1515	**scoliandra** (Meyrick, 1915) (Stenoma)	Guyana

Oecophorinae

Tribe Oecophorini

ACARTOPHILA Meyrick, 1932

1516	**microsacta** Meyrick, 1932	Costa Rica
1517	**stauromacha** Meyrick, 1932	Colombia

ALICIANA Clarke, 1978

1518	**albella** (Blanchard, 1852) (Epigraphia)	Chile
1519	**geminata** Clarke, 1978	Chile

ALTIURA Clarke, 1978

1520	**maculata** Clarke, 1978	Chile

ALYNDA Clarke, 1978

1521	**cinnamomea** Clarke, 1978	Chile
1522	**sarissa** Clarke, 1978	Chile
1523	**striata** Clarke, 1978	Chile

ANIUTA Clarke, 1978

1524	**melanoma** Clarke, 1978	Chile
1525	**ochroleuca** Clarke, 1978	Chile

ARCTOPODA Butler, 1883
    Polypseustis Dognin, 1908

1526	**maculosa** Butler, 1883	Chile
	cuprea (Dognin, 1908) (Polypseustis)	Chile

ATELOSTICHA Meyrick, 1883
    Crypsynarthra Lower, 1901

1527	**depressariella** (Walker, 1864) (Cryptolechia)	Brazil (Am)
1528	**ochrospora** Meyrick, 1928	Brazil (Am)
1529	**percnotoxa** Meyrick, 1920	Brazil (Am)

ATHA Clarke, 1978

1530	**trimacula** Clarke, 1978	Chile

ATOPOTORNA Meyrick, 1932

1531	**ptychoptila** Meyrick, 1932	Panama

BORKHAUSENIA Hübner, [1825]
    Amaurosetia Stephens, [1835]
    Pseudatemelia Rebel, 1910

1532	**bryotrophoides** (Zeller, 1877) (Oecophora)	Colombia
1533	**commixta** Meyrick, 1928	Peru
1534	**confarreatella** (Zeller, 1877) (Oecophora)	Colombia
1535	**crimnodes** Meyrick, 1912	Argentina
1536	**longipalpis** Meyrick, 1931	Argentina
1537	**minnetta** (Butler, 1883) (Oecophora)	Chile
1538	**praesul** Meyrick, 1931	Argentina
1539	**sensilis** Meyrick, 1926	Peru
1540	**syrmeutis** Meyrick, 1931	Argentina

BRYMBLIA Hodges, 1974

1541	**quadrimaculella** (Chambers, 1875) (Oecophora)	USA (Co)

CALLISTENOMA Butler, 1883

1542	**ustimacula** (Zeller, 1874) (Cryptolechia)	Chile
	zelleri Butler, 1883	Chile

CECIDOLECHIA Kieffer & Jörgensen, 1910
    Cecidolechia Strand, 1911

1543	**maculicostella** Kieffer & Jörgensen, 1910	Argentina
	maculicostella Strand, 1911	Argentina

CHEZALA Walker, 1864

1544	**lunularis** Meyrick, 1926	Colombia
	lunaris Gaede, 1938, missp.	

CORITA Clarke, 1978

1545	**amphichroma** Clarke, 1978	Chile

DECANTHA Busck, 1914

1546	**minuta** Busck, 1914	Panama

DEIA Clarke, 1978

1547	**lineola** Clarke, 1978	Chile

DESPINA Clarke, 1978

1548	**rhodosema** (Meyrick, 1931) (Borkhausenia)	Chile

DINOTROPA Meyrick, 1916

1549	**ochrocrossa** Meyrick, 1916	French Guiana

DITA Clarke, 1978

1550	**fasciatipedella** (Zeller, 1874) (Cryptolechia)	Chile
1551	**phococara** Clarke, 1978	Chile

DOLIOTECHNA Meyrick, 1914

1552	**bimarginata** Meyrick, 1929	Colombia
1553	**designata** Meyrick, 1914	Guyana
1554	**eucentra** Meyrick, 1920	Brazil (Pa)
1555	**hyalophaea** Meyrick, 1914	Guyana
1556	**integra** Meyrick, 1914	Guyana
1557	**orphnopis** Meyrick, 1914	Guyana
1558	**spilocrossa** Meyrick, 1920	Peru
1559	**triplacodes** Meyrick, 1920	Brazil (Am)
1560	**trissobathra** Meyrick, 1920	Peru
1561	**virginea** Meyrick, 1914	Guyana

DYSGNORIMA Zeller, 1877

1562	**subannulata** Zeller, 1877	Colombia

ENDROSIS Hübner, [1825]

1563	**sarcitrella** (Linnaeus, 1758) (Phalaena)	[Sweden?]
	sarcitella (Linnaeus, 1761) (Phalaena), missp.	
	fenestrella (Scopoli, 1763) (Phalaena)	[Yugoslavia]
	lactella ([Denis & Schiffermüller], 1775) (Tinea)	[Austria]
	betulinella (Hübner, [1818-19]) (Tinea)	Europe
	kennicottella Clemens, 1860	USA (Il)
	subditella (Walker, 1864) (Gelechia)	New Zealand
	lacteella [sic] var. antarctica Staudinger, 1899	Chile

EOMICHLA Meyrick, 1916
Astoxena Meyrick, 1930

1564	**argentidisca** (Dognin, 1905) (Borkhausenia)	Brazil (SP)
1565	**clotho** (Meyrick, 1930) (Astoxena)	French Guiana
1565.1	**hallwachsae** Clarke, 1983	Costa Rica
1566	**imperiella** (Busck, 1914) (Peleopoda)	Panama
1567	**irenella** (Busck, 1911) (Peleopoda)	French Guiana
1568	**leucoclista** Meyrick, 1930	Brazil (Pa)
1569	**maroniella** (Busck, 1911) (Peleopoda)	French Guiana
1570	**notandella** (Busck, 1911) (Peleopoda)	French Guiana
1571	**nummulata** Meyrick, 1916	French Guiana
1572	**regiella** (Busck, 1912) (Peleopoda)	Panama
1573	**thysiarcha** Meyrick, 1928	Bolivia
1574	**xystidota** Meyrick, 1918	French Guiana

ERAINA Clarke, 1978

1575	**thamnocephala** Clarke, 1978	Chile

EXOSPHRANTIS Meyrick, 1931

1576	**bibula** Meyrick, 1931	Argentina

HALIMARMARA Meyrick, 1931

1577	**atrivallata** Meyrick, 1931	Brazil (Am)

HARPELLA Schrank, 1802

1578	**semnodoxa** Meyrick, 1931	Chile

HELIOSTIBES Zeller, 1874

1579	**mathewi** Zeller, 1874	Chile

HOFMANNOPHILA Spuler, 1910

1580	**pseudospretella** (Stainton, 1849) (Oecophora)	UK (England)

HYPERSKELES Butler, 1883

1581	**choreutidea** Butler, 1883	Chile

IFEDA Hodges, 1966

1582	**perobtusa** (Meyrick, 1922) (Batrachedra)	Brazil (Am)

INGA Busck, 1908
Doxa Walsingham, 1912
Lysigrapha Meyrick, 1914
Pelomimas Meyrick, 1914
Orsimacha Meyrick, 1914
Siderograptis Meyrick, 1920
Phanerodoxa Meyrick, 1921
Epimoryctis Meyrick, 1930
Horomeristis Meyrick, 1931
Agriotorna Meyrick, 1931

1583	**analis** (Busck, 1914) (Cryptolechia)	Panama
1584	**ancorata** (Walsingham, 1912) (Cryptolechia)	Costa Rica
	capsaria (Meyrick, 1914) (Lysigrapha)	Guyana
1585	**brevisella** (Walker, 1864) (Cryptolechia)	Brazil (Am)
1586	**callierastis** (Meyrick, 1920) (Machimia)	Brazil (Am)
1587	**calycocentra** (Meyrick, 1931) (Horomeristis)	Peru
1588	**camelopis** (Meyrick, 1920) (Atelosticha)	Brazil (Am)
1589	**cancanodes** (Meyrick, 1918) (Machimia)	French Guiana
1590	**catasticta** (Meyrick, 1920) (Atelosticha)	Brazil (Pa)
1591	**caumatias** (Meyrick, 1929) (Machimia)	Colombia
1592	**cerophaea** (Meyrick, 1914) (Machimia)	Guyana
1593	**chlorochroa** (Meyrick, 1912) (Machimia)	Argentina
1594	**cnecodes** (Meyrick, 1920) (Atelosticha)	Brazil (Am)
1595	**concinna** (Meyrick, 1912) (Machimia)	Surinam
1596	**conserva** (Meyrick, 1914) (Machimia)	Guyana
1597	**corystes** (Meyrick, 1914) (Machimia)	Guyana
1598	**crossota** (Walsingham, 1912) (Cryptolechia)	Guatemala
1599	**crucifera** (Busck, 1914) (Psilocorsis)	Panama
1600	**cupidinea** (Meyrick, 1914) (Machimia)	Guyana
1601	**custodita** (Meyrick, 1928) (Machimia)	Brazil (Am)
1602	**cyclophthalma** (Meyrick, 1916) (Atelosticha)	French Guiana
	cyclophthalama Hodges, 1972, missp.	
1603	**deligata** (Meyrick, 1914) (Machimia)	Guyana
1604	**dilecta** (Meyrick, 1920) (Machimia)	Peru
1605	**distorta** (Meyrick, 1920) (Machimia)	Peru
1606	**elaphodes** (Meyrick, 1930) (Machimia)	Bolivia
1607	**empyrea** (Meyrick, 1920) (Machimia)	Brazil (Pa)
1608	**encamina** (Meyrick, 1912) (Machimia)	Venezuela
1609	**entaphrota** (Meyrick, 1915) (Machimia)	Colombia
1610	**erasicosma** (Meyrick, 1916) (Atelosticha)	French Guiana
1611	**eriocnista** (Meyrick, 1931) (Agriotorna)	Brazil (Pa)
1612	**erotias** (Meyrick, 1912) (Machimia)	Colombia
1613	**erythema** (Walsingham, 1912) (Cryptolechia)	Guatemala
	marcella (Busck, 1914) (Cryptolechia)	Panama
1614	**fervida** (Zeller, 1855) (Cryptolechia)	Brazil
	furvida (Walker, 1864) (Cryptolechia), missp.	
1615	**flava** (Zeller, 1839) (Depressaria)	Brazil
1616	**fundigera** (Meyrick, 1912) (Machimia)	Surinam
1617	**furva** (Meyrick, 1916) (Machimia)	French Guiana
1618	**genuina** (Meyrick, 1914) (Machimia)	Guyana
1619	**haemataula** (Meyrick, 1912) (Machimia)	Brazil (SP)
	haematula Hodges, 1972, missp.	
1620	**halosphora** (Meyrick, 1916) (Atelosticha)	French Guiana
1621	**helobia** (Meyrick, 1931) (Machimia)	Paraguay
1622	**hyperbolica** (Meyrick, 1928) (Machimia)	Brazil (Am)
1623	**icterota** (Meyrick, 1914) (Machimia)	Guyana
1624	**incensatella** (Walker, 1864) (Cryptolechia)	Brazil (Am)
1625	**inflammata** (Meyrick, 1916) (Machimia)	French Guiana
1626	**iracunda** (Meyrick, 1914) (Orsimacha)	Guyana
1627	**lacunata** (Meyrick, 1914) (Machimia)	Guyana
1628	**languida** (Meyrick, 1912) (Machimia), n. comb.	Venezuela
1629	**leptophragma** (Meyrick, 1920) (Siderograptis)	Brazil (Pa)
1630	**libidinosa** (Meyrick, 1926) (Machimia)	Colombia
1631	**loxobathra** (Meyrick, 1915) (Cryptolechia)	French Guiana
1632	**meliacta** (Meyrick, 1914) (Machimia)	Guyana
1633	**mercata** (Meyrick, 1914) (Machimia)	Guyana
1634	**mimobathra** (Meyrick, 1920) (Atelosticha)	Brazil (Am)
1635	**mixadelpha** (Meyrick, 1914) (Pelomimas)	Guyana
1636	**molifica** (Meyrick, 1914) (Machimia)	Peru
1637	**molybdopa** (Meyrick, 1920) (Siderograptis)	Brazil (Am)
1638	**mydopis** (Meyrick, 1914) (Machimia)	Guyana
1639	**neospila** (Meyrick, 1928) (Machimia)	Brazil (Am)
1640	**orthodoxa** (Meyrick, 1912) (Machimia)	Brazil (RJ)
1641	**orthophragma** (Meyrick, 1916) (Atelosticha)	French Guiana
1642	**pachybathra** (Meyrick, 1921) (Phanerodoxa)	Brazil (Am)
1643	**pagana** (Meyrick, 1916) (Machimia)	French Guiana
1644	**pagidotis** (Meyrick, 1918) (Machimia)	French Guiana
1645	**percnorma** (Meyrick, 1930) (Epimoryctis)	Brazil (Pe)
1646	**pericyclota** (Meyrick, 1920) (Machimia)	Peru
1647	**perioditis** (Meyrick, 1928) (Machimia)	Peru
1648	**petasodes** (Meyrick, 1914) (Orsimacha)	Guyana

1649	**phaeocrossa** (Meyrick, 1912) (Machimia)	Brazil (RJ)
1650	**plectanota** (Meyrick, 1918) (Machimia)	Guyana
1651	**porpotis** (Meyrick, 1914) (Machimia)	Guyana
1652	**pyrothyris** (Meyrick, 1916) (Machimia)	French Guiana
1653	**pyrrhoxantha** (Meyrick, 1931) (Machimia)	French Guiana
1654	**refuga** (Meyrick, 1916) (Machimia), **n. comb.**	French Guiana
1655	**rhodoclista** (Meyrick, 1916) (Atelosticha)	French Guiana
1656	**rosea** (Meyrick, 1920) (Machimia)	Brazil (Am)
1657	**roseomarginella** (Busck, 1911) (Cryptolechia)	French Guiana
1658	**ruricola** (Meyrick, 1914) (Machimia)	Venezuela
1659	**satura** (Meyrick, 1914) (Machimia)	Guyana
1660	**sciocrates** (Meyrick, 1929) (Machimia)	Colombia
1661	**sciotoxa** (Meyrick, 1914) (Machimia)	Guyana
1662	**semotella** (Walker, 1864) (Gelechia)	Brazil (Am)
1663	**separatella** (Walker, 1864) (Gelechia)	Brazil (Am)
1664	**signifera** (Meyrick, 1914) (Machimia)	Guyana
1665	**sodalis** (Walsingham, 1912) (Doxa)	Brazil (Pa)
	relicta (Meyrick, 1916) (Heliocausta)	Guyana
1666	**sparsiciliella** (Clemens, 1864) (Anesychia)	USA (Va)
	contrariella (Walker, 1864) (Cryptolechia)	"?"
	inscitella (Walker, 1864) (Cryptolechia)	"?"
	atropicta (Zeller, 1875) (Cryptolechia)	"N. Am."
	castigata (Meyrick, 1926) (Cryptolechia)	Costa Rica
1667	**speculatrix** (Meyrick, 1914) (Machimia)	Guyana
1668	**staphylitis** (Meyrick, 1916) (Machimia)	French Guiana
1669	**stativa** (Meyrick, 1920) (Atelosticha)	Peru
1670	**stereodesma** (Meyrick, 1916) (Machimia)	French Guiana
1671	**taboga** (Busck, 1914) (Epicallima)	Panama
1672	**textrina** (Meyrick, 1914) (Machimia)	Guyana
1673	**thermoxantha** (Meyrick, 1914) (Lysigrapha)	Peru
1674	**trailii** (Butler, 1877) (Cryptolechia)	Brazil (Am)
	corallina (Meyrick, 1914) (Machimia)	Guyana
	traili (Meyrick, 1922) (Machimia), missp.	
1675	**trifurcata** (Meyrick, 1918) (Machimia)	French Guiana
1676	**trygaula** (Meyrick, 1912) (Machimia)	Brazil (SP)
1677	**tubicen** (Meyrick, 1921) (Phanerodoxa)	Peru
1678	**versatilis** (Meyrick, 1921) (Phanerodoxa)	Brazil (Pa)
1679	**virginia** (Busck, 1914) (Doxa)	Panama
1680	**voluptaria** (Meyrick, 1914) (Machimia)	Guyana

IRENIA Clarke, 1978

1681	**curvula** Clarke, 1978	Chile
1682	**leucoxantha** Clarke, 1978	Chile

LELITA Clarke, 1978

1683	**acmaea** Clarke, 1978	Chile

LYGRONOMA Meyrick, 1913

1684	**cyanastra** (Meyrick, 1909) (Brenthia)	Bolivia
1685	**sporimaea** Meyrick, 1913	Guyana

MACAROCOSMA Meyrick, 1931

1686	**philochrysa** Meyrick, 1931	Chile

MATHILDANA Clarke, 1941

1687	**auricollis** (Walsingham, 1912) (Dasycera) [25]	Guatemala

MELOCHRYSIS Meyrick, 1916

1688	**heliaca** Meyrick, 1916	French Guiana

ODONNA Clarke, 1982

1689	**passiflorae** Clarke, 1982	Colombia
1690	**xenodora** Clarke, 1982	Colombia

PACHYPHOENIX Butler, 1883
  Tyriomorpha Meyrick, 1918
  Mattea Duckworth, 1966

1691	**phoenissa** (Butler, 1883) (Cryptolechia)	Chile
1692	**sanguinea** Butler, 1883	Chile

PHILOMUSAEA Meyrick, 1931
  Philomusea Clarke, 1978, missp.

1693	**brachyxista** Meyrick, 1931	Chile
1694	**craterias** Meyrick, 1931	Argentina
1695	**elissa** Meyrick, 1931	Argentina
1696	**incommoda** Meyrick, 1931	Argentina
1697	**meniscogramma** Clarke, 1978	Chile

PSEUDOECOPHORA Staudinger, 1899

1698	**vitellinella** Staudinger, 1899	Argentina

PYCNOTARSA Meyrick, 1920

1699	**hydrochroa** Meyrick, 1920	Brazil (Pa)

1700	**sulphurea** (Busck, 1914) (Cryptolechia)	Panama

PYRAMIDOBELA Braun, 1923
  Idioptila Meyrick, 1927

1701	**agyrtodes** (Meyrick, 1927) (Idioptila)	USA (Tx)
1702	**compulsa** Meyrick, 1931	Chile
1703	**epibryas** Meyrick, 1931	Brazil (SP)
1704	**ochrolepra** Powell, 1973	Mexico (Chia)
1705	**tetraphyta** Meyrick, 1931	Mexico (Hid)

RETHA Clarke, 1978

1706	**rustica** Clarke, 1978	Chile

REVONDA Clarke, 1978

1707	**eschara** Clarke, 1978	Chile

STASIXENA Meyrick, 1930

1708	**subagrestis** Meyrick, 1930	Brazil (RJ)

STRUTHOSCELIS Meyrick, 1913

1709	**acrobatica** Meyrick, 1913	Peru
1710	**semiotarsa** Meyrick, 1916	French Guiana

TERESITA Clarke, 1978

1711	**diffinis** (R. Felder & Rogenhofer, 1875) (Cryptolechia)	Chile
1712	**isaura** Clarke, 1978	Chile

THAUMATOLITA Walsingham, 1912

1713	**hamifera** Walsingham, 1912	Mexico (Mor)
1714	**stemonias** Meyrick, 1920	Peru

THEAMA E. M. Hering, 1958

1715	**argyrophorum** E. M. Hering, 1958	Argentina

UTILIA Clarke, 1978

1716	**falcata** Clarke, 1978	Chile
1717	**florinda** Clarke, 1978	Chile
1718	**luridella** (Zeller, 1874) (Cryptolechia)	Chile
1719	**ochracea** (Zeller, 1874) (Cryptolechia)	Chile

ZYMRINA Clarke, 1978

1720	**xanthosema** (Meyrick, 1931) (Borkhausenia)	Chile

Tribe Stathmopodini

ARAUZONA Walker, [1865]

1721	**basalis** Walker, [1865]	Brazil (Am)
1722	**moorei** Busck, 1913	Guyana

SNELLENIA Walsingham, 1889

1723	**flavipennis** (R. Felder & Rogenhofer, 1875) (Eretmocera)	S. Am.?
1724	**latipes** (Walker, [1865]) (Tinaegeria)	Brazil (Pa)

STATHMOPODA Herrich-Schäffer, 1853
  Boocara Butler, 1880
  Placostola Meyrick, 1887
  Erineda Busck, 1909
  Agrioscelis Meyrick, 1913
  Kakivoria Nagano, 1916

1725	**antischema** Meyrick, 1922	Peru
1726	**filicula** Clarke, 1978	Colombia

TINAEGERIA Walker, 1856

1727	**croconympha** Meyrick, 1921	Bolivia
1728	**fasciata** Walker, 1856	Brazil (Pa)
1729	**nephelozyga** Meyrick, 1930	"Brazil"
1730	**ochracea** Walker, 1856	Brazil (Pa)
	aeneiceps (R. Felder & Rogenhofer, 1875) (Eretmocera)	Colombia
1731	**pyromantis** Meyrick, 1921	Peru

Tribe Pleurotini

PLEUROTA Hübner, [1825]
  Eupleuris Hübner, [1825]
  Holoscolia Zeller, 1839
  Protasis Herrich-Schäffer, 1853

1732	**albastrigulella** (Kearfott, 1907) (Dorota [sic])	USA (Ca)

22. ELACHISTIDAE

by V. O. Becker

Elachistinae

DICRANOCTETES Braun, 1918
  Donacivola Busck, [1934]

1	**saccharella** (Busck, [1934]) (Donacivola)	Cuba

ELACHISTA Treitschke, 1833
  Aphelosetia Stephens, 1834
  Cycnodia Herrich-Schäffer, 1853
  Phigalia Chambers, 1875, preocc. (Duponchel, 1829
    [Geometridae])
  Atachia Wocke, 1876
  Neaera Chambers, 1880, preocc. (Robineau-Desvoidy,
    1830 [Diptera])
  Hecista Wallengren, 1881
  Aphigalia Dyar, 1903
  Irenicodes Meyrick, 1919
  Euproteodes Viette, 1954

2	**albisquamella** Zeller, 1877	Colombia
3	**luciliella** Zeller, 1877	Colombia
4	**tersectella** Zeller, 1877	Colombia

23. BLASTOBASIDAE

by V. O. Becker

Symmocinae

EUPOLELLA Fletcher, 1940, repl. name [26]
  Eupolis Meyrick, 1923, preocc. (Cambridge, 1900
    [Arachnida])

1	**stygnota** (Walsingham, 1911) (Glyphidocera)	Panama

GLYPHIDOCERA Walsingham, 1892
  Harpagandra Meyrick, 1918

2	**abiasta** Meyrick, 1936	Venezuela
3	**audax** Walsingham, 1892	St. Vincent
4	**capraria** Meyrick, 1929	Colombia
5	**carribea** Busck, 1910	Trinidad
	caribbea Meyrick, 1925, emend.	
6	**catectis** Meyrick, 1923	Ecuador
7	**cerochra** Meyrick, 1929	Colombia
8	**crocogramma** Meyrick, 1923	Brazil
9	**cryphiodes** (Meyrick, 1918) (Harpagandra)	Guyana
10	**dominicella** Walsingham, 1897	Dominica
11	**drosophaea** Meyrick, 1929	Colombia
12	**elpista** Walsingham, 1911	Panama
13	**eurrhipis** Meyrick, 1929	Colombia
14	**exsiccata** Meyrick, 1914	Guyana
15	**hypochloa** (Walsingham, 1911) (Dichomeris)	Mexico (Son)
16	**illiterata** Meyrick, 1929	Panama
17	**indocilis** Meyrick, 1930	Brazil (Pa)
18	**inurbana** Meyrick, 1914	Guyana
19	**lepidocyma** Meyrick, 1929	Colombia
20	**lophandra** Meyrick, 1929	Brazil (Pa)
21	**melithrepta** Meyrick, 1929	Colombia
22	**notolopha** Meyrick, 1929	Brazil (Pa)
23	**orthoctenis** Meyrick, 1923	Brazil (Pa)
24	**orthotenes** Meyrick, 1929	Colombia
25	**percnoleuca** Meyrick, 1923	Brazil (Am)
26	**perobscura** Walsingham, 1911	Mexico (Tab)
27	**psammolitha** Meyrick, 1923	Brazil
28	**ptilostoma** Meyrick, 1935	Argentina
29	**ptychocryptis** Meyrick, 1929	Colombia
30	**recticostella** Walsingham, 1897	Grenada
31	**reparabilis** Walsingham, 1912	Panama
32	**rhypara** Walsingham, 1911	Mexico (Gue)
33	**sagifera** Meyrick, 1929	Colombia
34	**salinae** Walsingham, 1911	Mexico (Oax)
35	**stenomorpha** Meyrick, 1923	Surinam
36	**thyrsogastra** Meyrick, 1929	Peru
37	**trachyacma** Meyrick, 1931	Paraguay
38	**umbrata** Walsingham, 1911	Guatemala
39	**vestita** Walsingham, 1911	Panama
40	**zophocrossa** Meyrick, 1929	Trinidad

OECIA Walsingham, 1897
  Macroceras Staudinger, 1876, preocc. (Semper,
    1870 [Mollusca])

41	**oecophila** (Staudinger, 1876) (Macroceras)	Italy
	maculata Walsingham, 1897	Brazil (Pa)

SCEPTEA Walsingham, 1911
  Sceptia McDunnough, 1939, missp.

42	**decedens** Walsingham, 1911	Mexico (Tab)

SYMMOCA Hübner, [1825]
  Simoca Weiler, 1877, missp.
  Parasymmoca Rebel, 1903
  Asarista Meyrick, 1935
  Conquassata Gozmány, 1957
  Symmoletria Gozmány, 1963

43	**vetusta** Meyrick, 1931	Brazil (Ba)

Blastobasinae

Tribe Blastobasini

AUXIMOBASIS Walsingham, 1892

44	**administra** Meyrick, 1922	Peru
45	**agrestis** Meyrick, 1922	Brazil (Pa)
46	**angusta** Meyrick, 1922	Colombia
47	**brevipalpella** Walsingham, 1897	Grenada
48	**coffeaella** Busck, 1925	Brazil (SP)
49	**constans** Walsingham, 1897	Virgin Is. (St. Thomas)
50	**flavicillata** Walsingham, 1897	Virgin Is. (St. Thomas)
51	**flavida** Meyrick, 1922	Brazil (Pa)
52	**incretata** Meyrick, 1931	Peru
53	**insularis** Walsingham, 1897	Virgin Is. (St. Thomas)
54	**liberatella** (Walker, 1864) (Gelechia)	Brazil (Am)
55	**normalis** Meyrick, 1918	Ecuador
56	**obstricta** Meyrick, 1918	Guyana
57	**persimilella** Walsingham, 1892	St. Vincent
58	**prolixa** Meyrick, 1922	Peru
59	**variolata** Walsingham, 1897	Virgin Is. (St. Thomas)

BLASTOBASIS Zeller, 1855
  Epistetus Walsingham, 1894
  Prostesis Walsingham, 1908

60	**aequivoca** Meyrick, 1922	Guyana
61	**anachasta** Meyrick, 1928	Brazil (Am)
62	**aphilodes** Meyrick, 1918	Colombia
63	**argillacea** Walsingham, 1897	Virgin Is. (St. Thomas)
64	**atmosema** Meyrick, 1930	Brazil (Pa)
65	**atmozona** Meyrick, 1939	Argentina
66	**candidata** Meyrick, 1922	Peru
67	**commendata** Meyrick, 1922	Brazil (Am)
68	**cophodes** Meyrick, 1918	Peru
69	**crotospila** Meyrick, 1926	Galapagos Is.
70	**ergastulella** Zeller, 1877	Colombia
71	**gracilis** Walsingham, 1897	Grenada
72	**grenadensis** Walsingham, 1897	Grenada
73	**guilandinae** Busck, 1900	USA (Fl)
74	**lecaniella** Busck, 1913	Guyana
75	**leucogona** Zeller, 1877	Colombia
76	**leucozyga** Meyrick, 1936	Venezuela
77	**neozona** (Meyrick, 1918) (Exinotis)	Guyana
78	**ochrobathra** Meyrick, 1921	Guyana
79	**pacalis** Meyrick, 1922	Brazil (Pa)
80	**subolivacea** Walsingham, 1897	Virgin Is. (St. Thomas)
81	**triangularis** Walsingham, 1897	Virgin Is. (St. Thomas)

HOLCOCERA Clemens, 1863
  Catacrypsis Walsingham, 1907
  Cynotes Walsingham, 1907
  Prosodica Walsingham, 1907

82	**adjutrix** Meyrick, 1918	Guyana
83	**aphanes** (Zeller, 1877) (Blastobasis)	Colombia
84	**basiplagata** Walsingham, 1912	Guatemala
85	**chloropeda** Meyrick, 1922	Brazil (Pa)
86	**controversella** (Zeller, 1877) (Blastobasis)	Colombia
87	**cylindroma** Meyrick, 1918	Colombia
88	**digesta** Meyrick, 1922	Brazil (Pa)
89	**eusaris** Meyrick, 1922	Peru
90	**hemiteles** Walsingham, 1912	Guatemala
91	**homochromatica** Walsingham, 1912	Mexico (Gue)
92	**limicola** Meyrick, 1918	Ecuador
*93	**nana** Dietz, 1910	USA (Tx)
94	**nephalia** Walsingham, 1907	Guatemala
95	**orthophrontis** Meyrick, 1932	Bolivia
96	**percnoscia** Meyrick, 1932	Brazil (SC)
97	**pinae** Amsel, 1962	Guatemala
98	**proagorella** (Zeller, 1877) (Blastobasis)	Colombia
99	**pugionaria** Meyrick, 1918	Guyana
*100	**pusilla** Dietz, 1910	USA (Tx)
101	**supletella** (Zeller, 1877) (Blastobasis)	Colombia
102	**sympasta** Meyrick, 1918	Peru
103	**titanica** Walsingham, 1912	Mexico (Oax)

ICONISMA Walsingham, 1897

104	**macrocera** Walsingham, 1897	Virgin Is. (St. Thomas)

METALLOCRATES Meyrick, 1930

105	**transformata** Meyrick, 1930	Brazil (Am)

VALENTINIA Walsingham, 1907

106	**bromeliae** Walsingham, 1912	Mexico (Ver)
107	**neptes** Walsingham, 1912	Mexico (Gue)
108	**tarachodes** Walsingham, 1912	Mexico (Gue)

Tribe Pigritiini

PIGRITIA Clemens, 1860

109	**astuta** Meyrick, 1918	Colombia
110	**biatomella** Walsingham, 1897	Virgin Is. (St. Thomas)
111	**mediocris** Walsingham, 1897	Virgin Is. (St. Thomas)
112	**troctis** Meyrick, 1922	Barbados

24. COLEOPHORIDAE

by V. O. Becker

Coleophorinae

COLEOPHORA Hübner, [1822]
  Eupista Hübner, [1825]
  Apista Hübner, [1825]
  Haploptilia Hübner, [1825]
  Porrectaria Haworth, 1828
  Damophila Curtis, 1832
  Astyages Stephens, 1834
  Metallosetia Stephens, 1834
  Casas Wallengren, 1881
  Casigneta Wallengren, 1881
  Corythangela Meyrick, 1897
  Casignetella Strand, 1928
  Calaritania Mariani, 1943
  Heringiella Börner, 1944
  Tolleophora Căpuse, 1971
  Ionescumia Căpuse, 1971
  Stollia Căpuse, 1971
  Razowskia Căpuse, 1971
  Orghidania Căpuse, 1971
  Frederickoenigia Căpuse, 1971
  Suireia Căpuse, 1971
  Zagulajevia Căpuse, 1971
  Amseliphora Căpuse, 1971
  Nemesia Căpuse, 1971
  Zangheriphora Căpuse, 1971
  Bourgogneja Căpuse, 1971
  Aureliania Căpuse, 1971
  Bacescuia Căpuse, 1971
  Klinzigedia Căpuse, 1971
  Vladdelia Căpuse, 1971
  Klimeschja Căpuse, 1971
  Glaseria Căpuse, 1971
  Valvulongia Căpuse, 1971
  Falkovitshia Căpuse, 1972
  Helopharea Falkovitsh, 1972
  Cricotechna Falkovitsh, 1972
  Plegmidia Falkovitsh, 1972
  Agapalsa Falkovitsh, 1972
  Phylloschema Falkovitsh, 1972
  Bima Falkovitsh, 1972
  Systrophoeca Falkovitsh, 1972
  Aporiptura Falkovitsh, 1972
  Symphypoda Falkovitsh, 1972
  Oedicaula Falkovitsh, 1972
  Argyractinia Falkovitsh, 1972
  Chnoocera Falkovitsh, 1972
  Orthographis Falkovitsh, 1972
  Phagolamia Falkovitsh, 1972
  Monotemachia Falkovitsh, 1972
  Corethropoea Falkovitsh, 1972
  Characia Falkovitsh, 1972
  Perygra Falkovitsh, 1972
  Perygridia Falkovitsh, 1972
  Luzulina Falkovitsh, 1972
  Carpochena Falkovitsh, 1972
  Abaraschia Căpuse, 1973
  Amselghia Căpuse, 1973
  Ardania Căpuse, 1973
  Ascleriductia Căpuse, 1973
  Baraschia Căpuse, 1973
  Benanderpia Căpuse, 1973
  Calcomarginia Căpuse, 1973
  Caleophora Căpuse, 1973, missp.
  Corothropoea Căpuse, 1973
  Cornulivalvulia Căpuse, 1973, missp.
  Dumitrescumia Căpuse, 1973
  Ecebalia Căpuse, 1973
  Globulia Căpuse, 1973
  Hamuliella Căpuse, 1973
  Helvalbia Căpuse, 1973
  Ionnemesia Căpuse, 1973
  Kasyfia Căpuse, 1973
  Kuznetzovvlia Căpuse, 1973

Latisacculia Căpuse, 1973
Longibacillia Căpuse, 1973
Lucidaesia Căpuse, 1973
Lvaria Căpuse, 1973
Membrania Căpuse, 1973
Metapista Căpuse, 1973
Multicoloria Căpuse, 1973
Neugenvia Căpuse, 1973
Nosyrislia Căpuse, 1973
Ortohgraphis Căpuse, 1973, missp.
Oudejansia Căpuse, 1973
Paravalvulia Căpuse, 1973
Patzakia Căpuse, 1973
Postvinculia Căpuse, 1973
Proglaseria Căpuse, 1973
Quadratia Căpuse, 1973
Rhamnia Căpuse, 1973
Sacculia Căpuse, 1973
Scleriductia Căpuse, 1973
Tollsia Căpuse, 1973
Tuberculia Căpuse, 1973
Ulna Căpuse, 1973
Ductispira Căpuse, 1973
Klimeschjosefia Căpuse, 1975

1	**anisota** Meyrick, 1927	Bermuda
2	**breyeri** Pastrana, 1963	Argentina
3	**chiarelliae** Pastrana, 1963	Argentina
4	**decipiens** Walsingham, 1914	Mexico (Gue)
5	**exarga** Meyrick, 1917	Colombia
6	**haywardi** Pastrana, 1963	Argentina
7	**intexta** Meyrick, 1917	Peru
8	**lepyropis** Meyrick, 1921	Brazil (Pa)
9	**pelinopis** Meyrick, 1933	Mexico (DF)
10	**picticornis** Walsingham, 1897[27]	(West Indies)
11	**pulchricornis** Walsingham, 1897	Virgin Is. (St. Croix/St. Thomas)
12	**zymotica** Meyrick, 1931	Paraguay

TOCASTA Busck, 1912

13	**priscella** Busck, 1912	Panama

Batrachedrinae

AMBLYTENES Meyrick, 1930

14	**lunatica** Meyrick, 1930	Brazil (Pa)

BATRACHEDRA Herrich-Schäffer, 1853
  Eustaintonia Spuler, 1910

15	**calator** Hodges, 1966	Mexico (Hid)
16	**comosae** Hodges, 1966	Puerto Rico
*17	**concitata** Meyrick, 1928	USA (NM)
18	**conspersa** Meyrick, 1916	Ecuador
19	**copia** Clarke, 1957	Mexico (Mex)
20	**daduchus** Hodges, 1966	Jamaica
*21	**decoctor** Hodges, 1966, extralim.	
	a) **bermudensis** Hodges, 1966	Bermuda
22	**dolichoscia** Meyrick, 1928	Colombia
*23	**enormis** Meyrick, 1928	USA (NM)
24	**knabi** (Walsingham, 1909) (Homaledra)	Mexico (Ver)
25	**linaria** Clarke, 1957	Mexico (BCS)
26	**meator** Hodges, 1966	St. Lucia
27	**nuciferae** Hodges, 1966	Brazil (Ba)
28	**paritor** Hodges, 1966	Jamaica
29	**repertor** Hodges, 1966	Panama
30	**rixator** Hodges, 1966	Colombia
31	**theca** Clarke, 1957	Mexico (DF)

CHEDRA Hodges, 1966

32	**delector** Hodges, 1966	Chile

25. MOMPHIDAE

by V. O. Becker

ANCHIMOMPHA Clarke, 1965

1	**melaleuca** Clarke, 1965	Chile (Juan Fernandez Is.)

MOMPHA Hübner, [1825]
  Laverna Curtis, 1839
  Lophoptilus Sircom, 1848
  Cyphophora Herrich-Schäffer, 1853
  Psacaphora Herrich-Schäffer, 1853
  Anybia Stainton, 1854
  Wilsonia Clemens, 1864
  Leucophryne Chambers, 1875

2	**agonistes** (Walsingham, 1909) (Laverna)	Guatemala
3	**basisignella** (Zeller, 1877) (Laverna)	Colombia
4	**conspersa** (Walsingham, 1892) (Anybia)	St. Vincent

5	**constellaris** Meyrick, 1931	Chile
6	**conviva** Meyrick, 1931	Argentina
7	**crassinodis** (Zeller, 1877) (Laverna)	Colombia
8	**eloisella** (Clemens, 1860) (Laverna)	USA [Pa?]
	magnatella (Zeller, 1873) (Phyllocnistis)	USA (Ma)
	oenotheraeella (Chambers, 1875) (Laverna), repl. name	USA (Tx)
	lyonetiella (Chambers, 1875) (Laverna)	USA (Tx)
9	**exodias** Meyrick, 1931	Argentina
10	**exsultans** (Zeller, 1877) (Laverna)	Colombia
11	**farinacea** (Walsingham, 1909) (Laverna)	Mexico (Gue)
12	**heterolychna** Meyrick, 1922	Peru
13	**laevinella** (Zeller, 1877) (Laverna)	Colombia
14	**leucochrysis** Meyrick, 1935	Argentina
15	**metallifera** (Walsingham, 1897) (Anybia)	Jamaica
16	**musota** Meyrick, 1917	Peru
17	**obsessa** (Walsingham, 1909) (Laverna)	Mexico (Gue)
18	**ochrosemia** (Zeller, 1877) (Laverna)	Colombia
19	**permota** Meyrick, 1917	Colombia
20	**pernota** (Walsingham, 1909) (Laverna)	Mexico (Gue)
21	**phalaropis** Meyrick, 1922	Peru
22	**piperatella** (Walsingham, 1897) (Anybia)	Virgin Is. (St. Thomas)
23	**polygoni** (Zeller, 1877) (Asychna)	Colombia
24	**sectifera** Meyrick, 1922	Brazil (Am)
25	**sympotica** Meyrick, 1931	Chile
26	**tripunctata** (Walsingham, 1897) (Anybia)	Virgin Is. (St. Thomas)
27	**trithalama** Meyrick, 1922	Brazil (Am)
28	**verruculella** (Zeller, 1877) (Laverna)	Colombia

SYNALLAGMA Busck, 1907

| 29 | **busckiella** Engel, 1907 | USA (Pa) |
| | busckellum Walsingham, 1909, emend. | |

26. AGONOXENIDAE

by V. O. Becker

Blastodacninae

HOMOEOPREPES Walsingham, 1909
   Homeoprepes Hodges, 1978, missp.

1	**felisae** Clarke, 1962	Colombia
2	**sympatrica** Clarke, 1962	Colombia
3	**trochiloides** Walsingham, 1909	Costa Rica

MICROCOLONA Meyrick, 1897

| 4 | **transennata** Meyrick, 1922 | Brazil (Pa) |

NANODACNA Clarke, 1964

5	**ancora** Clarke, 1964	Chile (Juan Fernandez Is.)
6	**indiscriminata** Clarke, 1965	Chile (Juan Fernandez Is.)
7	**logistica** (Meyrick, 1931) (Colonophora)	Argentina
8	**vinacea** (Meyrick, 1922) (Homaledra)	Peru

NICANTHES Meyrick, 1928

| 9 | **rhodoclea** Meyrick, 1928 | Guyana |

PAMMECES Zeller, 1863

10	**albivitella** Zeller, 1863	Venezuela
11	**citraula** Meyrick, 1922	Peru
12	**crocoxysta** Meyrick, 1922	Brazil (Am)
13	**lithochroma** Walsingham, 1897	Dom. Rep.
14	**pallida** Walsingham, 1897	Virgin Is. (St. Thomas)
15	**phlogophora** Walsingham, 1909	Panama
16	**problema** Walsingham, 1915	Colombia

PANCLINTIS Meyrick, 1929

| 17 | **socia** Meyrick, 1929 | Colombia |

PROCHOLA Meyrick, 1915

18	**aedilis** Meyrick, 1915	Guyana
19	**agypsota** Meyrick, 1922	Brazil (Am)
20	**basichlora** Meyrick, 1922	Brazil (Pa)
21	**catacentra** Meyrick, 1922	Brazil (Am)
22	**catholica** Meyrick, 1917	Guyana
23	**chloropis** Meyrick, 1922	Brazil (Am)
24	**euclina** Meyrick, 1922	Peru
25	**fuscula** Forbes, 1931	Puerto Rico
26	**holomorpha** Meyrick, 1931	Paraguay
27	**obstructa** Meyrick, 1915	Ecuador
28	**ochromicta** Meyrick, 1922	Brazil (Am)
29	**oppidana** Meyrick, 1915	Guyana
30	**orphnopa** Meyrick, 1922	Brazil (Am)
31	**orthobasis** Meyrick, 1922	Brazil (Pa)
32	**pervallata** Meyrick, 1922	Brazil (Pa)
33	**prasophanes** Meyrick, 1922	Brazil (Pa)

34	**revecta** Meyrick, 1922	Guyana
35	**semialbata** Meyrick, 1922	Brazil (Am)
36	**sollers** Meyrick, 1917	Guyana
37	**subtincta** (Meyrick, 1922) (Syntetrernis)	Brazil (Am)

SYNTETRERNIS Meyrick, 1922

| 38 | **neocompsa** Meyrick, 1933 | Argentina |
| 39 | **xiphodes** Meyrick, 1922 | Peru |

ZARATHA Walker, 1864

40	**macrocera** R. Felder & Rogenhofer, 1875	Brazil (Am)
41	**mesonyctia** Meyrick, 1909	Bolivia
42	**pterodactylella** Walker, 1864	Brazil (Am)
	niveiventris R. Felder & Rogenhofer, 1875	Colombia

27. COSMOPTERIGIDAE

by V. O. Becker

Antequerinae

ANTEQUERA Clarke, 1941

| 1 | **acertella** (Busck, 1913) (Semioscopis) | USA (Ca) |
| | extimulata (Meyrick, 1928) (Borkhausenia) | Mexico (BCN) |

EUCLEMENSIA Grote, 1878, repl. name
   Hamadryas Clemens, 1864, preocc. (Hübner, 1806 [Nymphalidae])

| *2 | **schwarziella** Busck, [1901] | USA (Az) |

Cosmopteriginae

ANONCIA Clarke, 1941

| 3 | **chordostoma** (Meyrick, 1912) (Cryptolechia) | Argentina |
| 4 | **diveni** (Heinrich, 1921) (Borkhausenia) | USA (Tx) |

APHANOSARA Forbes, 1931

| 5 | **planistes** Forbes, 1931 | Puerto Rico |

COSMOPTERIX Hübner, [1825]
   Cosmopteryx Zeller, 1839, emend.

6	**abnormalis** Walsingham, 1897 (Cosmopteryx [sic])	Haiti
7	**apiculata** Meyrick, 1922 (Cosmopteryx [sic])	Brazil (Am)
8	**astrapias** Walsingham, 1909 (Cosmopteryx [sic])	Mexico (Tab)
9	**attenuatella** (Walker, 1864) (Gelechia)	Jamaica
	flavofasciata Wollaston, 1879 (Cosmopteryx [sic])	St. Helena
	mimetis Meyrick, 1897 (Cosmopteryx [sic])	Australia
	antillia Forbes, 1931 (Cosmopteryx [sic])	Puerto Rico
	superba Gozmány, 1960 (Cosmopteryx [sic])	Egypt
10	**callichalca** Meyrick, 1922 (Cosmopteryx [sic])	Brazil (Am)
11	**citrinopa** Meyrick, 1915 (Cosmopteryx [sic])	Peru
12	**diaphora** Walsingham, 1909 (Cosmopteryx [sic])	Mexico (Gue)
13	**erasmia** Meyrick, 1915 (Cosmopteryx [sic])	Guyana
*14	**facunda** Hodges, 1962	USA (Tx)
15	**inaugurata** Meyrick, 1922 (Cosmopteryx [sic])	Brazil (Pa)
16	**interfracta** Meyrick, 1922 (Cosmopteryx [sic])	Brazil (Pa)
17	**irrubricata** Walsingham, 1909 (Cosmopteryx [sic])	Mexico (Ver)
18	**isotoma** Meyrick, 1915 (Cosmopteryx [sic])	Guyana
19	**nyctiphanes** Meyrick, 1915 (Cosmopteryx [sic])	Ecuador
20	**ochleria** Walsingham, 1909 (Cosmopteryx [sic])	Mexico (Tab)
21	**pentachorda** Meyrick, 1915 (Cosmopteryx [sic])	Peru
22	**pyrozela** Meyrick, 1922 (Cosmopteryx [sic])	Brazil (Pa)
23	**sanctivicentii** Walsingham, 1892 (Cosmopteryx [sic])	St. Vincent
24	**similis** Walsingham, 1897 (Cosmopteryx [sic])	Virgin Is. (St. Croix)
25	**teligera** Meyrick, 1915 (Cosmopteryx [sic])	Colombia
26	**tenax** Meyrick, 1915 (Cosmopteryx [sic])	Colombia
27	**tetragramma** Meyrick, 1915 (Cosmopteryx [sic])	Guyana
28	**thrasyzela** Meyrick, 1915 (Cosmopteryx [sic])	Guyana
29	**venefica** Meyrick, 1915 (Cosmopteryx [sic])	Peru
30	**violenta** (Meyrick, 1916) (Batrachedra)	Guyana
31	**xanthura** Walsingham, 1909 (Cosmopteryx [sic])	Mexico (Tab)

DROMIAULIS Meyrick, 1922

| 32 | **excitata** Meyrick, 1922 | Peru |

ECBALLOGONIA Walsingham, 1912

| 33 | **bimetallica** Walsingham, 1912 | Mexico (Ver) |

HARPOGRAPTIS Meyrick, 1925

| 34 | **eucharacta** (Meyrick, 1922) (Stomopteryx) | Brazil (Pa) |

LABDIA Walker, 1864

| 35 | **cyanocoma** Meyrick, 1922 | Peru |

LIMNAECIA Stainton, 1851
   Lymnaecia Kimball, 1965, missp.

| 36 | **dasytricha** Meyrick, 1917 | Guyana |

MORILOMA Busck, 1912

| 37 | **pardella** Busck, 1912 | Panama |

PYRODERCES Herrich-Schäffer, 1853
   Anatrachyntis Meyrick, 1915
   Sathrobrota Hodges, 1962

38	**albistrigella** (Möschler, 1890) (Batrachedra)	Puerto Rico
39	**rileyi** (Walsingham, 1882) (Batrachedra)	USA (Ga)
	stigmatophora (Walsingham, 1897)	Virgin Is. (St. Thomas)
	(Batrachedra)	

"SCAEOSOPHA" auct. (not Meyrick, 1914)

| 40 | **albicellata** Meyrick, 1931, mispl. | Brazil |
| 41 | **citrocarpa** Meyrick, 1931, mispl. | Brazil (ES) |

SEMATOPTIS Meyrick, 1931

| 42 | **amphilychna** Meyrick, 1931 | Chile |

TELADOMA Busck, 1932

| *43 | **murina** Hodges, 1962 | USA (Az) |
| *44 | **nebula** Hodges, 1978 | USA (Az) |

TRICLONELLA Busck, 1901
   Anorcota Meyrick, 1920
   Pharmacoptis Meyrick, 1932

45	**aglaogramma** Meyrick, 1931	Brazil (ES)
46	**antidectis** (Meyrick, 1914) (Epicallima)	USA (Az)
*47	**bicoloripennis** Hodges, 1962	USA (Tx)
48	**calyptrodes** Meyrick, 1922	Brazil (Am)
49	**chionozona** Meyrick, 1931	Brazil
50	**cruciformis** Meyrick, 1931	Brazil (ES)
51	**diglypta** Meyrick, 1931	Brazil
52	**elliptica** Meyrick, 1916	French Guiana
53	**etearcha** Meyrick, 1920	Peru
54	**euzosta** Walsingham, 1912	Panama
55	**iphicleia** Meyrick, 1924	Costa Rica
56	**philantha** Meyrick, 1920	Brazil (Am)
57	**pictoria** Meyrick, 1916	French Guiana
58	**platyxantha** (Meyrick, 1909) (Promalactis)	Bolivia
59	**rhabdophora** Forbes, 1930	Puerto Rico
	breviramis (Meyrick, 1932) (Pharmacoptis)	?
60	**sequella** Busck, 1914	Panama
61	**trachyxyla** Meyrick, 1920	Brazil (Pa)
62	**triargyra** Meyrick, 1920	Peru
63	**turbinalis** Meyrick, 1933	Argentina
64	**umbrigera** Meyrick, 1929	Panama
65	**xanthota** Walsingham, 1912	Mexico (Gue)

URANGELA Busck, 1912

| 66 | **pygmaea** Busck, 1912 | Panama |

Chrysopeleiinae

ASCALENIA Wocke, 1876

67	**pancrypta** (Meyrick, 1915) (Cholotis)	Cuba
68	**plumbata** (Meyrick, 1915) (Cholotis)	Guyana
69	**praediata** Meyrick, 1922	Peru
70	**revelata** Meyrick, 1922	Brazil (Pa)
71	**vadata** Meyrick, 1922	Peru

ERITARBES Walsingham, 1909

| 72 | **guttata** Busck, 1914 | Panama |
| 73 | **otiosa** Walsingham, 1909 | Mexico (Tab) |

ITHOME Chambers, 1875
   Eriphia Chambers, 1875, preocc. (Latreille, 1817
   [Crustacea])

74	**anthraceuta** (Meyrick, 1915) (Cholotis)	Peru
75	**cathidrota** (Meyrick, 1915) (Cholotis)	Guyana
76	**chersota** (Meyrick, 1915) (Cholotis)	Colombia
77	**concolorella** (Chambers, 1875) (Eriphia)	USA (Tx)
	unimaculella Chambers, 1875	USA (Tx)
	unomaculella auct., missp.	
78	**curvipunctella** (Walsingham, 1892) (Anybia)	St. Vincent
	quinquepunctata (Forbes, 1931) (Eriphia)	Puerto Rico
*79	**edax** Hodges, 1962	USA (Tx)
80	**iresiarcha** (Meyrick, 1915) (Cholotis)	Guyana

81	**pelasta** (Meyrick, 1915) (Cholotis)	Guyana
82	**pernigrella** (Forbes, 1931) (Eriphia)	Puerto Rico
83	**pignerata** (Meyrick, 1922) (Ascalenia)	Peru

LEPTOZESTIS Meyrick, 1924

| 84 | **chalcoptila** (Meyrick, 1922) (Syntomactis) | Peru |

OBITHOME Hodges, 1964

| *85 | **punctiferella** (Busck, 1906) (Mompha) | USA (Tx) |

PERIMEDE Chambers, 1874

86	**annulata** Busck, 1914	Panama
*87	**battis** Hodges, 1962	USA (Az)
88	**catapasta** Walsingham, 1909	Mexico (Gue)
89	**purpurascens** Forbes, 1931	Puerto Rico

PERIPLOCA Braun, 1919

| *90 | **juniperi** Hodges, 1978 | USA (Wy) |

STILBOSIS Clemens, 1860
   Aeaea Chambers, 1874
   Amaurogramma Braun, 1919

91	**alcyonis** Meyrick, 1917	French Guiana
92	**alsocoma** Meyrick, 1917	Guyana
93	**amphibola** Walsingham, 1909	Mexico (Gue)
94	**argyritis** Meyrick, 1922	Brazil (Pa)
95	**chrysorrhabda** Meyrick, 1922	Peru
96	**condylota** Meyrick, 1917	Guyana
97	**devoluta** Meyrick, 1917	Guyana
98	**gnomonica** Meyrick, 1917	Guyana
99	**hypanthes** Meyrick, 1917	Guyana
100	**incincta** Walsingham, 1909	Mexico (Tab)
101	**juvenis** Walsingham, 1909	Mexico (Gue)
102	**lonchocarpella** Busck, [1934]	Cuba
103	**phaeoptera** Forbes, 1931	Puerto Rico
104	**symphracta** Meyrick, 1917	Guyana
105	**tesquella** Clemens, 1860	USA [Pa?]
	tesquatella Chambers, 1878, missp.	
	quinquicristatella (Chambers, 1881) (Laverna)	USA (Ma)

WALSHIA Clemens, 1864

106	**albicornella** Busck, 1914	Panama
107	**calcarata** Walsingham, 1909	Mexico (Tab)
108	**detracta** Walsingham, 1909	Mexico (Gue)
109	**pentapyrga** (Meyrick, 1922) (Mompha)	Brazil

28. SCYTHRIDIDAE

by V. O. Becker

SCYTHRIS Hübner, [1825]
   Butalis Treitschke, 1833, preocc. (Boie, 1826 [Aves])
   Arotrura Walsingham, 1888
   Colinita Busck, 1907

1	**depressa** Meyrick, 1931	Paraguay
2	**dimota** Meyrick, 1931	Paraguay
3	**dividua** Meyrick, 1916	Peru
4	**ejiciens** Meyrick, 1928	Peru
5	**fluvialis** Meyrick, 1916	Colombia
6	**immunis** Meyrick, 1916	Peru
7	**inanima** Meyrick, 1916	Peru
8	**medullata** Meyrick, 1916	Peru
9	**notorrhoa** Meyrick, 1921	Brazil (Am)
10	**plocogastra** Meyrick, 1931	Paraguay
11	**tibicina** Meyrick, 1916	Peru
12	**zeugmatica** Meyrick, 1931	Brazil (Pa)

29. GELECHIIDAE [28]

by V. O. Becker

Anomologinae

ISOPHRICTIS Meyrick, 1917

| *1 | **actiella** Barnes & Busck, 1920 | USA (Ca) |

MEGACRASPEDUS Zeller, 1839
   Neda Chambers, 1874, preocc. (Mulsant, 1850
   [Coleoptera])
   Pycnobathra Lower, 1901
   Autoneda Busck, 1903, repl. name
   Toxoceras Chrétien, 1915
   Megacraspedas Barnes & McDunnough, 1917, missp.

| 2 | **exilis** Walsingham, 1909 | Mexico (Gue) |

MONOCHROA Heinemann, 1870
    Catabrachmia Rebel, 1909

3	**absconditella** (Walker, 1864) (Gelechia)	"N. Am."
	palpiannulella (Chambers, 1872) (Gelechia)	USA (Ky)

NEALYDA Dietz, 1900

4	**accincta** Meyrick, 1923	Brazil (Pa)
5	**bicolor** (Walsingham, 1892)(Didactylota)	St. Vincent
6	**bougainvilleae** E. M. Hering, 1955	Argentina
7	**leucozostra** Meyrick, 1923	Brazil (Pa)
8	**pisoniae** Busck, 1900	USA (Fl)

Gelechiinae

ARISTOTELIA Hübner, [1825]
    Ergatis Heinemann, 1870, preocc. (Blackwell, 1841
      [Arachnida])
    Isochasta Meyrick, 1886
    Eucatoptus Walsingham, 1897

9	**aphiltra** Meyrick, 1917	Peru
10	**argyractis** Meyrick, 1923	Brazil (Am)
11	**calculatrix** Meyrick, 1923	Brazil (Pa)
12	**chalybeichroa** (Walsingham, 1897) (Euca- toptus)	Virgin Is. (St. Thomas)
	chalybochroa Meyrick, 1925, emend.	
13	**chloroneura** Meyrick, 1923	Brazil (Pa)
14	**corallina** Walsingham, 1909	Mexico (Gue)
15	**cosmographa** Meyrick, 1917	Peru
16	**crassicornis** Walsingham, 1897	Virgin Is. (St. John)
17	**cynthia** Meyrick, 1917	Peru
18	**cytheraea** Meyrick, 1917	Colombia
19	**dasypoda** Walsingham, 1910	Mexico (Tab)
20	**diolcella** Forbes, 1931	Puerto Rico
21	**elachistella** (Zeller, 1877) (Gelechia)	Colombia
22	**ephoria** Meyrick, 1917	Peru
23	**erycina** Meyrick, 1917	Ecuador
24	**eupatoriella** Busck, [1934]	Cuba
25	**hieroglyphica** Walsingham, 1909	Mexico (Tab)
26	**howardi** Walsingham, 1909	Mexico (Son)
27	**lignicolora** Forbes, 1931	Puerto Rico
28	**naxia** Meyrick, 1926	Galapagos Is.
29	**oribatis** Meyrick, 1917	Peru
30	**pantalaena** (Walsingham, 1911) (Untomia)	Mexico (Tab)
31	**paphia** Meyrick, 1917	Peru
32	**parephoria** Clarke, 1951	Argentina
33	**paterata** Meyrick, 1914	Guyana
34	**penicillata** (Walsingham, 1897) (Eucatoptus)	Haiti
35	**perfossa** Meyrick, 1917	Ecuador
	radicata Meyrick, 1917	Colombia
36	**perplexa** Clarke, 1951	Argentina
37	**probolopis** Meyrick, 1923	Brazil (Am)
38	**pudibundella** (Zeller, 1873) (Gelechia)	USA (Tx)
	intermediella (Chambers, 1878) (Gelechia)	USA (Tx)
39	**pulicella** Walsingham, 1897	Virgin Is. (St. Thomas)
40	**pyrodercia** Walsingham, 1910	Mexico (Gue)
41	**roseosuffusella** (Clemens, 1860) (Gelechia)	USA [Pa?]
	belella (Walker, 1864) (Gelechia)	"N. Am."
42	**rubidella** (Clemens, 1860) (Gelechia)	USA [Pa?]
	rubensella (Chambers, 1872) (Gelechia)	USA (Ky)
	pudibundella.-(Chambers, 1877) (Gelechia), misid. (not Zeller, 1873)	
43	**sarcodes** Walsingham, 1910	Panama
44	**saturnina** Meyrick, 1917	Peru
45	**squamigera** Walsingham, 1909	Mexico (Gue)
46	**subrosea** Meyrick, 1914	Guyana
47	**trossulella** Walsingham, 1897	Dom. Rep.
48	**vagabundella** Forbes, 1931	Puerto Rico
49	**veteranella** (Zeller, 1877) (Tachyptilia)	Brazil
50	**vicana** Meyrick, 1917	Peru

ARLA Clarke, 1942

*51	**diversella** (Busck, 1916) (Gelechia)	USA (Ca)

AROGA Busck, 1914
    Aruga Janse, 1958, missp.

52	**paraplutella** (Busck, 1910) (Gelechia)	USA (Ca)

AROGALEA Walsingham, 1910

53	**albilingua** Walsingham, 1911	Mexico (Gue)
54	**archaea** Walsingham, 1911	Mexico (Gue)
55	**crocipunctella** (Walsingham, 1892) (Lita)	St. Vincent
56	**melitoptila** (Meyrick, 1923) (Telphusa)	Brazil (Am)
57	**senecta** Walsingham, 1911	Mexico (Gue)
58	**soronella** Busck, 1914	Panama

BARTICEJA Povolný, 1967

59	**epitricha** (Meyrick, 1917) (Phthorimaea)	Guyana
	ochrotoma (Meyrick, 1923) (Telphusa)	Brazil (Pa)

CALLIPRORA Meyrick, 1914

60	**centrocrossa** Meyrick, 1922	Brazil (Am)
61	**clistogramma** Meyrick, 1926	Brazil (SP)
62	**erethistis** Meyrick, 1922	Peru
63	**eurydelta** Meyrick, 1922	Peru
64	**pentagramma** Meyrick, 1914	Guyana
65	**peritura** Meyrick, 1922	Brazil (Pa)
66	**platyxipha** Meyrick, 1922	Brazil (Am)
67	**rhodogramma** Meyrick, 1922	Brazil (Am)
68	**tetraplecta** Meyrick, 1922	Peru
69	**trigramma** Meyrick, 1914	Guyana

CHIONODES Hübner, [1825]
    Chionoda Hübner, [1826], missp.

70	**argosema** (Meyrick, 1917) (Gelechia)	Ecuador
71	**consona** (Meyrick, 1917) (Gelechia)	Peru
72	**dryobathra** (Meyrick, 1917) (Gelechia)	Colombia
73	**eburata** (Meyrick, 1917) (Gelechia)	Colombia
74	**icriodes** (Meyrick, 1931) (Gelechia)	Chile
75	**lacticoma** (Meyrick, 1917) (Gelechia)	Peru
76	**litigiosa** (Meyrick, 1917) (Gelechia)	Ecuador
77	**mediofuscella** (Clemens, 1863) (Gelechia)	USA [Pa?]
	vagella (Walker, 1864) (Gelechia)	Canada
	fuscoochrella (Chambers, 1872) (Depressaria)	USA (Ky)
	liturosella (Zeller, 1873) (Gelechia)	USA (Tx)
	rhedaria (Meyrick, 1923) (Gelechia)	Canada
78	**pentadora** (Meyrick, 1917) (Gelechia)	French Guiana
79	**perissosema** (Meyrick, 1932) (Gelechia)	Argentina
80	**salva** (Meyrick, 1925) (Phthorimaea), repl. name	Virgin Is. (St. Thomas)
	leucocephala (Walsingham, 1897) (Gelechia), preocc. (not Lower, 1893)	Virgin Is. (St. Thomas)
81	**spiridoxa** (Meyrick, 1931) (Gelechia)	Paraguay

COLEOTECHNITES Chambers, 1880
    Evagora Clemens, 1860, preocc. (Péron & Lesueur,
      1810 [Coelenterata])
    Eidothea Chambers, 1873, preocc. (Risso, 1826
      [Mollusca])
    Eidothoa Chambers, 1873, missp.
    Coleotechnistes Riley, 1891, missp.
    Eucordylea Dietz, 1900
    Pulicalvaria T. N. Freeman, 1963

*82	**elucidella** (Barnes & Busck, 1920) (Eucordylea)	USA (Ca)
83	**vagatioella** (Chambers, 1873) (Eidothoa [sic])	USA (Ky)
	dorsivittella (Zeller, 1873) (Gelechia)	USA (Tx)

DELTOPHORA Janse, 1950

84	**caymana** Sattler, 1979	Cayman Is.
85	**duplicata** Sattler, 1979	Cayman Is.
86	**flavocincta** Sattler, 1979	Colombia
87	**glandiferella** (Zeller, 1873) (Gelechia)	USA (Tx)
88	**lanceella** Sattler, 1979	Guyana
89	**minuta** Sattler, 1979	Brazil (Pa)
90	**suffusella** Sattler, 1979	Paraguay

EPHYSTERIS Meyrick, 1908
    Microcraspedus Janse, 1958
    Ephystereris Janse, 1960, missp.
    Echinoglossa Clarke, 1965

91	**trinota** (Clarke, 1965) (Echinoglossa)	Chile (Juan Fernandez Is.)

EUCHIONODES Clarke, 1950

92	**traditionis** Clarke, 1950	Argentina

EUDACTYLOTA Walsingham, 1911

93	**barberella** (Busck, 1903) (Neodactylota)	USA (Az)
94	**iobapta** (Meyrick, 1927) (Stomopteryx)	USA (Tx)

EVIPPE Chambers, 1873
    Phaetusa Chambers, 1875, preocc. (Wagler, 1832 [Aves])
    Tholerostola Meyrick, 1917

95	**aequorea** (Meyrick, 1917) (Recurvaria)	Peru
96	**aulonota** (Meyrick, 1917) (Aristotelia)	Ecuador
97	**evippella** (Forbes, 1931) (Tholerostola)	Puerto Rico
98	**leuconota** (Zeller, 1873) (Gelechia)	USA (Tx)
	plutella (Chambers, 1875) (Phaetusa)	USA (Tx)
99	**omphalopa** (Meyrick, 1917) (Tholerostola)	Ecuador
100	**plumata** (Meyrick, 1917) (Aristotelia)	Guyana

EXOTELEIA Wallengren, 1881
    Paralechia Busck, 1903
    Heringia Spuler, 1910, preocc. (Rondani, 1856 [Diptera])
    Heringiola Strand, 1917, repl. name

101	**ithycosma** (Meyrick, 1914) (Strobisia)	Guyana

FACULTA Busck, 1939

102	**inaequalis** (Busck, 1910) (Gelechia)	USA (NM)
	inaequalis (Walsingham, 1911) (Gelechia), preocc. (not Busck, 1910)	Mexico (Son)
	anisectis (Meyrick, 1923) (Gelechia), repl. name	Mexico (Son)
	clistrodoma (Meyrick, 1923) (Gelechia)	USA (Az)

FASCISTA Busck, 1939

103	**albipectus** (Walsingham, 1911) (Gelechia)	Mexico (Son)
104	**vorax** (Meyrick, 1939) (Phthorimaea)	Argentina

FRISERIA Busck, 1939

105	**acaciella** (Busck, 1906) (Telphusa)	USA (Tx)
*106	**caieta** Hodges, 1966	USA (Az)
107	**cockerelli** (Busck, 1903) (Gelechia)	USA (NM)
	lindenella (Busck, 1903) (Gelechia)	USA (Tx)
	malindella (Busck, 1910) (Gelechia)	USA (NM)
	sarcochlora (Meyrick, 1929) (Gelechia)	USA (NM)
108	**infracta** (Walsingham, 1911) (Gelechia)	Mexico (Gue)
109	**lacticaput** (Walsingham, 1911) (Gelechia),	Mexico (Gue)
	lacticeps (Meyrick, 1925), emend.	
*110	**nona** Hodges, 1966	USA (Az)
111	**paphlactis** (Meyrick, 1912) (Gelechia)	Brazil (SP)
112	**repentina** (Walsingham, 1911) (Gelechia)	Mexico (Gue)

GELECHIA Hübner, [1825]
Guenea Bruand, 1850
Galechia Desmarest, [1857], missp.
Gelschia Nowicki, 1865, missp.
Cirrha Chambers, 1872
Oeseis Chambers, 1875
Gelecia Watt, 1920, missp.

113	**bathrochlora** Meyrick, 1932	Brazil (SC)
114	**bufo** Walsingham, 1911	Mexico (Gue)
115	**cacoderma** Walsingham, 1911	Mexico (Gue)
116	**caespitella** Zeller, 1877	Colombia
117	**cerussata** Walsingham, 1911	Mexico (Ver)
118	**chlorocephala** Meyrick, 1932	Mexico (DF)
119	**clopica** Meyrick, 1931	Argentina
120	**concinna** Walsingham, 1911	Mexico (Gue)
121	**creberrima** Walsingham, 1911	Mexico (Gue)
122	**cuneifera** Walsingham, 1911	Mexico (Gue)
123	**delapsa** Meyrick, 1931	Brazil (ES)
124	**diacmota** Meyrick, 1932	Brazil (SC)
125	**dolbyi** (Walsingham, 1911) (Dichomeris)	Panama
126	**elephantopis** Meyrick, 1936	Venezuela
127	**exclarella** Möschler, 1890	Puerto Rico
128	**flammulella** Walsingham, 1897	Virgin Is. (St. Thomas)
129	**gnathodoxa** Meyrick, 1926	Galapagos Is.
130	**goniospila** Meyrick, 1931	Argentina
131	**hetaeria** Walsingham, 1911	Mexico (Ver)
132	**impurgata** Walsingham, 1911	Mexico (Gue)
133	**lapidescens** Meyrick, 1916, repl. name	Mexico (Son)
	lithodes Walsingham, 1911, preocc. (not Meyrick, 1886)	Mexico (Son)
134	**leptospora** Meyrick, 1932	Costa Rica
135	**nephelophracta** Meyrick, 1932	Costa Rica
136	**neptica** Walsingham, 1911	Mexico (Gue)
137	**nigripectus** Walsingham, 1911	Mexico (Son)
138	**nucifer** Walsingham, 1911	Mexico (Son)
	nucifera Walsingham, 1925, emend.	
139	**ophiaula** Meyrick, 1931	Argentina
140	**ophiomorpha** Meyrick, 1935	Argentina
141	**pertinens** Meyrick, 1931	Paraguay
142	**petraea** Walsingham, 1911	Guatemala
143	**picrogramma** Meyrick, 1929	Guyana
144	**platydoxa** Meyrick, 1923	Guyana
145	**pleroma** Walsingham, 1911	Mexico (Gue)
146	**protozona** Meyrick, 1926	Galapagos Is.
147	**rhypodes** Walsingham, 1911	Mexico (Son)
148	**scotodes** Walsingham, 1911	Mexico (Gue)
149	**sonorensis** Walsingham, 1911	Mexico (Son)
150	**suspensa** Meyrick, 1923	Brazil (Am)
151	**synthetica** Walsingham, 1911	Mexico (Gue)
*152	**thymiata** (Meyrick, 1929) (Nothris)	USA (Az)
153	**traducella** Busck, 1914	Panama
154	**veneranda** Walsingham, 1911	Mexico (Son)
155	**xylobathra** Meyrick, 1936	Venezuela

GNORIMOSCHEMA Busck, 1900
Gnorimochema Dyar, [1903], missp.
Tuta Kieffer & Jörgensen, 1910
Gonorimoschema Deurs, 1954, missp.
Gnorrimoschema Hartig, 1964, missp.
Lerupsia Riedl, 1965
Neoschema Povolný, 1967
Larupsia Soffner, 1967, missp.

156	**atriplicella** Kieffer & Jörgensen, 1910	Argentina

157	**borsaniella** Köhler, 1939	Argentina
158	**cestrivora** Clarke, 1950	Argentina
	cestivora Hayward, 1969, missp.	
159	**dudiella** Busck, 1903	USA (Az)
160	**euchthonia** (Meyrick, 1939) (Phthorimaea)	Argentina
161	**exacta** (Meyrick, 1917) (Phthorimaea)	Guyana
162	**involuta** (Meyrick, 1917) (Phthorimaea)	Guyana
163	**motasi** Povolný, 1976	Colombia
164	**perfidiosa** (Meyrick, 1917) (Phthorimaea)	Colombia
165	**saphirinella** (Chambers, 1875) (Gelechia)	USA (Co)
166	**urosema** (Meyrick, 1917) (Phthorimaea)	Peru

HAPALOSARIS Meyrick, 1917

167	**petulans** Meyrick, 1917	Colombia

KEIFERIA Busck, 1939

168	**brunnea** Povolný, 1973	Virgin Is. (St. Thomas)
169	**colombiana** Povolný, 1975	Colombia
170	**lycopersicella** (Walsingham, 1897) (Eucatoptus)	Virgin Is. (St. Croix)
	lenta (Meyrick, 1917) (Phthorimaea)	Peru
	lycopersicella (Busck, 1928) (Phthorimaea)	USA (Hi)
	elmorei (Keifer, 1936) (Gnorimoschema)	USA (Ca)
171	**rusposoria** Povolný, 1970	Grenada

LITA Treitschke, 1833

172	**dialis** Hodges, 1966	USA (Az)
*173	**pagella** Hodges, 1966	USA (Az)
*174	**princeps** (Busck, 1910) (Gnorimoschema)	USA (Ca)
*175	**recens** Hodges, 1966	USA (Ca)
*176	**rectistrigella** (Barnes & Busck, 1920) (Gelechia)	USA (Ca)
176.1	**texanella** (Chambers, 1880) (Anesychia)	USA (Tx)
	chambersella Dyar, [1903]	USA (Tx)
177	**variabilis** (Busck, 1903) (Gelechia)	USA (Ca/Co)

LOCHARCA Meyrick, 1923

178	**emicans** Meyrick, 1923	Peru

NEODACTYLOTA Busck, 1903

*179	**basilica** Hodges, 1966	USA (Az)

ORSOTRICHA Meyrick, 1914

180	**venosa** (Butler, 1883) (Topeutis)	Chile

PHTHORIMAEA Meyrick, 1902
Phtyorimaea Turner, 1919, missp.
Phthorimoea Povolný & Zakopal, 1951, missp.
Pthorimaea Issiki, 1957, missp.
Phthorimea Diakonoff, [1968], missp.
Phtorimea Oei-Dharma, 1969, missp.

181	**ferella** (Berg, 1875) (Gelechia)	Argentina
182	**impudica** Walsingham, 1911	Panama
183	**interjuncta** Meyrick, 1931	Brazil
184	**jamaicensis** (Walsingham, 1897) (Gelechia)	Jamaica
185	**operculella** (Zeller, 1873) (Gelechia)	USA (Tx)
	terrella (Walker, 1864) (Gelechia), preocc. (not Denis & Schiffermüller, 1775)	Australia
	solanella (Boisduval, 1875) (Bryotropha)	?
	tabacella (Ragonot, 1879) (Gelechia)	?
	sedata (Butler, 1880) (Gelechia)	New Zealand
	epicentra Meyrick, 1909	South Africa
186	**sphenophora** (Walsingham, 1897) (Gelechia)	Grenada

POLYHYMNO Chambers, 1874
Copocercia Zeller, 1877
Oegoconiodes Matsumura, 1931
Oegoconides Neave, 1940, missp.

187	**charigramma** Meyrick, 1929	Brazil (Am)
188	**colleta** Walsingham, 1911	Mexico (Gue)
189	**conflicta** Meyrick, 1917	Peru
190	**convergens** Walsingham, 1911	Mexico (Gue)
191	**crambinella** (Zeller, 1877) (Copocercia)	Colombia
192	**gladiata** Meyrick, 1917	Colombia
193	**leucocras** Walsingham, 1911	Mexico (Son)
194	**luteostrigella** Chambers, 1874	USA (Tx)
	fuscostrigella Chambers, 1876	USA (Tx)
195	**subaequalis** Walsingham, 1911	Mexico (Gue)

PSEUDARLA Clarke, 1965

196	**miranda** Clarke, 1965	Chile (Juan Fernandez Is.)

PTYCERATA Ely, 1910
Scrobipalpopsis Povolný, 1967
Subgenus **Ptycerata** Ely, 1910

197	**chili** (Povolný, 1967) (Scrobipalpopsis)	Chile
198	**solanivora** (Povolný, 1973) (Scrobipalpopsis)	Costa Rica

Subgenus **Scrobischema** Povolný, 1979

199	**vergarai** (Povolný, 1979) (Scrobipalpopsis)	Colombia

RECURVARIA Haworth, 1828
Telea Stephens, 1834, preocc. (Hübner, 1819
[Saturniidae])
Aphanaula Meyrick, 1895
Hinnebergia Spuler, 1910
Microlechia Turati, 1924

200	**annulicornis** (Walsingham, 1897) (Aristotelia)	Virgin Is. (St. Thomas)
201	**eromene** (Walsingham, 1897) (Aristotelia)	Virgin Is. (St. Thomas)
202	**febriculella** (Zeller, 1877) (Teleia)	Colombia
203	**filicornis** (Zeller, 1877) (Teleia)	Colombia
204	**flagellifer** Walsingham, 1910	Mexico (Gue)
	flagellifera Meyrick, 1925, missp.	
205	**insequens** Meyrick, 1931	Brazil (SP)
206	**intermissella** (Zeller, 1877) (Teleia)	Colombia
207	**kitella** (Walsingham, 1897) (Aristotelia)	Haiti
208	**melanostictella** (Zeller, 1877) (Teleia)	Colombia
209	**merismatella** (Zeller, 1877) (Teleia)	Colombia
210	**nothostigma** Meyrick, 1914	Guyana
211	**ornatipalpella** (Walsingham, 1897) (Aristotelia)	Grenada
212	**ostariella** (Walsingham, 1897) (Aristotelia)	Virgin Is. (St. Thomas)
213	**penetrans** Meyrick, 1923	Brazil (Am)
214	**picula** Walsingham, 1910	Mexico (Ver)
215	**pleurosaris** Meyrick, 1923	Brazil (Am)
216	**putella** Busck, 1914	Panama
217	**rhicnota** Walsingham, 1910	Guatemala
218	**rhombophorella** (Zeller, 1877) (Teleia)	Colombia
219	**sartor** Walsingham, 1910	Mexico (Gue)
220	**saxea** Meyrick, 1923	Brazil (Pa)
221	**senariella** (Zeller, 1877) (Teleia)	Colombia
222	**sticta** Walsingham, 1910	Mexico (Gue)
223	**syncstia** Meyrick, 1939	Argentina
224	**thiodes** Meyrick, 1917	Colombia
225	**thysanota** Walsingham, 1910	Mexico (Gue)
226	**trigonophorella** (Zeller, 1877) (Teleia)	Colombia
227	**xanthotricha** Meyrick, 1917	Peru

SCHISTOPHILA Chrétien, 1899

228	**fuscella** Forbes, 1931	Puerto Rico

SCROBIPALPULA Povolný, 1964
Subgenus **Eurysacca** Povolný, 1967

229	**melanocampa** (Meyrick, 1917) (Phthorimaea)	Peru

Subgenus **Magnifascia** Povolný, 1967

230	**aulorrhoa** (Meyrick, 1935) (Phthorimaea)	Argentina

Subgenus **Scrobipalpula** Povolný, 1964

231	**absoluta** (Meyrick, 1917) (Phthorimaea)	Peru
232	**chiquitella** (Busck, 1909) (Gnorimoschema)	USA (NM)
233	**conifera** (Meyrick, 1916) (Chelaria)	Ecuador
234	**crustaria** (Meyrick, 1917) (Phthorimaea)	Peru
235	**daturae** (Zeller, 1877) (Doryphora)	Colombia
236	**densata** (Meyrick, 1917) (Phthorimaea)	Peru
	laciniosa (Meyrick, 1931) (Phthorimaea)	Argentina
237	**gregalis** (Meyrick, 1917) (Phthorimaea)	Peru
238	**gregariella** (Zeller, 1877) (Lita)	Colombia
239	**hemilitha** (Clarke, 1965) (Gnorimoschema)	Chile (Juan Fernandez Is.)
240	**henshawiella** (Busck, 1903) (Gnorimoschema)	USA (Co)
	ochreostrigella (Chambers, 1877) (Gelechia), preocc. (not Chambers, 1875)	USA (Co)
241	**ilyella** (Zeller, 1877) (Lita)	Colombia
242	**isochlora** (Meyrick, 1931) (Phthorimaea)	Paraguay
243	**melanolepis** (Clarke, 1965) (Gnorimoschema)	Chile (Juan Fernandez Is.)
244	**motasi** Povolný, 1976	Colombia
245	**patagonica** Povolný, 1977	Argentina
246	**stirodes** (Meyrick, 1931) (Phthorimaea)	Argentina
247	**trichinaspis** (Meyrick, 1917) (Phthorimaea)	Peru

SOPHRONIA Hübner, [1825]

248	**mediatrix** Zeller, 1877	Colombia

STEGASTA Meyrick, 1904

249	**biniveipunctata** (Walsingham, 1897) (Gelechia)	Grenada
250	**bosqueella** (Chambers, 1875) (Oecophora)	USA
	basqueella (Chambers, 1875) (Oecophora), missp.	
	bosquella (Chambers, 1878) (Gelechia), emend.	
	costipunctella (Möschler, 1890) (Gelechia)	Puerto Rico

251	**capitella** (Fabricius, 1794) (Alucita)	"West Indies"
	capitatus (Fabricius, 1798) (Ypsolophus), repl. name	"West Indies"
	robustella (Walker, 1864) (Gelechia)	Dom. Rep.
	rivulella (Möschler, 1890) (Gelechia)	Puerto Rico
252	**comissata** Meyrick, 1923	Brazil (Am)
253	**donatella** (Walker, 1864) (Gelechia)	Jamaica
254	**phalacra** (Walsingham, 1911) (Gelechia)	Mexico (Son)
255	**postpallescens** (Walsingham, 1897) (Gelechia)	Grenada
256	**scoteropis** Meyrick, 1931	Paraguay
257	**zygotoma** Meyrick, 1917	Ecuador

STOMOPTERYX Heinemann, 1870
Inotica Meyrick, 1913
Instica Sharp, 1915, missp.
Acraeologa Meyrick, 1921
Stomopterix Turati, 1922, missp.
Stromopteryx Pierce & Metcalfe, 1935, missp.

258	**phaeopa** Meyrick, 1918	Peru

SYMMETRISCHEMA Povolný, 1967

259	**altisona** (Meyrick, 1917) (Phthorimaea)	Peru
260	**aquilina** (Meyrick, 1917) (Phthorimaea)	Peru
261	**ardeola** (Meyrick, 1931) (Phthorimaea)	Paraguay
262	**atrifascis** (Meyrick, 1917) (Phthorimaea)	Peru
263	**capsica** (Bradley & Povolný, 1965) (Gnorimoschema)	Montserrat
264	**capsivorum** Povolný, 1973	Peru
265	**loquax** (Meyrick, 1917) (Phthorimaea)	Peru
266	**plaesiosema** (Turner, 1919) (Phthorimaea)	Australia
	melanoplintha (Meyrick, 1926) (Phthorimaea)	New Zealand
	tuberosella (Busck, 1931) (Gnorimoschema)	Peru
267	**striatella** (Murtfeldt, 1900) (Eucatoptus)	USA (Mo)
268	**ventratella** (Zeller, 1877) (Gelechia)	Colombia

TAYGETE Chambers, 1873

269	**altivola** (Meyrick, 1929) (Epithectis)	Peru
270	**balsamopa** (Meyrick, 1923) (Epithectis)	Peru
271	**barydelta** (Meyrick, 1923) (Epithectis)	Brazil (Pa)
272	**citranthes** (Meyrick, 1923) (Epithectis)	Brazil (Pa)
273	**consociata** (Meyrick, 1914) (Epithectis)	Guyana
274	**critica** (Walsingham, 1910) (Epithectis)	Mexico (Gue)
275	**ignavella** (Zeller, 1877) (Gelechia)	Colombia
276	**lasciva** (Walsingham, 1910) (Epithectis)	Panama
277	**notospila** (Meyrick, 1923) (Epithectis)	Brazil (Am)
278	**parvella** (Fabricius, 1794) (Alucita)	"West Indies"
	xylochroa (Meyrick, 1926) (Epithectis)	Bermuda
279	**platysoma** (Walsingham, 1910) (Epithectis)	Mexico (Ver/Dur)
280	**sphecophila** (Meyrick, 1936) (Epithectis)	Trinidad

TELPHUSA Chambers, 1872
Adrasteia Chambers, 1872
Adrastia Kirby, 1874, emend.
Geniadophora Walsingham, 1897
Telephusa Beirne, 1938, missp.

281	**auxoptila** Meyrick, 1926	Colombia
282	**callitechna** Meyrick, 1914	Guyana
	praefinita (Meyrick, 1917) (Mompha)	French Guiana
283	**delatrix** Meyrick, 1923	Peru
284	**distictella** Forbes, 1931	Puerto Rico
285	**extranea** (Walsingham, 1892) (Poecilia)	St. Vincent
286	**hemicycla** Meyrick, 1932	Paraguay
287	**latebricola** Meyrick, 1932	Virgin Is. (St. Thomas)
288	**medulella** Busck, 1914	Panama
289	**melanoleuca** Walsingham, 1911	Mexico (Gue)
290	**obligata** Busck, 1914	Panama
291	**ochrifoliata** Walsingham, 1911	Mexico (Ver)
292	**orgilopis** Meyrick, 1923	Brazil (Pa)
293	**penetratrix** Meyrick, 1931	Paraguay
294	**perspicua** (Walsingham, 1897) (Gelechia)	Haiti
295	**quinquedentata** (Walsingham, 1911) (Gelechia)	Mexico (Gue)
296	**ripula** Walsingham, 1911	Guatemala
297	**smaragdopis** Meyrick, 1926	Costa Rica
298	**translucida** (Walsingham, 1892) (Bryotropha)	Dominica

THIOTRICHA Meyrick, 1886
Reuttia O. Hofmann, 1898
Thiotrica Inoue, 1954, missp.
Thiothricha Hartig, 1956, missp.

299	**argoxantha** Meyrick, 1914	Guyana
300	**aucupatrix** Meyrick, 1929	Peru
301	**cleodorella** (Zeller, 1877) (Gelechia)	Colombia
302	**godmani** (Walsingham, 1892) (Polyhymno)	St. Vincent
303	**laterestriata** (Walsingham, 1897) (Polyhymno)	Virgin Is. (St. Thomas)
304	**sciurella** (Walsingham, 1897) (Polyhymno)	Virgin Is. (St. Thomas)

TILDENIA Povolný, 1967

305	**gudmannella** (Walsingham, 1897) (Gelechia)	Virgin Is. (St. Thomas)

Anacampsinae

ANACAMPSIS Curtis, 1827
    Anacompsis Desmarest, [1857], missp.
    Tachyptilia Heinemann, 1870
    Tuchyptilia Kirby, 1871, missp.
    Tachiptilia Chambers, 1878, missp.
    Tachyptilix Hartmann, 1880, missp.
    Agriastis Meyrick, 1914
    Agriastsi Busck, 1919, missp.
    Tachoptilia Daltry, 1926, missp.
    Trachyphilia Le Marchand, 1947, missp.
    Trachyptilia Le Marchand, 1947, missp.

306	**aedificata** Meyrick, 1929	Brazil (Pa)
307	**capyrodes** Meyrick, 1922	Brazil (Am)
308	**cenelpis** (Walsingham, 1911) (Untomia)	Mexico (Tab)
309	**cistulata** Meyrick, 1914	Guyana
310	**conistica** Walsingham, 1910	Mexico (Son)
311	**considerata** Meyrick, 1922	Brazil (Am)
312	**cornifer** Walsingham, 1897	Virgin Is. (St. Croix)
	cornifera Meyrick, 1925, missp.	
313	**crypticopa** (Meyrick, 1931) (Gelechia)	Argentina
314	**diplodelta** Meyrick, 1922	Brazil (Pa)
315	**flexiloqua** Meyrick, 1922	Peru
316	**humilis** Hodges, 1970	Uruguay
317	**idiocentra** Meyrick, 1922	Brazil (Pa)
318	**inquieta** (Meyrick, 1914) (Agriastis)	Guyana
319	**insularis** Walsingham, 1897	Virgin Is. (St. Thomas)
320	**languens** Meyrick, 1918	Ecuador
321	**lapidella** Walsingham, 1897	Grenada
322	**lithodelta** Meyrick, 1922	Peru
323	**meibomiella** Forbes, 1931	Puerto Rico
324	**multinotata** (Meyrick, 1918) (Gelechia)	Guyana
325	**nocturna** (Meyrick, 1914) (Agriastis)	Guyana
326	**peloptila** (Meyrick, 1914) (Agriastis)	Guyana
327	**perquisita** Meyrick, 1922	Brazil (Pa)
328	**petrographa** Meyrick, 1922	Brazil (Pa)
329	**phytomiella** Busck, 1914	Panama
330	**poliombra** Meyrick, 1922	Brazil (Am)
331	**pomaceella** (Walker, 1864) (Gelechia)	Brazil (Am)
332	**prasina** Meyrick, 1914	Guyana
333	**primigenia** Meyrick, 1918	Ecuador
324	**quinquepunctella** Walsingham, 1897	Grenada
335	**rhabdodes** Walsingham, 1910	Mexico (Tab)
336	**scalata** (Meyrick, 1914) (Agriastis)	Guyana
	caneodes Meyrick, 1922	Brazil (Pa)
337	**subactella** (Walker, 1864) (Gelechia)	"Australia" [S. Am.?]
338	**tridentella** (Walsingham, 1910) (Cathegesis)	Mexico (Gue)
339	**ursula** Walsingham, 1910	Mexico (Mor)
340	**viretella** (Zeller, 1877) (Gelechia)	Brazil

BATTARISTIS Meyrick, 1914
    Duvita Busck, 1916

341	**acroglypta** Meyrick, 1929	Colombia
342	**amphiscolia** Meyrick, 1914	Guyana
343	**ardiophora** Meyrick, 1914	Guyana
344	**atelesta** Meyrick, 1914	Guyana
345	**bistrigella** (Busck, 1914) (Anacampsis)	Panama
346	**concisa** Meyrick, 1929	Cuba
347	**coniosema** Meyrick, 1922	Peru
348	**curtella** (Busck, 1914) (Anacampsis)	Panama
349	**emissurella** (Walker, 1864) (Gelechia)	Brazil (Am)
	severella (Walker, 1864) (Cryptolechia)	Brazil (Am)
	fuliginosa (R. Felder & Rogenhofer, 1875) (Gelechia)	Brazil (Am)
	dorsalis (Busck, 1914) (Anacampsis)	Panama
	astroconis (Meyrick, 1918) (Compsolechia)	Surinam
350	**ichnota** Meyrick, 1914	Guyana
351	**melanamba** Meyrick, 1914	Guyana
352	**nigratomella** (Clemens, 1863) (Gelechia)	USA [Pa?]
	apicilinella (Clemens, 1863) (Gelechia)	USA [Pa?]
	apicistrigella (Chambers, 1872) (Parasia)	USA (Ky)
353	**orthocampta** Meyrick, 1914	Guyana
354	**parazela** Meyrick, 1929	Brazil (Pa)
355	**perinaeta** (Walsingham, 1910) (Anacampsis)	Mexico (Gue)
356	**prismatopa** Meyrick, 1922	Guyana
357	**rhythmodes** Meyrick, 1929	Brazil (Pa)
358	**sphenodelta** Meyrick, 1922	Brazil (Am)
359	**stereogramma** Meyrick, 1914	Guyana
360	**symphora** (Walsingham, 1911) (Untomia)	Mexico (Tab)
361	**syngraphopa** Meyrick, 1922	Peru
	syngraphora Meyrick, 1925, missp.	
362	**synocha** Meyrick, 1922	Peru
363	**tricentrota** Meyrick, 1931	Brazil
364	**unistrigella** (Busck, 1914) (Anacampsis)	Panama

COMPSOLECHIA Meyrick, 1918

365	**abolitella** (Walker, 1864) (Gelechia)	"Australia" [S. Am.]
366	**abruptella** (Walker, 1864) (Gelechia)	Brazil (Am)
	sectella (Walker, 1864) (Gelechia)	Brazil (Am)
367	**accinctella** (Walker, 1864) (Gelechia)	Brazil (Am)

368	**acosmeta** (Walsingham, 1910) (Anacampsis)	Mexico (Son)
369	**aequilibris** Meyrick, 1931	Brazil (RJ)
370	**amaurota** (Meyrick, 1914) (Anacampsis)	Guyana
371	**amazonica** Meyrick, 1918, repl. name	Brazil (Am)
	suffusella (Walker, 1864) (Gelechia), preocc. (not Douglas, 1850)	Brazil (Am)
372	**ambusta** (Walsingham, 1910) (Anacampsis)	Mexico (Tab)
	brochospila (Meyrick, 1914) (Anacampsis)	Guyana
373	**anthracura** (Meyrick, 1914) (Anacampsis)	Guyana
374	**antiplaca** Meyrick, 1922	Peru
375	**argyracma** Meyrick, 1922	Brazil (Pa)
376	**atmastra** Meyrick, 1929	Peru
377	**balia** (Walsingham, 1910) (Anacampsis)	Mexico (Gue)
378	**binotatella** (Walker, 1864) (Gelechia)	Brazil (Am)
379	**blepharopa** (Meyrick, 1914) (Anacampsis)	Guyana
380	**compalea** (Walsingham, 1910) (Anacampsis)	Mexico (Gue)
381	**canofusella** (Walker, 1864) (Gelechia)	Brazil (Am)
382	**caryoterma** Meyrick, 1922	Brazil (Am)
383	**cassidata** (Meyrick, 1914) (Anacampsis)	Guyana
384	**chelidonia** Meyrick, 1922	Brazil (Pa)
385	**chrysoplaca** (Meyrick, 1912) (Anacampsis)	Venezuela
386	**cognatella** (Walker, 1864) (Gelechia)	Brazil (Am)
387	**corymbas** Meyrick, 1918	Guyana
388	**crocodilopa** Meyrick, 1922	Peru
389	**desectella** (Zeller, 1877) (Tachyptilia)	Cuba
390	**dicax** (Meyrick, 1914) (Anacampsis)	Guyana
391	**diortha** (Meyrick, 1914) (Anacampsis)	Guyana
392	**diplolychna** Meyrick, 1922	Brazil (Pa)
393	**drachmaea** Meyrick, 1922	Brazil (Am)
394	**dryochrossa** Meyrick, 1922	Brazil (Am)
395	**elephas** (Walsingham, 1910) (Anacampsis)	Mexico (Gue)
396	**epibola** (Walsingham, 1910) (Anacampsis)	Mexico (Ver)
397	**erebodelta** Meyrick, 1922	Peru
398	**eupecta** (Meyrick, 1914) (Anacampsis)	Guyana
399	**eurygypsa** Meyrick, 1922	Peru
400	**fasciella** (R. Felder & Rogenhofer, 1875) (Gelechia)	Brazil
401	**ferreata** (Meyrick, 1914) (Anacampsis)	Guyana
402	**glaphyra** (Walsingham, 1910) (Anacampsis)	Mexico (Ver)
403	**halmyra** (Meyrick, 1914) (Anacampsis)	Guyana
404	**hemileucas** Meyrick, 1922	Brazil (Am)
405	**incurva** (Meyrick, 1914) (Anacampsis)	Guyana
406	**inusta** (Meyrick, 1914) (Anacampsis)	Guyana
407	**ischnoptera** Meyrick, 1922	Brazil (Pa)
408	**lagunculariella** (Busck, 1900) (Anacampsis)	USA (Fl)
409	**leucorrhapta** (Meyrick, 1914) (Anacampsis)	Guyana
410	**lingulata** Meyrick, 1918	Colombia
411	**lithomorpha** (Meyrick, 1914) (Anacampsis)	Guyana
412	**loxogramma** Meyrick, 1922	Brazil (Am)
413	**mangelivora** (Walsingham, 1897) (Anacampsis)	Virgin Is. (St. Thomas)
414	**melanophaea** (Forbes, 1931) (Anacampsis)	Puerto Rico
415	**mesodelta** Meyrick, 1922	Brazil (Am)
416	**metadupa** (Walsingham, 1910) (Anacampsis)	Mexico (Tab)
417	**mniocosma** Meyrick, 1922	Peru
418	**molybdina** (Walsingham, 1910) (Anacampsis)	Mexico (Gue)
419	**monochromella** (Walker, 1864) (Gelechia)	Brazil (Am)
	displicitella (Walker, 1864) (Gelechia)	Brazil (Am)
420	**neurophora** Meyrick, 1922	Brazil (Am)
421	**niobella** (R. Felder & Rogenhofer, 1875) (Gelechia)	Brazil (Am)
	tornoptila Meyrick, 1922	Brazil (Am)
422	**niphocentra** Meyrick, 1922	Peru
423	**nuptella** (R. Felder & Rogenhofer, 1875) (Gelechia)	Brazil (Am)
	scholias Meyrick, 1922	Peru
424	**ocelligera** (Butler, 1883) (Gelechia)	Chile
425	**orthophracta** (Meyrick, 1914) (Anacampsis)	Guyana
426	**parmata** Meyrick, 1918	Colombia
427	**peculella** (Busck, 1914) (Anacampsis)	Panama
428	**pentastra** Meyrick, 1922	Brazil (Pa)
429	**percnospila** (Meyrick, 1914) (Anacampsis)	Guyana
430	**perlatella** (Walker, 1864) (Gelechia)	Brazil (Am)
	secundella (Walker, 1864) (Gelechia)	Brazil (Am)
	smaragdulella (Walker, 1864) (Gelechia)	Brazil (Pa)
431	**petromorpha** Meyrick, 1922	Peru
432	**phaeotoxa** Meyrick, 1922	Brazil (Pa)
433	**phepsalitis** Meyrick, 1922	Brazil (Pa)
434	**picticornis** (Walsingham, 1897) (Aristotelia)	Virgin Is. (St. Croix)
435	**platiastis** Meyrick, 1922	Brazil (Pa)
436	**plumbeolata** (Walsingham, 1897) (Anacampsis)	Virgin Is. (St. Croix)
437	**praenivea** (Meyrick, 1914) (Anacampsis)	Guyana
438	**ptochogramma** Meyrick, 1922	Brazil (Pa)
439	**pungens** Meyrick, 1922	Peru
440	**quadrifascia** (Walker, 1864) (Gelechia)	Brazil (Am)
	superella (Walker, 1864) (Gelechia)	Brazil (Am)
441	**recta** Meyrick, 1922	Brazil (Am)
442	**refracta** (Meyrick, 1914) (Anacampsis)	Guyana
443	**religata** Meyrick, 1922	Peru
444	**repandella** (Walker, 1864) (Gelechia)	Brazil (Am)
	subscriptella (Walker, 1864) (Gelechia)	Brazil (Am)
	episema (Walsingham, 1910) (Anacampsis)	Mexico (Ver)
445	**rhombica** Meyrick, 1922	Peru
446	**salebrosa** Meyrick, 1918	Colombia
447	**sciomima** Meyrick, 1922	Brazil (Am)
448	**scitella** (Walker, 1864) (Gelechia)	Brazil (Am)

449   scopulata (Meyrick, 1914) (Anacampsis)   Guyana
450   scutella (Zeller, 1877) (Gelechia)   [S. Am?]
451   secretella (Walker, 1864) (Gelechia)   Brazil (Am)
            plejadella (R. Felder & Rogenhofer, 1875)   Brazil (Am)
            (Gelechia)
            trimolybda (Meyrick, 1914) (Anacampsis)   Guyana
452   seductella (Walker, 1864) (Gelechia)   Brazil (Am)
453   sesamodes Meyrick, 1922   Peru
454   siderophaea (Walsingham, 1910) (Anacampsis)   Mexico (Ver)
455   solidella (Walker, 1864) (Gelechia)   Brazil (Am)
456   sporozona (Meyrick, 1914) (Anacampsis)   Guyana
457   stasigastra Meyrick, 1922   Brazil (Pa)
458   stelliferella (Walker, 1864) (Gelechia)   Brazil (Am)
            speciosella (Walker, 1864) (Gelechia)   Brazil (Am)
459   stillata Meyrick, 1922   Peru
460   subapicalis (Walker, 1864) (Gelechia)   Brazil (Am)
461   succincta (Walsingham, 1910) (Anacampsis)   Mexico (Tab)
462   suffectella (Walker, 1864) (Gelechia)   Brazil (Am)
463   superfusella (Walker, 1864) (Gelechia)   Brazil (Am)
464   susceptella (Walker, 1864) (Gelechia)   Brazil (Am)
465   suspectella (Walker, 1864) (Gelechia)   Brazil (Am)
466   tardella (Walker, 1864) (Gelechia)   Brazil (Am)
            collocatella (Walker, 1864) (Gelechia)   Brazil (Am)
            sublatella (Walker, 1864) (Gelechia)   Brazil (Am)
467   terrenella (Busck, 1914) (Anacampsis)   Panama
468   tetrortha Meyrick, 1922   Brazil (Am)
469   thysanora (Meyrick, 1914) (Anacampsis)   Peru
470   titanota (Walsingham, 1910) (Anacampsis)   Guatemala
471   trachycnemis Meyrick, 1922   Peru
472   trajectella (Walker, 1864) (Gelechia)   Brazil (Am)
            diazeucta Meyrick, 1918, repl. name   Brazil (Am)
473   transjectella (Walker, 1864) (Gelechia)   Brazil (Am)
474   trapezias Meyrick, 1922   Brazil (Pa)
475   trochilea (Walsingham, 1910) (Anacampsis)   Mexico (Tab)
476   versatella (Walker, 1864) (Gelechia)   Brazil (Am)
477   volubilis Meyrick, 1922   Peru
478   zebrina (Walsingham, 1910) (Anacampsis)   Mexico (Tab)

HOLOPHYSIS Walsingham, 1910
     Hoplophysis McDunnough, 1939, missp.

479   anoma Walsingham, 1910   Mexico (Ver/Tab)
480   autodesma (Meyrick, 1918) (Zalithia)   Colombia
481   auxiliaris (Meyrick, 1918) (Zalithia)   Colombia
482   barydesma (Meyrick, 1918) (Zalithia)   Ecuador
483   quadrimaculata Walsingham, 1910   Mexico (Tab)
484   stagmatophoria Walsingham, 1910   Mexico (Gue)
485   tentatella (Walker, 1864) (Gelechia)   Brazil (Am)
486   xanthostoma Walsingham, 1910   Mexico (Gue)

STROBISIA Clemens, 1860
     Systasiota Walsingham, 1910

487   argentifrons Walsingham, 1911   Mexico (Tab)
488   iridipennella Clemens, 1860   USA [Pa?]
            aphroditeella Chambers, 1872   USA (Ky)
            aphroditella Frey, 1878, missp.
            irridipennella Busck, [1903], missp.
489   leucura (Walsingham, 1910) (Systasiota)   Mexico (Ver)
490   sapphiritis Meyrick, 1914   Guyana
491   spintheropis Meyrick, 1922   Brazil
492   stellaris (R. Felder & Rogenhofer, 1875) (Pancalia)   Colombia

UNTOMIA Busck, 1906

493   acicularis Meyrick, 1918   Ecuador
494   alticolens Walsingham, 1911   Mexico (Gue)
            alticolans Meyrick, 1925, emend.
495   horista Walsingham, 1911   Mexico (Tab)
496   juventella (Walsingham, 1897) (Ypsolophus)   Jamaica
497   latistriga Walsingham, 1911   Mexico (Mor)
498   melanobathra Meyrick, 1918   Ecuador
499   rotundata Walsingham, 1911   Mexico (Tab)

Chelariinae

CRASIMORPHA Meyrick, 1923

500   infuscata Hodges, 1963   USA (Hi)
501   peragrapta Meyrick, 1923   French Guiana

EPILECHIA Busck, 1939

502   catalinella (Busck, 1907) (Gelechia)   USA (Ca)
            tehuacana (Busck, 1913) (Gelechia)   Mexico (DF)

HAPLOCHELA Meyrick, 1923

503   trigonota (Walsingham, 1911) (Psoricoptera)   Mexico (Tab)
            mundana (Meyrick, 1914) (Chelaria)   Guyana

HYPATIMA Hübner, [1825]
     Chelaria Haworth, 1828
     Hypatina Stephens, 1835, missp.

     Allocota Meyrick, 1904, preocc. (Motschulsky, 1860
          [Coleoptera])
     Cynestomorpha Meyrick, 1904
     Deuteroptila Meyrick, 1904
     Cymatomorpha Meyrick, 1904
     Semodictis Meyrick, 1909
     Allocotaniana Strand, 1913, repl. name
     Episacta Turner, 1919
     Cellaria Neave, 1939, missp.
     Cheleria Lhomme, [1948], missp.

504   euchorda (Meyrick, 1923) (Chelaria)   Brazil (Pa)
505   hora (Busck, 1914) (Psoricoptera)   Panama

PECTINOPHORA Busck, 1917

506   gossypiella (Saunders, 1844) (Depressaria)   India
            umbripennis (Walsingham, 1885) (Gelechia)   India

PESSOGRAPTIS Meyrick, 1923

507   cancellata (Meyrick, 1914) (Chelaria)   Guyana
508   cyanactis Meyrick, 1930   Brazil (Pa)
509   thalamias Meyrick, 1923   Brazil (Am)

PORPODRYAS Meyrick, 1920

510   prasinantha Meyrick, 1920   French Guiana

PROSTOMEUS Busck, 1903

511   brunneus Busck, 1903   USA (Fl)

SEMOPHYLAX Meyrick, 1932

512   apicepuncta (Busck, 1911) (Psoricoptera)   French Guiana
            praesignis (Meyrick, 1913) (Anisoplaca)   Peru
            apicipuncta (Meyrick, 1925) (Chelaria), emend.

SITOTROGA Heinemann, 1870
     Silotroga Kirby, 1871, missp.
     Nesolechia Meyrick, 1921
     Syngenomictis Meyrick, 1927
     Sitotrogus Matsumura, 1931, missp.
     Sitotrega Borg, 1932, missp.
     Sititroga Costa Lima, 1945, missp.

513   cerealella (Olivier, 1789) (Alucita)   [Europe]
            hordei (Kirby, 1815) (Tinea)   ?
            arctella (Walker, 1864) (Gelechia)   Sri Lanka
            melanarthra (Lower, 1900) (Gelechia)   Australia
            palearis (Meyrick, 1913) (Epithectis)   India
            ochrescens (Meyrick, 1938) (Aristotelia)   China
514   coarctella Zeller, 1877   Colombia

Dichomeridinae

ACOMPSIA Hübner, [1825]
     Acampsia Westwood, 1840, missp.
     Accompsia Bruand, 1850, missp.
     Brachycrossata Heinemann, 1870
     Brachicrossata Hartmann, 1880, missp.
     Cathegesis Walsingham, 1910

515   angulifera Walsingham, 1897   Grenada
516   psoricopterella (Walsingham, 1892) (Brachycrossata)   Grenada
517   vinitincta (Walsingham, 1910) (Cathegesis)   Guatemala

ANORTHOSIA Clemens, 1860
     Sagaritis Chambers, 1872, preocc. (Billberg, 1820
          [Crustacea])
     Anorthodisca Gaede, 1937, missp.

518   capillata Walsingham, 1911   Guatemala
519   punctipennella Clemens, 1860   USA [Pa?]
            gracilella (Chambers, 1872) (Sagaritis)   USA (Ky)

BRACHMIA Hübner, [1825]
     Braclunia Stephens, 1834, missp.
     Cladodes Heinemann, 1870, preocc. (Solier, 1849
          [Coleoptera])
     Ceratophora Heinemann, 1870, preocc. (Gray,
          [1835] [Reptilia])
     Eudodacles Snellen, 1889, repl. name
     Aulacomima Meyrick, 1904
     Apethistis Meyrick, 1908

520   chambersella (Murtfeldt, 1874) (Gelechia)   USA (Mo)
            subalbusella (Chambers, 1874) (Gelechia)   USA (Tx)
            parvipulvella (Chambers, 1874) (Gelechia)   USA (Tx)
            inaequepulvella (Chambers, 1875) (Gelechia)   USA (Mo)
            subalbella (Walsingham, 1911) (Dichomeris), emend.
            subalbella Meyrick, 1925, emend.
521   convolvuli (Walsingham, 1908) (Trichotaphe)   Canary Is.

	crypsilychna Meyrick, 1914	India
	dryadopa Meyrick, 1918	South Africa
	effera (Meyrick, 1918) (Lecithocera)	India
	emigrans (Meyrick, 1921) (Lecithocera)	Barbados
522	lyrella (Walsingham, 1911) (Dichomeris)	Guatemala
523	virescens Walsingham, 1911	Mexico (Gue)

BRACHYACMA Meyrick, 1886
    Lathontogenus Walsingham, 1897
    Paraspistes Meyrick, 1905
    Lipatia Busck, 1910
    Paraspistis Busck, 1914, missp.
    Brachyaema Povolný, 1964, missp.
    Lathontogonus Diakonoff, [1968], missp.
    Brachiacma Common, 1970, missp.
    Lathontogenes Hodges, 1983, missp.

524	palpigera (Walsingham, 1891) (Gelechia)	Mozambique
	adustipennis (Walsingham, 1897)	Virgin Is. (St. Croix)
	(Lathontogenus)	
	ioloncha (Meyrick, 1905) (Paraspistes)	Sri Lanka
	crotalariella (Busck, 1910) (Lipatia)	Trinidad
	epichorda Turner, 1919	Australia

COTYLOSCIA Meyrick, 1923

525	caustonota (Meyrick, 1914) (Trichotaphe)	Guyana
526	terracocta (Walsingham, 1911) (Anorthosia)	Panama
527	triplagella (Walker, 1864) (Gelechia)	Brazil (Am)

CYMOTRICHA Meyrick, 1923
    Oxysactis Meyrick, 1923

528	abortiva (Walsingham, 1911) (Dichomeris)	Guatemala
529	amauropis Meyrick, 1923	Peru
530	amphicosma Meyrick, 1930	Brazil (Pa)
531	cinctella (Walker, 1864) (Gelechia)	Brazil (Am)
	subrutila Meyrick, 1923	Brazil (Am)
532	cinnamicostella (Zeller, 1877) (Epicorthylis)	Panama
533	cyclospila (Meyrick, 1918) (Trichotaphe)	French Guiana
534	designatella (Walker, 1864) (Gelechia)	Brazil (Am)
535	directa (Meyrick, 1912) (Trichotaphe)	Venezuela
536	ellipsias (Meyrick, 1922) (Dichomeris)	Peru
537	exallacta Meyrick, 1923	Peru
538	fluctuans Meyrick, 1923	Peru
539	inspiciens Meyrick, 1931	Paraguay
540	intentella (Walker, 1864) (Gelechia)	Brazil (Am)
541	jugata (Walsingham, 1911) (Dichomeris)	Mexico (Tab)
542	melanota (Walsingham, 1911) (Dichomeris)	Mexico (Ver)
543	melissia (Walsingham, 1911) (Dichomeris)	Panama
544	memnonia (Meyrick, 1913) (Trichotaphe)	Brazil (RJ)
545	miltophragma (Meyrick, 1922) (Dichomeris)	Brazil (Pa)
546	mochlopis Meyrick, 1923	Brazil (Am)
547	ostensella (Walker, 1864) (Gelechia)	Brazil (Am)
548	pectinella (Forbes, 1931) (Trichotaphe), n. comb.	Puerto Rico
549	permundella (Walker, 1864) (Gelechia)	Brazil (Am)
	tactella (Walker, 1864) (Gelechia)	Brazil (Am)
550	procyphodes (Meyrick, 1922) (Dichomeris)	Brazil (Am)
551	ptilocompa (Meyrick, 1922) (Dichomeris)	Peru
552	rubiginosella (Walker, 1864) (Cryptolechia)	Brazil (Am)
	ochropyga (Walsingham, 1911) (Dichomeris)	Mexico (Tab)
553	serena (Meyrick, 1909) (Trichotaphe)	Bolivia
554	sphyrocopa Meyrick, 1918	French Guiana
555	subdentata (Meyrick, 1922) (Dichomeris)	Brazil (Pa)
556	sumptella (Walker, 1864) (Gelechia)	Brazil (Am)
557	thalamopa (Meyrick, 1922) (Dichomeris)	Brazil (Am)
558	thalpodes (Meyrick, 1922) (Dichomeris)	Peru
559	themelia (Meyrick, 1913) (Trichotaphe)	Brazil (SP)
560	thesmiopa (Meyrick, 1922) (Dichomeris)	Brazil (Pa)
561	trigonella (Walsingham, 1892) (Trichotaphe)	St. Vincent
562	tristicta (Busck, 1914) (Dichomeris)	Panama
563	turrita (Meyrick, 1914) (Trichotaphe)	Guyana

DICHOMERIS Hübner, [1818]
    Oxybelia Hübner, [1825]
    Rhinosia Treitschke, 1833
    Rhobonda Walker, 1864, preocc. (Walker, 1863 [Choreutidae])
    Carna Walker, 1864, repl. name; preocc. (Gistel, 1848 [Echinodermata])
    Gaeza Walker, 1864
    Eurysara Turner, 1919
    Euryzancla Turner, 1919
    Macrozancla Turner, 1919
    Brochometis Meyrick, 1923

564	acrolychna Meyrick, 1922	Brazil (Pa)
565	acuminata (Staudinger, 1876) (Mesophleps)	Italy
	ianthes (Meyrick, 1887) (Hypsolophus [sic])	Reunion
	rusticus (Walsingham, 1892) (Ypsolophus)	St. Vincent
	ammoxanthus (Meyrick, 1904) (Ypsolophus)	Australia
	ochrophanes (Meyrick, 1907) (Ypsolophus)	Sri Lanka
566	aequata Meyrick, 1914	Guyana
567	antisticha Meyrick, 1926	Costa Rica

568	argentinella (Berg, 1885) (Ypsolophus)	Argentina
569	baccata Meyrick, 1923	Brazil (Am)
570	brachymetra Meyrick, 1923	Peru
571	cachrydias Meyrick, 1914	Guyana
572	caryaefoliella (Chambers, 1872) (Ypsolophus)	USA (Ky)
	caryifoliella Meyrick, 1925, emend.	
573	crossospila Meyrick, 1933	Costa Rica
574	dignella Walsingham, 1911	Mexico (Gue)
575	famulata Meyrick, 1914	Guyana
	granivora Meyrick, 1932	Trinidad
576	fida Meyrick, 1923	Brazil (Pa)
577	formulata (Meyrick, 1922) (Trichotaphe)	Brazil (Am)
578	hemichrysella (Walker, 1863) (Psecadia)	Brazil (Am)
	excisorella (Walker, 1864) (Sophronia)	Brazil (Am)
	macroptera Meyrick, 1914	Guyana
579	heteracma Meyrick, 1923	Brazil (Am)
580	hexasticta Walsingham, 1911	Mexico (Gue)
581	horiodes Meyrick, 1923	Brazil (Am)
582	horocompsa Meyrick, 1933	Bolivia
583	indigna (Walsingham, 1892) (Ypsolophus)	St. Vincent
584	ingloria Meyrick, 1923	Peru
585	instans Meyrick, 1923	Peru
586	ligulella Hübner, [1818]	USA
	contubernatella (Fitch, 1853) (Chaetochilus)	USA (NY)
	pometella (Harris, 1853) (Rhinocera)	USA (Ma)
	malifoliella (Fitch, 1854) (Chaetochilus)	USA (NY)
	flavivittella (Clemens, 1864) (Ypsolophus)	USA (Ky)
	quercipominella (Chambers, 1872) (Ypsolophus)	USA (Ky)
	reedella (Chambers, 1872) (Ypsolophus)	USA (Ky)
	ruderella (Chambers, 1878) (Ypsolophus), missp.	
587	lucrifuga Meyrick, 1923	Brazil (Pa)
588	lypetica Walsingham, 1911	Mexico (Gue)
589	mexicana Walsingham, 1911	Mexico (Son)
590	percnopholis Walsingham, 1911	Guatemala
591	piperata (Walsingham, 1892) (Ypsolophus)	St. Vincent
592	plexigramma Meyrick, 1922	Guyana
593	prensans Meyrick, 1922	Brazil (Pa)
594	punctatella (Walker, 1864) (Rhobonda)	Brazil (Am)
595	saturata Meyrick, 1923	Brazil (Pa)
596	sciastes Walsingham, 1911	Mexico (Ver)
597	squalens Meyrick, 1914	Guyana
598	stratella (Walsingham, 1897) (Ypsolophus)	Trinidad
599	stratigera Meyrick, 1922	Brazil (Am)
600	substratella Walsingham, 1911	Mexico (Tab)
601	tactica Meyrick, 1918	Ecuador
602	thermodryas Meyrick, 1923	Peru
603	zomias Meyrick, 1914	Guyana

EUNEBRISTIS Meyrick, 1923

604	cinclidias (Meyrick, 1918) (Noeza)	French Guiana
605	gyralea (Meyrick, 1922) (Noeza)	Brazil (Am)
606	oncotera (Walsingham, 1911) (Noeza)	Mexico (Gue)
607	zachroa (Meyrick, 1914) (Noeza)	Guyana
608	zingarella (Walsingham, 1897) (Malacotricha)	Virgin Is. (St. Croix)

HAPALONOMA Meyrick, 1914

| 609 | argyracta Meyrick, 1914 | Guyana |
| 610 | sublustricella (Walker, 1864) (Gelechia) | Brazil (Am) |

ILINGIOTIS Meyrick, 1914
    Sirogenes Meyrick, 1923

611	hemeropa Meyrick, 1923	Brazil (Am)
612	sevectella (Walker, 1864) (Gelechia)	Brazil (Am)
613	thermophaea (Meyrick, 1923) (Sirogenes)	Brazil (Am)
614	thrasynta Meyrick, 1914	Guyana
615	vigilans Meyrick, 1914	Guyana

MYROPHILA Meyrick, 1923

616	carycina (Meyrick, 1914) (Trichotaphe)	Guyana
617	caryophragma Meyrick, 1923	Guyana
618	diacnista Meyrick, 1923	Guyana

ONEBALA Walker, 1864
    Helcystogramma Zeller, 1877
    Dectobathra Meyrick, 1904

619	adaequata (Meyrick, 1914) (Helcystogramma)	Guyana
	adequata Clarke, 1969, missp.	
620	anisopa (Meyrick, 1918) (Anacampsis)	Colombia
621	archigrapha Meyrick, 1929	Colombia
622	carycastis (Meyrick, 1922) (Helcystogramma)	Brazil (Pa)
623	cerinura (Meyrick, 1923) (Brachmia)	Brazil (Pa)
624	chalyburga (Meyrick, 1922) (Helcystogramma)	Brazil (Pa)
625	daedalea (Walsingham, 1911) (Dichomeris)	Mexico (Tab)
626	elliptica (Forbes, 1931) (Trichotaphe), n. comb.	Puerto Rico
627	meconitis (Meyrick, 1913) (Trichotaphe)	Argentina
628	ribeella (Zeller, 1877) (Helcystogramma)	Panama
629	rusticella (Walker, 1864) (Gelechia)	Brazil (Am)
630	sertigera (Meyrick, 1923) (Helcystogramma)	Peru
631	stellatella Busck, 1914 (Dichomeris)	Panama

632	**symbolica** (Meyrick, 1914) (Helcystogramma)	Guyana
633	**tegulella** (Walsingham, 1897) (Trichotaphe)	Grenada
	servilis (Walsingham, 1911) (Dichomeris)	Panama
634	**trichocyma** (Meyrick, 1923) (Brachmia)	Brazil (Am)

PACHYSARIS Meyrick, 1914

635	**collina** Meyrick, 1914	Peru
636	**contrita** Meyrick, 1922	Brazil (Pa)
637	**paenitens** Meyrick, 1923	Brazil (Am)
638	**rurigena** Meyrick, 1914	Guyana

PARANOEA Walsingham, 1911

639	**latescens** Walsingham, 1911	Mexico (Tab)

PLOCAMOSARIS Meyrick, 1912
  Noeza Walker, 1866, preocc. (Meigen, 1800 [Diptera])
  Neochrista Meyrick, 1923

640	**auritogata** (Walsingham, 1911) (Noeza)	Panama
641	**pandora** Meyrick, 1912	Brazil (RJ)
642	**pyropis** (Meyrick, 1918) (Noeza)	French Guiana
643	**telegraphella** (Walker, 1866) (Noeza)	Brazil (Am)

PROPHORAULA Meyrick, 1922

644	**pyrrhopis** Meyrick, 1922	Brazil (Am)

SEMIOMERIS Meyrick, 1923

645	**pyretodes** Meyrick, 1914	Guyana

TEUCHOPHANES Meyrick, 1914

646	**cornuta** (Busck, 1914) (Dichomeris)	Panama
647	**luminosa** (Busck, 1914) (Dichomeris)	Panama
	leucopleura Meyrick, 1914	Guyana
648	**perceptella** (Busck, 1914) (Dichomeris)	Panama

TRICHOTAPHE Clemens, 1860
  Begoe Chambers, 1862
  Malacotricha Zeller, 1873
  Tricotaphe Riley, 1891, missp.
  Malacotriche Busck, 1903, missp.
  Malachotriche Busck, 1903, missp.

649	**aenigmatica** Clarke, 1962	Mexico (Ver)
650	**anticrates** Meyrick, 1931	Brazil
651	**ardesiella** (Walsingham, 1911) (Dichomeris)	Mexico (Ver)
652	**argigastra** (Walsingham, 1911) (Dichomeris)	Mexico (Ver)
	argogastra Meyrick, 1925, missp.	
653	**arotrosema** (Walsingham, 1911) (Dichomeris)	Mexico (Ver)
654	**atriguttata** Meyrick, 1931	Paraguay
655	**aurisulcata** Meyrick, 1922	Brazil (Am)
656	**carinella** (Walsingham, 1911) (Dichomeris)	Mexico (Gue)
657	**chalinopis** Meyrick, 1935	Argentina
*658	**condaliavorella** Busck, 1900, mispl.	USA (Fl)
659	**cyanoneura** Meyrick, 1922	Brazil (Pa)
660	**euparypha** Meyrick, 1922	Peru
661	**evitata** (Walsingham, 1911) (Dichomeris)	Panama
662	**excavata** (Busck, 1914) (Dichomeris)	Panama
663	**fulvicilia** Meyrick, 1922	Brazil (Am)
664	**habrochitona** (Walsingham, 1911) (Dichomeris)	Panama
665	**manella** (Möschler, 1890) (Ypsolophus)	Puerto Rico
666	**mistipalpis** (Walsingham, 1911) (Dichomeris)	Panama
	violaria Meyrick, 1914	Guyana
667	**nessica** (Walsingham, 1911) (Dichomeris)	Panama
668	**opsonoma** (Meyrick, 1914) (Dichomeris)	Guyana
669	**porphyrogramma** Meyrick, 1914	Guyana
670	**renascens** (Walsingham, 1911) (Dichomeris)	Mexico (Tab)
671	**retracta** Meyrick, 1922	Brazil (Pa)
672	**semicuprata** Meyrick, 1922	Peru
673	**syringota** Meyrick, 1926	Peru
674	**tangolias** Gyen, 1913	Chile
675	**thrypsandra** Meyrick, 1923	Ecuador
676	**varronia** (Busck, 1913) (Dichomeris)	Guyana
677	**vetustella** (Walker, 1864) (Paepia)	Brazil (Am)
678	**xerodes** (Walsingham, 1911) (Dichomeris)	Mexico (Tab)
679	**xuthostola** (Walsingham, 1911) (Dichomeris)	Mexico (Tab)

VAZUGADA Walker, 1864

680	**abscessella** (Walker, 1863) (Psecadia)	Brazil (Am)
	strigiplenella Walker, 1864	Brazil (Am)
	zonostoma (Meyrick, 1914) (Dichomeris)	Guyana
681	**amphicoma** (Meyrick, 1912) (Dichomeris)	Brazil (SP)
682	**costalis** (Busck, 1914) (Dichomeris)	Panama
683	**leucostena** (Walsingham, 1911) (Dichomeris)	Mexico (Gue)
684	**macrosphena** (Meyrick, 1913) (Trichotaphe)	Brazil (SP)
685	**percnacma** Meyrick, 1923	Peru
686	**phoenogramma** Meyrick, 1930	Brazil (Pa)

Lecithocerinae

DEOCLONA Busck, 1903
  Proclesis Walsingham, 1911
  Lioclepta Meyrick, 1922
  Deoclana Fletcher, 1929, missp.

687	**complanata** (Meyrick, 1922) (Lioclepta), **n. comb.**	Peru
688	**eriobotryae** Busck, 1939	Argentina
689	**xanthoselene** (Walsingham, 1911) (Proclesis)	Guyana
	xanthoselena Meyrick, 1925, emend.	

Unplaced Genera

ACROPHILETES Meyrick, 1932

690	**cosmocrossa** Meyrick, 1932	Bolivia

ADULLAMITIS Meyrick, 1932
  Adullanitis Gaede, 1937, missp.

691	**emancipata** Meyrick, 1932	Brazil (Pa)

AEROTYPIA Walsingham, 1911

692	**pleurotella** Walsingham, 1911	Mexico (Mor)

AGATHACTIS Meyrick, 1929

693	**toxocosma** Meyrick, 1929	Guyana

AGELIARCHES Meyrick, 1923

694	**rhizogramma** Meyrick, 1923	Brazil (Pa)

ALSODRYAS Meyrick, 1914

695	**deltochlora** Meyrick, 1922	Brazil (Pa)
696	**lactaria** Meyrick, 1914	Guyana
697	**prasinoptila** Meyrick, 1922	Brazil (Pa)

ANOMOXENA Meyrick, 1917

698	**spinigera** Meyrick, 1917	Colombia
699	**tetraxoa** Meyrick, 1917	Ecuador

ANTHINORA Meyrick, 1914

700	**xanthophanes** Meyrick, 1914	Guyana

ANTHISTARCHA Meyrick, 1925
  Antistarcha Costa Lima, 1945, missp.

701	**binocularis** Meyrick, 1929	Brazil (Ba)
702	**geniatella** (Busck, 1914) (Gelechia)	Panama

APOTACTIS Meyrick, 1918

703	**citroptila** Meyrick, 1933	Costa Rica

APOTHETOECA Meyrick, 1922

704	**synaphrista** Meyrick, 1922	Chile (Juan Fernandez Is.)

AROTROMIMA Meyrick, 1929

705	**politica** Meyrick, 1929	Guyana

BELTHECA Busck, 1914
  Anterethista Meyrick, 1914
  Antherethista Gaede, 1937, missp.

706	**phosphoropa** (Meyrick, 1922) (Anterethista)	Peru
707	**picolella** Busck, 1914	Panama
	heteractis (Meyrick, 1914) (Anterethista)	Guyana

BESCIVA Busck, 1914

708	**longitudinella** Busck, 1914	Panama

BRACHYPSALTIS Meyrick, 1931

709	**subalbata** Meyrick, 1931	Argentina

BRUCHIANA Jörgensen, 1916

710	**cassiaella** Jörgensen, 1916	Argentina

CATALEXIS Walsingham, 1909

711	**tapinota** Walsingham, 1909	Guatemala

CATOPTRISTIS Meyrick, 1925

712	**trissoxantha** (Meyrick, 1922) (Strobisia)	Peru

CERYCANGELA Meyrick, 1925

713	**sacricola** (Meyrick, 1922) (Zalithia)	Brazil (Am)

CHALCOMIMA Meyrick, 1929

714	**hoplodoxa** Meyrick, 1929	Peru

CHARISTICA Meyrick, 1925

715	**caeligena** (Meyrick, 1922) (Zalithia)	Brazil (Am)
716	**callichroma** (Meyrick, 1914) (Zalithia)	Guyana
717	**exteriorella** (Walker, 1864) (Gelechia)	Brazil (Am)
718	**ioploca** (Meyrick, 1922) (Zalithia)	Brazil (Am)
719	**iriantha** (Meyrick, 1914) (Zalithia)	Guyana
720	**porphyraspis** (Meyrick, 1909) (Strobisia)	Bolivia
721	**rhodopetala** (Meyrick, 1922) (Zalithia)	Brazil (Am)
722	**sandaracota** (Meyrick, 1914) (Zalithia)	Guyana
723	**walkeri** (Walsingham, 1911) (Strobisia)	Panama
	euphracta (Meyrick, 1914) (Zalithia)	Guyana

CLISTOTHYRIS Zeller, 1877

724	**villosula** Zeller, 1877	Colombia

COLEOSTOMA Meyrick, 1922

725	**entryphopa** Meyrick, 1922	Brazil (Pa)

COLONANTHES Meyrick, 1923

726	**plectanopa** Meyrick, 1923	Peru

COMMATICA Meyrick, 1909
    Apopira Walsingham, 1911

727	**acropelta** Meyrick, 1914	Guyana
728	**bifuscella** (Forbes, 1931) (Anacampsis)	Puerto Rico
729	**chionura** Meyrick, 1914	Guyana
730	**crossotorna** Meyrick, 1929	Guyana
731	**cryptina** (Walsingham, 1911) (Untomia)	Mexico (Tab)
732	**cyanorrhoa** Meyrick, 1914	Guyana
733	**emplasta** Meyrick, 1914	Guyana
734	**eremna** Meyrick, 1909	Bolivia
735	**extremella** (Walker, 1864) (Gelechia)	Brazil (Am)
736	**falcatella** (Walker, 1864) (Gelechia)	Brazil (Am)
	rostella (R. Felder & Rogenhofer, 1875)	Brazil (RJ)
	(Gelechia)	
737	**hexacentra** Meyrick, 1922	Brazil (Am)
738	**lupata** Meyrick, 1914	Guyana
739	**metochra** Meyrick, 1914	Guyana
740	**nerterodes** Meyrick, 1914	Guyana
741	**palirrhoa** Meyrick, 1922	Brazil (Am)
742	**parmulata** Meyrick, 1914	Guyana
743	**phanocrossa** Meyrick, 1922	Brazil (Am)
744	**placoterma** Meyrick, 1918	Colombia
745	**pterygota** Meyrick, 1929	Brazil (Pa)
746	**servula** Meyrick, 1922	Peru
747	**stygia** Meyrick, 1922	Brazil (Am)
748	**xanthocarpa** Meyrick, 1922	Peru

COMPSOSARIS Meyrick, 1914
    Gompsosaris Gaede, 1937, missp.

749	**flavidella** (Busck, 1914) (Recurvaria)	Panama
750	**testacea** Meyrick, 1914	Guyana

COPTICOSTOLA Meyrick, 1929

751	**acuminata** (Walsingham, 1911) (Untomia)	Mexico (Tab)

CRAMBODOXA Meyrick, 1913[29]

752	**platyaula** Meyrick, 1913	Colombia

DARLIA Clarke, 1950

753	**praetexta** Clarke, 1950	Argentina

DIASTALTICA Walsingham, 1910

754	**separabilis** Walsingham, 1910	Guatemala

DISSOPTILA Meyrick, 1914

755	**asphaltitis** Meyrick, 1914	Guyana
756	**crocodora** Meyrick, 1922	Peru
757	**disrupta** Meyrick, 1914	Guyana
758	**mutabilis** Meyrick, 1914	Guyana
759	**prozona** Meyrick, 1914	Guyana

DREPANOTERMA Walsingham, 1897

760	**lacticaudellum** Walsingham, 1897	Grenada

ELASIPRORA Meyrick, 1914

761	**rostrigera** Meyrick, 1914	Guyana

EMPEDAULA Meyrick, 1918

762	**phanerozona** Meyrick, 1922	Brazil (Pa)
763	**rhodocosma** (Meyrick, 1914) (Aristotelia)	Guyana

ERIPNURA Meyrick, 1914

764	**criodes** Meyrick, 1914	Guyana

ERISTHENODES Meyrick, 1935

765	**tetrapetra** Meyrick, 1935	Argentina

ERYTHRIASTIS Meyrick, 1925

766	**rhodocrossa** (Meyrick, 1914) (Pachnistis)	Guyana
767	**rubentula** (Meyrick, 1914) (Pachnistis)	Guyana

ETHIROSTOMA Meyrick, 1914

768	**interpolata** Meyrick, 1922	Peru
769	**semiacma** Meyrick, 1914	Guyana

EUNOMARCHA Meyrick, 1923
    Atoponeura Busck, 1914, preocc. (Szépligeti, 1905
    [Hymenoptera])

770	**violacea** (Busck, 1914) (Atoponeura)	Panama
	glycinopis Meyrick, 1923	Brazil (Am)

EUZONOMACHA Meyrick, 1925

771	**subjectella** (Walker, 1864) (Gelechia)	Brazil (Am)

FORTINEA Busck, 1914

772	**auriciliella** Busck, 1914	Panama

GALTICA Busck, 1914

773	**venosa** Busck, 1914	Panama

GLAUCACNA Forbes, 1931
    Glaucagna Gaede, 1937, missp.

774	**iridea** Forbes, 1931	Puerto Rico

ILARCHES Meyrick, 1933

775	**notaula** Meyrick, 1933	Bolivia

IPHIMACHAERA Meyrick, 1931[30]

776	**picticollis** (Walsingham, 1912) (Mnesichara)	Panama
	decapitata Meyrick, 1931	Brazil (SP)

ISEMBOLA Meyrick, 1926

777	**diasticta** Meyrick, 1926	Ecuador

LEISTOGENES Meyrick, 1927[31]

778	**rebellis** Meyrick, 1927	Peru

LOGISIS Walsingham, 1909

779	**achroea** Walsingham, 1909	Costa Rica

LOPHAEOLA Meyrick, 1932

780	**inquinata** Meyrick, 1932	Brazil (Am)

MERIDORMA Meyrick, 1925

781	**thrombodes** (Meyrick, 1914) (Ptocheuusa)	Guyana

METABOLAEA Meyrick, 1923

782	**chlorophthalma** Meyrick, 1923	Brazil (Am)

METOPLEURA Busck, 1912

783	**potosi** Busck, 1912	Mexico (SLP)

MOLOPOSTOLA Meyrick, 1920

784	**calumnians** Meyrick, 1926	Brazil (Pa)
785	**rufitecta** Meyrick, 1920	French Guiana

OESTOMORPHA Walsingham, 1911

786	**alloea** Walsingham, 1911	Mexico (Ver)

OXYCRYPTIS Meyrick, 1912

787	**attonita** Meyrick, 1912	Colombia

OXYLECHIA Meyrick, 1917

788	**confirmata** Meyrick, 1917	Colombia

PACHYGENEIA Meyrick, 1923

789	**clitellaria** Meyrick, 1923	Brazil (Am)
	oxyloba (Meyrick, 1929) (Gnorimoschema)	Brazil (Am)

PARASTEGA Meyrick, 1912

790	**chionostigma** (Walsingham, 1911) (Psoricoptera)	Panama
	ochropis Meyrick, 1914	Ecuador/Surinam
791	**hemisigna** Clarke, 1951	Argentina
792	**niveisignella** (Zeller, 1877) (Psoricoptera)	Panama
	curvatella Busck, 1914	Panama
793	**trichella** Busck, 1914	Panama

PARELECTROIDES Clarke, 1952, repl. name
    Parelectra Meyrick, 1925, preocc. (Dognin, 1914
    [Noctuidae])

794	**chalybea** (R. Felder & Rogenhofer, 1875) (Simaethis),	Brazil (Am)
	n. comb.	
795	**helicopis** (Meyrick, 1922) (Strobisia)	Brazil (Pa)
796	**scintillula** (Walsingham, 1911) (Strobisia)	Mexico (Tab)
797	**selectella** (Walker, 1864) (Gelechia)	Brazil (Am)
798	**subvectella** (Walker, 1864) (Gelechia)	Brazil (Am)

PAVOLECHIA Busck, 1914
    Desmaucha Meyrick, 1918

799	**argentea** Busck, 1914	Panama
	chrysostoma (Meyrick, 1918) (Desmaucha)	Guyana

PELOCNISTIS Meyrick, 1932

800	**xylozona** Meyrick, 1932	Brazil (Am)

PERIORISTICA Walsingham, 1910

801	**chalcopera** Walsingham, 1910	Mexico (Gue)

PHYLOPATRIS Meyrick, 1923

802	**terpnodes** Meyrick, 1923	Brazil (Pa)

PROMOLOPICA Meyrick, 1925

803	**epiphantha** Meyrick, 1925	Brazil (Pa)

PTILOSTONYCHIA Walsingham, 1911 [32]
    Ptilonostychia Fletcher, 1929, missp.

804	**plicata** Walsingham, 1911	Panama

RHYNCHOTONA Meyrick, 1923

805	**phaeostrota** Meyrick, 1923	Peru

SATRAPODOXA Meyrick, 1925

806	**regia** (Meyrick, 1914) (Strobisia)	Guyana

SCLEROGRAPTIS Meyrick, 1923

807	**oxytypa** Meyrick, 1923	Guyana

SIMONEURA Walsingham, 1911

808	**ophitis** Walsingham, 1911	Mexico (Tab)

SOROTACTA Meyrick, 1914

809	**bryochlora** Meyrick, 1922	Brazil (Am)
810	**viridans** Meyrick, 1914	Guyana

STACHYOSTOMA Meyrick, 1923

811	**psilodoxa** Meyrick, 1923	Ecuador

STAGMATURGIS Meyrick, 1923

812	**catharosema** Meyrick, 1923	Brazil (Am)

STEREMNIODES Meyrick, 1923

813	**sciactis** Meyrick, 1923	Brazil (Am)

STEREODMETA Meyrick, 1931

814	**xylodeta** Meyrick, 1931	Brazil (ES)

STIBARENCHES Meyrick, 1930

815	**bifissa** Meyrick, 1930	Brazil (Pa)

SYMPHANACTIS Meyrick, 1925

816	**hetaera** (Meyrick, 1914) (Ptocheuusa)	Guyana

SYNACTIAS Meyrick, 1931

817	**micranthis** Meyrick, 1931	Paraguay

TABERNILLAIA Walsingham, 1911
    Tabernillaea Meyrick, 1925, emend.

818	**ephialtes** Walsingham, 1911	Panama

TAPHROSARIS Meyrick, 1922

819	**malthacopa** Meyrick, 1922	Brazil (Am)

TECIA Kieffer & Jörgensen, 1910
    Fapua Kieffer & Jörgensen, 1910
    Lata Kieffer & Jörgensen, 1910

820	**albinervella** (Kieffer & Jörgensen, 1910) (Fapua)	Argentina
821	**kiefferi** Kieffer & Jörgensen, 1910	Argentina
822	**mendozella** Kieffer & Jörgensen, 1910	Argentina

THRYPSIGENES Meyrick, 1914
    Thripsigenes Clarke, 1955, missp.

823	**colluta** Meyrick, 1914	Guyana
824	**furvescens** Meyrick, 1914	Guyana

THYRSOMNESTIS Meyrick, 1929

825	**ceramoxantha** Meyrick, 1929	Colombia

TOCMIA Walker, 1864

826	**versicolorella** Walker, 1864	Brazil (Am)

TRICHEMBOLA Meyrick, 1918

827	**idiarcha** Meyrick, 1931	Paraguay

ZELOSYNE Walsingham, 1911

828	**olga** Meyrick, 1915	Guyana
829	**poecilosoma** Walsingham, 1911	Panama

Unplaced Species

830	"Gaea" **lilloi** Köhler, 1941, mispl. [33]	Argentina

Copromorphoidea [34]

30. COPROMORPHIDAE

by J. B. Heppner

CATHELOTIS Meyrick, 1926

1	**sanidopa** Meyrick, 1926	Colombia

ENDOTHAMNA Meyrick, 1922

2	**marmarocyma** Meyrick, 1922	Chile

LOTISMA Busck, 1909

3	**trigonana** (Walsingham, 1879) (Sciaphila)	USA (Ca)
	kincaidiella (Busck, 1904) (Hemerophila)	USA (Wa)
4	**vulcanicola** Meyrick, 1932	Costa Rica

NEOPHYLARCHA Meyrick, 1926

5	**helicosema** Meyrick, 1926	Guyana/French Guiana

ORDRUPIA Busck, 1911
    Ordupia Busck, 1911, missp.
    Ardrupia Busck, 1911, missp.

6	**fanniella** Busck, 1912	Panama
7	**friserella** Busck, 1911	French Guiana
	fabricata Meyrick, 1915	Guyana
	dasyleuca Meyrick, 1926	Peru
8	**macroctenis** Meyrick, 1926	Peru

PHYCOMORPHA Meyrick, 1914

9   **escharitis** Meyrick, 1916   Colombia

RHOPALOSETIA Meyrick, 1926

10   **phlyctaenopa** Meyrick, 1926   French Guiana

SARIDACMA Meyrick, 1930

11   **ilyopis** Meyrick, 1930   Brazil (Pa)

SYNCAMARIS Meyrick, 1932

12   **argophthalma** Meyrick, 1932   Brazil (SC)

31. ALUCITIDAE

by S. E. Miller

ALINGUATA Fleming, 1948

1   **neblina** Fleming, 1948   Venezuela

ALUCITA Linnaeus, 1758
     Orneodes Latreille, 1796, suppr. (ICZN Op. 450)
     Euchiradia Hübner, [1825]
     Alucitina Heydenreich, 1851

2   **acalles** (Walsingham, 1915) (Orneodes), n. comb.   Costa Rica
3   **ancalopa** (Meyrick, 1921) (Orneodes), n. comb.   Brazil/Guyana
4   **arriguttii** (Pastrana, 1960) (Orneodes), n. comb.   Bolivia
5   **bridarollii** (Pastrana, 1951) (Orneodes), n. comb.   Paraguay
6   **brunnea** (Fletcher, 1926) (Orneodes), n. comb.   Brazil (Am)
7   **eudactyla** R. Felder & Rogenhofer, 1875   Colombia/Brazil
8   **flavicincta** (Walsingham, 1915) (Orneodes), n. comb. Jamaica
9   **jujuyensis** (Pastrana, 1953) (Orneodes), n. comb.   Argentina
10   **mulciber** (Meyrick, 1932) (Orneodes), n. comb.   Costa Rica
11   **nasuta** Zeller, 1877   Colombia
12   **nubifera** (Meyrick, 1921) (Orneodes), n. comb.   Colombia
13   **panolbia** (Walsingham, 1915) (Orneodes), n. comb.   Guatemala
14   **patria** (Meyrick, 1921) (Orneodes), n. comb.   Guyana
15   **proseni** (Pastrana, 1951) (Orneodes), n. comb.   Argentina
16   **punctiferella** Walker, 1866   Belize
17   **riggii** (Orfila, 1949) (Orneodes), n. comb.   Argentina
18   **sertifera** (Meyrick, 1921) (Orneodes), n. comb.   French Guiana
19   **stephanopsis** (Meyrick, 1921) (Orneodes), n. comb.   Brazil (Am)
20   **tandilensis** (Pastrana, 1960) (Orneodes), n. comb.   Argentina
21   **trachydesma** (Meyrick, 1929) (Orneodes), n. comb.   Bolivia

HEXERETMIS Meyrick, 1929

22   **argo** Meyrick, 1929   Peru
23   **willineri** Pastrana, 1953   Bolivia

PAELIA Walker, 1866

24   **lunuligera** Walker, 1866   Brazil (Am)

32. CARPOSINIDAE

by J. B. Heppner

CARPOSINA Herrich-Schäffer, 1853
  Subgenus **Carposina** Herrich-Schäffer, 1853

1   **engalactis** Meyrick, 1932   Brazil (SC)
2   **phycitana** Walsingham, 1914   Panama

  Subgenus **Trepsitypa** Meyrick, 1913

3   **cardinata** (Meyrick, 1911) (Trepsitypa)   Guyana

  Subgenus **Dipremna** Davis, 1969

4   **cretata** Davis, 1969   Puerto Rico

  Subgenus **Epipremna** Davis, 1969

5   **dominicae** Davis, 1969   Dominica

  Subgenus **Hypopremna** Davis, 1969

6   **bullata** Meyrick, 1913   Guyana

ATOPOSEA Davis, 1969

7   **maxima** (Meyrick, 1912) (Carposina)   Colombia

33. EPERMENIIDAE

by J. B. Heppner

Ochromolopinae

PAROCHROMOLOPIS Gaedike, 1977

1   **parishi** Gaedike, 1977   Peru
2   **psittacanthus** Heppner, 1980   Costa Rica
3   **syncrata** (Meyrick, 1921) (Epermenia)   Peru

34. GLYPHIPTERIGIDAE

by J. B. Heppner

COTAENA Walker, [1865]

1   **mediana** Walker, [1865]   Brazil (Pa)
2   **plenella** (Busck, 1914) (Glyphipteryx [sic])   Panama

MYRSILA Boisduval, [1875]

3   **auripennis** Boisduval, [1875]   Brazil (Pa)

PHALERARCHA Meyrick, 1913

4   **chrysorma** Meyrick, 1913   Guyana
5   **eumitrella** (Busck, 1914) (Ussara)   Panama

CRONICOMBRA Meyrick, 1920

6   **deltodes** (Walsingham, 1914) (Glyphipteryx [sic])   Mexico (Tab)
7   **essedaria** Meyrick, 1926   Peru
8   **granulata** Meyrick, 1920   Brazil (Pa)
9   **lamella** (Busck, 1914) (Porpe)   Panama
10   **palpella** (Walsingham, 1914) (Glyphipteryx [sic])   Mexico (Tab)
11   **phaeobathra** (Meyrick, 1932) (Ussara)   Brazil (Go)
12   **porphyrospila** (Meyrick, 1926) (Machlotica)   Peru

TAENIOSTOLELLA Fletcher, 1940, repl. name
     Taeniostola Meyrick, 1920, preocc. (Bezzi, 1913
     [Diptera])

13   **celophora** (Meyrick, 1920) (Taeniostola)   Brazil (Pa)
14   **litura** (R. Felder & Rogenhofer, 1875) (Oecophora)   Brazil (Am)

MACHLOTICA Meyrick, 1909
     Maclotica Busck, 1915, missp.

15   **chrysodeta** Meyrick, 1909   Bolivia
16   **eurymolybda** Meyrick, 1926   Peru

NEOMACHLOTICA Heppner, 1981

17   **actinota** (Walsingham, 1914) (Glyphipteryx [sic])   Mexico (Tab)
18   **atractias** (Meyrick, 1909) (Machlotica)   Bolivia
19   **nebras** (Meyrick, 1909) (Machlotica)   Bolivia

TRAPEZIOPHORA Walsingham, 1892

20   **gemmula** Walsingham, 1892   St. Vincent

RHABDOCRATES Meyrick, 1931

21   **sporomantis** Meyrick, 1931   Peru

USSARA Walker, 1864
     Setiostoma R. Felder & Rogenhofer, 1875 (not
     Zeller, 1875 [Oecophoridae])
     Usara Busck, [1934], missp.

22   **ancobathra** Meyrick, 1932   Brazil (Am)
23   **ancyristis** Meyrick, 1920   Brazil (Pa)
24   **arquata** Meyrick, 1926   Colombia
25   **chalcodesma** Meyrick, 1913   Guyana
26   **chrysangela** Meyrick, 1922   Peru
27   **decoratella** Walker, 1864   Brazil (Am)
28   **eurythmiella** Busck, 1914   Panama
29   **flaviceps** (R. Felder & Rogenhofer, 1875) (Setiostoma) Brazil (Am)
30   **olyranta** Meyrick, 1931   Brazil (RS)

SERICOSTOLA Meyrick, 1927[35]

31   **rhodanopa** Meyrick, 1927   Colombia

GLYPHIPTERIX Hübner, [1825]
     Heribeia Stephens, 1829
     Aechmia Treitschke, 1833
     Aecimia Boisduval, 1836, missp.
     Glyphipteryx Zeller, 1839, emend. (not Curtis, 1827
     [Cosmopterigidae])
     Glyphiteryx Fischer von Röslerstamm, 1841, missp.
     Anacampsoides Bruand, 1850, nom. oblit.
     Glypipteryx Stainton, 1854, missp.
     Glyphopteryx Herrich-Schäffer, 1854, emend.
     Glyphiptoryx Mann & Rogenhofer, 1878, missp.
     Glyphptieryx Turati, 1879, missp.
     Glyphipterys Christoph, 1882, missp.
     Glyphyteryx Hampson, 1918, missp.
     Glyphteryx Watt, 1920, missp.

32	**atelura** Meyrick, 1920 (Glyphipteryx [sic])	Guyana
33	**callidelta** Meyrick, 1922 (Glyphipteryx [sic])	Peru
34	**caudatella** Walsingham, 1897 (Glyphipteryx [sic])	Grenada
35	**cestrota** Meyrick, 1915 (Glyphipteryx [sic])	Peru
36	**chrysallacta** Meyrick, 1922 (Glyphipteryx [sic])	Peru
37	**colorata** Meyrick, 1913 (Glyphipteryx [sic])	Guyana
38	**columnaris** Meyrick, 1913 (Glyphipteryx [sic])	Guyana
39	**conosema** Meyrick, 1913 (Glyphipteryx [sic])	Guyana
40	**cornigerella** Zeller, 1877 (Glyphipteryx [sic])	Colombia
41	**crinita** Meyrick, 1913 (Glyphipteryx [sic])	Guyana
42	**epastra** Meyrick, 1922 (Glyphipteryx [sic])	Peru
43	**expurgata** Meyrick, 1922 (Glyphipteryx [sic])	Peru
44	**falcigera** Meyrick, 1913 (Glyphipteryx [sic])	Guyana
45	**hologramma** Meyrick, 1920 (Glyphipteryx [sic])	Brazil (Pa)
46	**indomita** Meyrick, 1922 (Glyphipteryx [sic])	Brazil (Am)
47	**invicta** Meyrick, 1920 (Glyphipteryx [sic])	Brazil (Pa)
	iinvicta Clarke, 1969, missp.	
48	**ioclista** Meyrick, 1913 (Glyphipteryx [sic])	Guyana
49	**leptocona** Meyrick, 1922 (Glyphipteryx [sic])	Peru
50	**neochorda** Meyrick, 1922 (Glyphipteryx [sic])	Peru
51	**nugella** R. Felder & Rogenhofer, 1875 (Glyphipteryx [sic])	Colombia
52	**oligastra** Meyrick, 1926 (Glyphipteryx [sic])	Colombia
53	**orthodeta** Meyrick, 1922 (Glyphipteryx [sic])	Brazil (Am)
54	**oxyglypta** (Meyrick, 1929) (Acrolepia)	Panama
55	**paradisea** Walsingham, 1897 (Glyphipteryx [sic])	Grenada
56	**perfracta** Meyrick, 1922 (Glyphipteryx [sic])	Peru
57	**platyochra** Meyrick, 1920 (Glyphipteryx [sic])	Brazil (Pa)
58	**pseudostoma** Meyrick, 1922 (Glyphipteryx [sic])	Guyana
59	**refractella** Zeller, 1877 (Glyphipteryx [sic])	Colombia
60	**repletana** (Walker, 1864) (Ussara)	Brazil (Am)
61	**septemstrigella** Zeller, 1877 (Glyphipteryx [sic])	Colombia
62	**speculans** Meyrick, 1922 (Glyphipteryx [sic])	Brazil (Am)
63	**stasichlora** Meyrick, 1931 (Glyphipteryx [sic])	Peru
64	**syndecta** Meyrick, 1915 (Glyphipteryx [sic])	Peru
65	**synorista** Meyrick, 1922 (Glyphipteryx [sic])	Brazil (Am)
66	**uncta** Meyrick, 1913 (Glyphipteryx [sic])	Guyana
67	**unguifera** Meyrick, 1922 (Glyphipteryx [sic])	Peru
68	**variata** Meyrick, 1913 (Glyphipteryx [sic])	Guyana
69	**versicolor** Meyrick, 1913 (Glyphipteryx [sic])	Guyana
70	**voluptella** R. Felder & Rogenhofer, 1875 (Glyphipteryx [sic])	Brazil (Am)
71	**xanthoplecta** Meyrick, 1922 (Glyphipteryx [sic])	Peru
72	**zalodisca** Meyrick, 1920 (Glyphipteryx [sic])	Brazil (Pa)

DIPLOSCHIZIA Heppner, 1981

73	**glaucophanes** (Meyrick, 1922) (Glyphipteryx [sic])	Brazil (Am)
74	**tetratoma** (Meyrick, 1913) (Glyphipteryx [sic])	Guyana
75	**urophora** (Walsingham, 1914) (Glyphipteryx [sic])	Guatemala

Yponomeutoidea

35. PLUTELLIDAE

by J. B. Heppner

Plutellinae

CALLIATHLA Meyrick, 1931

| 1 | **peplophanes** Meyrick, 1931 | Argentina |

EUCALLIATHLA Clarke, 1967

| 2 | **candidella** (Blanchard, 1852) (Aecophora [sic]) | Chile |

EUCERATIA Walsingham, 1881

| 3 | **argentea** (Busck, 1912) (Calantica) | Mexico (Ver) |

EUDOLICHURA Clarke, 1965

| 4 | **exuta** Clarke, 1965 | Chile |

LEUROPERNA Clarke, 1965

| 5 | **leioptera** Clarke, 1965 | Chile (Juan Fernandez Is.) |

ORTHENCHES Meyrick, 1886

6	**clytandra** Meyrick, 1931	Chile
7	**dissimulatrix** Meyrick, 1931	Chile
8	**osteacma** Meyrick, 1931	Argentina
9	**semicretata** Meyrick, 1931	Argentina

PHILAUSTERA Meyrick, 1927

| 10 | **signifera** Meyrick, 1927 | Colombia |

PLUTELLA Schrank, 1802
Anadetia Hübner, [1825]
Euota Hübner, [1825]
Cerostoma.- Stephens, 1834 (not Latreille, 1802)
Creagria Sodoffsky, 1837

11	**acrodelta** Meyrick, 1931	Argentina
12	**culminata** Meyrick, 1931	Argentina
13	**deltodoma** Meyrick, 1931	Chile
14	**diluta** Meyrick, 1931	Chile
15	**nephelaegis** Meyrick, 1931	Argentina
16	**porrectella** (Linnaeus, 1758) (Phalaena Tinea)	[Sweden?]
	hesperidella (Hübner, 1796) (Tinea)	[Europe]
	vigilaciella Clemens, 1860	USA [Pa?]
17	**rectivittella** Zeller, 1877	Colombia
18	**xylostella** (Linnaeus, 1758) (Phalaena Tinea)	[Sweden?]
	cinerea (Geoffroy, 1785) (Tinea)	[Europe]
	maculipennis (Curtis, 1832) (Cerostoma)	UK (England)
	annulatellus Wood, 1839	UK (England)
	cruciferarum Zeller, 1843	Germany
	brassicella Fitch, 1856	USA (NY)
	limbipennella Clemens, 1860	USA [Pa?]
	mollipedella Clemens, 1860	USA [Pa?]
	cicerella (Rondani, 1876) (Gelechia)	Italy
	dubiosella (Beutenmüller, 1889) (Cerostoma)	USA (Ak)
	albovenosa Walsingham, 1907	USA (Hi)
	dudiosella (Moriuti, 1977) (Cerostoma), missp.	

THALASSONYMPHA Meyrick, 1931

| 19 | **mysteriodes** Meyrick, 1931 | Argentina |

YPSOLOPHA Latreille, 1796
Ypsolophus Fabricius, 1798
Cerostoma Latreille, 1802
Hypsolopha Billberg, 1820
Theristis Hübner, [1825]
Harpipterix Hübner, [1825]
Abebaea Hübner, [1825]
Harpipteryx Treitschke, 1833, emend.
Chaetochilus Stephens, 1834
Harpepteryx Sodoffsky, 1837, emend.
Hypolepia Guenée, 1845, nom. nud.
Pteroxia Guenée, 1845, nom. nud.
Harpopteryx Agassiz, 1846, emend.
Hypsilophus Agassiz, 1846, emend.
Periclymenobius Wallengren, 1880
Credemnon Wallengren, 1880
Trachoma Wallengren, 1880
Pluteloptera Chambers, 1880
Plutelloptera Walsingham, 1881, missp.
Alapa Kieffer & Jörgensen, 1910
Mapa Strand, 1911, redesc. [for Alapa]
Pycnopogon Chrétien, 1922
Credemna Forbes, 1923, missp.
Melitonympha Meyrick, 1927
Chalconympha Meyrick, 1931
Credemon Moriuti, 1977, missp.

20	**cordillerella** (Kieffer & Jörgensen, 1910) (Alapa)	Argentina
	cordillerella (Strand, 1911) (Mapa), redesc.	Argentina
21	**crispulella** (Berg, 1875) (Cerostoma)	Argentina
22	**eurypepla** (Meyrick, 1931) (Chalconympha)	Argentina
23	**hydraea** (Meyrick, 1919) (Cerostoma)	Peru
24	**malacodoxa** (Meyrick, 1932) (Cerostoma)	Argentina
25	**telluris** (Clarke, 1965) (Melitonympha), n. comb.	Chile (Juan Fernandez Is.)

Praydinae

ATEMELIA Herrich-Schäffer, 1853

| 26 | **contrariella** Zeller, 1877 | Colombia |

PRAYS Hübner, [1825]

| 27 | **stratella** Zeller, 1877 | Colombia |

36. YPONOMEUTIDAE

by J. B. Heppner

Attevinae [36]

ATTEVA Walker, 1854
Poeciloptera Clemens, 1860, preocc. (Latreille, 1829 [Hemiptera])
Amblothridia Wallengren, 1861
Corinea Walker, 1863
Oeta Grote, 1865
Carthara Walker, 1866, preocc. (Walker, 1865 [Pyralidae])
Synadia Walker, 1866, repl. name
Scintilla Guenée, 1879, preocc. (Deshayes, 1856 [Mollusca])
Syblis Guenée, 1879

1	**cosmogona** Meyrick, 1931	Brazil (RJ)
2	**ergatica** Walsingham, 1914	Belize
3	**flavivitta** (Walker, 1866) (Carthara)	Colombia

4	**fulviguttata** (Zeller, 1873) (Oeta)	Surinam
	glaucopidella (Guenée, 1879) (Syblis)	Jamaica
5	**hysginiella** (Wallengren, 1861) (Amblothridia)	Panama
	sylpharis (Butler, 1877) (Cydosia)	Galapagos Is.
6	**microsticta** Walsingham, 1914	Mexico (Pue)
7	**numeratrix** Meyrick, 1930	Brazil (Pa)
8	**punctella** (Cramer, 1781) (Phalaena)	Surinam
	pastulella (Fabricius, 1787) (Tinea)	Surinam
	pustulella (Fabricius, 1794) (Tinea)	?
	subtilis (Hübner, [1819]) (Crameria)	Surinam
	aurea (Fitch, 1856) (Deiopeia)	USA (Ga)
	compta (Clemens, 1860) (Poeciloptera)	USA (Tx)
	gemmata (Grote, 1873) (Oeta)	Cuba
	aurera (Stretch, 1873) (Oeta), missp.	
	fastuosa (Zeller, 1877) (Oeta)	Cuba
	compta var. floridana (Neumoegen, 1891) (Oeta)	USA (Fl)
	edithella Busck, 1908	USA (Tx)
	exquisita Busck, 1912	Mexico (Coa)
9	**siderea** (Walsingham, 1892) (Oeta)	Virgin Is. (St. Thomas)

LACTURA Walker, 1854
  Dianasa Walker, 1854
  Mieza Walker, 1854
  Sarbena Walker, 1865, preocc. (Walker, 1862 [Noctuidae])
  Themiscyra Walker, 1865
  Cyptasia Walker, 1866
  Buxeta Walker, 1866
  Enaemia Zeller, 1872
  Pseudotalara Druce, 1885
  Pseudocaprima Walsingham, 1900
  Epidictica Turner, 1903
  Hedycharis Turner, 1903
  Eriopyrrha Meyrick, 1913

10	**atrolinea** (Barnes & McDunnough, 1913) (Mieza)	USA (Tx)
11	**basistriga** (Barnes & McDunnough, 1913) (Mieza)	USA (Tx)
12	**chrysippa** (Druce, 1885) (Pseudotalara)	Guatemala
13	**citrina** (Busck, 1913) (Mieza)	Costa Rica
14	**euthoracica** (Schaus, 1912) (Pseudotalara)	Costa Rica
15	**irrorata** (Busck, 1913) (Mieza)	Costa Rica
16	**lateralis** (Dyar, 1912) (Pseudotalara)	Mexico (Ver)
17	**pseudophile** (Dyar, 1912) (Pseudotalara)	Mexico (Oax)
18	**rutila** (Bartlett-Calvert, 1893) (Hyponomeuta [sic])	Chile
19	**schausia** Busck, 1920	Guatemala
20	**schenoxantha** (Schaus, 1912) (Pseudotalara)	Costa Rica
21	**spatula** (Busck, 1913) (Mieza)	Costa Rica
22	**subfervens** (Walker, 1854) (Mieza)	"N. Am."
	subferreus (Herrich-Schäffer, 1866) (Mieza), missp.	

PYGMOCRATES Meyrick, 1932

| 23 | **lissopeda** Meyrick, 1932 | Brazil (SC) |

Yponomeutinae

ANCHIMACHETA Walsingham, 1914

24	**capnodes** Walsingham, 1914	Mexico (Gue)
25	**iodes** Walsingham, 1914	Mexico (Gue)
26	**tolmetes** Walsingham, 1914	Mexico (Gue)

DITRIGONOPHORA Walsingham, 1897

| 27 | **marmoreipennis** Walsingham, 1897 | Grenada |

EUARNE Möschler, 1890

| 28 | **obligatella** Möschler, 1890 | Puerto Rico |

ITHUTOMUS Butler, 1883
  Ithytomus Meyrick, 1914, emend.

| 29 | **formosus** Butler, 1883 | Chile |

SPILADARCHA Meyrick, 1913

| 30 | **derelicta** Meyrick, 1913 | Guyana |

SYNCERASTIS Meyrick, 1931 37

| 31 | **ptisanopa** Meyrick, 1931 | Chile |

TEINOPTILA Sauber, 1902

| 32 | **calcarata** (Meyrick, 1924) (Hyponomeuta [sic]) | Bermuda |

TOECORHYCHIA Butler, 1883

| 33 | **cinerea** Butler, 1883 | Chile |

URODUS Herrich-Schäffer, 1854 38
  Trichostibas Zeller, 1863
  Pexicnemidia Möschler, 1890, n. syn. 39
  Paratiquadra Walsingham, 1897

34	**amphilocha** Meyrick, 1924	Brazil (Pa)
35	**aphanoptis** Meyrick, 1930	Brazil (Pa)
36	**auchmera** Walsingham, 1914	Guatemala
37	**brachyanches** Meyrick, 1931	Brazil (RJ)
38	**calligera** (Zeller, 1877) (Trichostibas)	Cuba
39	**carabopa** Meyrick, 1925	Peru
40	**chiquita** (Busck, 1910) (Trichostibas)	Costa Rica
41	**chrysoconis** Meyrick, 1932	Peru
42	**costaricae** (Busck, 1910) (Trichostibas)	Costa Rica
43	**cumulata** Walsingham, 1914	Mexico (Ver)
44	**cyanombra** (Meyrick, 1913) (Trichostibas)	Argentina
45	**cyclopica** Meyrick, 1930	Brazil (Pa)
46	**decens** Meyrick, 1925	Costa Rica
47	**distincta** (Strand, 1911) (Trichostibas)	Panama
48	**favigera** (Meyrick, 1913) (Trichostibas)	Peru
49	**fonteboae** (Strand, 1911) (Trichostibas)	Brazil (Am)
50	**forficulella** (Walsingham, 1897) (Paratiquadra)	Jamaica
51	**fulminalis** Meyrick, 1931	Brazil (MG)
52	**fumosa** (Zeller, 1863) (Trichostibas)	Venezuela
53	**hephaestiella** (Zeller, 1877) (Trichostibas)	Panama
54	**hexacentris** Meyrick, 1931	Brazil (MG)
55	**hypsicrates** Meyrick, 1925	Colombia
56	**imitans** (R. Felder & Rogenhofer, 1875) (Trichostibas)	Colombia
57	**imitata** Druce, 1884	Guatemala
58	**iophlebia** (Zeller, 1877) (Trichostibas)	"Antilles"
59	**isoxesta** Meyrick, 1932	Bolivia/Costa Rica
60	**isthmiella** (Busck, 1910) (Trichostibas)	Panama
61	**lithophaea** (Meyrick, 1913) (Trichostibas)	Guyana
62	**marantica** Walsingham, 1914	Panama
63	**merida** (Strand, 1911) (Trichostibas)	Venezuela
64	**mirella** (Möschler, 1890) (Pexicnemidia), n. comb.	Puerto Rico
65	**modesta** Druce, 1884	Guatemala
66	**monura** Herrich-Schäffer, 1854	Venezuela
67	**niphatma** Meyrick, 1925	Colombia
68	**opticosema** Meyrick, 1930	Brazil (Pa)
69	**ovata** (Zeller, 1877) (Trichostibas)	Cuba
70	**pallidicostella** (Walsingham, 1897) (Trichostibas)	Jamaica
71	**perischias** Meyrick, 1925	Brazil (Am)
72	**porphyrina** Meyrick, 1932	Colombia/Costa Rica
73	**praetextata** (Meyrick, 1913) (Trichostibas)	Peru
74	**pulvinata** Meyrick, 1924	Peru
75	**sanctipaulensis** (Strand, 1911) (Trichostibas)	Brazil (Am)
76	**scythrochalca** Meyrick, 1932	Costa Rica
77	**sordidata** (Zeller, 1877) (Trichostibas)	Puerto Rico
78	**spumescens** Meyrick, 1925	Peru
79	**staphylina** Meyrick, 1932	Brazil (SC)
80	**sympiestis** Meyrick, 1925	Brazil (Am)
81	**transverseguttata** (Zeller, 1877) (Trichostibas)	Panama
	transversiguttata Walsingham, 1914, emend.	
82	**triancycla** Meyrick, 1931	Paraguay
83	**venatella** (Busck, 1910) (Trichostibas)	Brazil (Pr)
84	**xiphura** Meyrick, 1931	Brazil (SC)

XYROSARIS Meyrick, 1907
  Xyrosaria Kearfott, [1903], missp.

| 85 | **mnesicentra** Meyrick, 1913 | Guyana |

YPONOMEUTA Latreille, 1796
  Hyphantes Hübner, [1806], suppr. (ICZN Op. 97)
  Erminea Haworth, [1811]
  Hyponomeuta Billberg, 1820, emend.
  Coenyphantes Hübner, [1822]
  Nygmia Hübner, [1825], preocc. (Hübner, [1820]
    [Lymantriidae])
  Hyponomeuta Sodoffsky, 1837, emend.
  Hyponomenta Turner, 1898, missp.

| 86 | **eusoma** Walsingham, 1914 | Mexico (Ver) |
| 878 | **triangularis** Möschler, 1890 | Puerto Rico |

ZELLERIA Stainton, 1849

88	**cirrhoscia** Meyrick, 1931	Argentina
89	**leucoschista** Meyrick, 1931	Argentina
90	**pistopis** Meyrick, 1931	Argentina

37. ARGYRESTHIIDAE

by J. B. Heppner

ARGYRESTHIA Hübner, [1825] 40
  Argyrosetia Stephens, 1829
  Oligos Treitschke, 1830
  Ederesa Curtis, 1833
  Ismene Stephens, 1834
  Blastotere Ratzeburg, 1840
  Argyrestia MacKay, 1972, missp.

1	**biruptella** Zeller, 1877	Colombia
2	**carcinomatella** Zeller, 1877	Colombia
3	**conspersa** Butler, 1883	Chile
4	**diffractella** Zeller, 1877	Colombia
5	**melitaula** Meyrick, 1918	Colombia

6	ochridorsis Zeller, 1877	Colombia
7	percussella Zeller, 1877	Colombia

38. DOUGLASIIDAE

by J. B. Heppner

PROTONYCTIA Meyrick, 1931

1	originalis Meyrick, 1931	Ecuador

39. ACROLEPIIDAE

by J. B. Heppner

ACROLEPIOPSIS Gaedike, 1970
  Argiope Chambers, 1873, preocc. (Audouin, 1827 [Arachnida])

1	aureella (Blanchard, 1852) (Elachista), n. comb.	Chile
2	bythodes (Meyrick, 1919) (Acrolepia), n. comb.	Peru
3	cestrella (Busck, [1934]) (Acrolepia), n. comb.	Cuba
4	chalcolampra (Meyrick, 1931) (Acrolepia), n. comb.	Chile
5	chariphanes (Meyrick, 1931) (Acrolepia), n. comb.	Chile
6	elaphrodes (Meyrick, 1919) (Acrolepia), n. comb.	Peru
7	halosema (Meyrick, 1931) (Acrolepia), n. comb.	Argentina
8	jaspidata (Meyrick, 1919) (Acrolepia), n. comb.	Peru
9	maculella (Blanchard, 1852) (Elachista), n. comb.	Chile
10	marmaropis (Meyrick, 1919) (Acrolepia), n. comb.	Colombia
11	mixotypa (Meyrick, 1931) (Acrolepia), n. comb.	Chile
12	niphosperma (Meyrick, 1931) (Acrolepia), n. comb.	Argentina
13	poliopis (Meyrick, 1919) (Acrolepia), n. comb.	Peru
14	prasinaula (Meyrick, 1927) (Acrolepia), n. comb.	Colombia
15	seraphica (Meyrick, 1931) (Acrolepia), n. comb.	Argentina
16	syrphacopis (Meyrick, 1919) (Acrolepia), n. comb.	Peru
17	tharsalea (Walsingham, 1914) (Acrolepia), n. comb.	Guatemala
18	xiphias (Meyrick, 1931) (Acrolepia), n. comb.	Chile

ANTISPASTIS Meyrick, 1926

19	clarkei Pastrana, 1952	Argentina
20	selectella (Walker, 1863) (Adela) [41]	Brazil (Am)
21	xylophragma Meyrick, 1926	Peru

40. HELIODINIDAE

by J. B. Heppner

Schreckensteiniinae [42]

SCHRECKENSTEINIA Hübner, [1825]
  Chrysocorys Curtis, 1833

1	inferiorella Zeller, 1877	Colombia
2	jocularis Walsingham, 1914	Guatemala

Heliodininae

AMPHICLADA Meyrick, 1912

3	fervescens Meyrick, 1912	Grenada

COPOCENTRA Meyrick, 1909

4	calliscelis Meyrick, 1909	Bolivia
5	notopyrsa Meyrick, 1936	Peru
6	porphyropis Meyrick, 1922	Brazil (Am)
7	saltatoria Meyrick, 1922	Brazil (Pa)
8	submetallica Meyrick, 1922	Peru

CREMBALASTIS Meyrick, 1915

9	erythrorma Meyrick, 1915	Peru

CYCLOPLASIS Clemens, 1864

10	basiplagata Walsingham, 1897	Virgin Is. (St. Thomas)
	basiplagiata Forbes, 1930, missp.	
11	gnathodes Meyrick, 1917	Ecuador
12	habrarcha Meyrick, 1917	Ecuador
		Ecuador

HELIODINES Stainton, 1854 [43]
  Aetole Chambers, 1875
  Aetola Frey, 1884, missp.
  Heliodinides Turner, 1941, missp.

*13	albaciliella Busck, 1910	USA (NM)
	albiciliella Meyrick, 1913, emend.	
14	aureoflamma Walsingham, 1897	Virgin Is. (St. Thomas)
*15	bella (Chambers, 1875) (Aetole)	USA (Tx)
16	choneuta Meyrick, 1915	Colombia
*17	ciccella Barnes & Busck, 1920	USA (Az)
18	demarcha Meyrick, 1917	Peru
19	isoleura Meyrick, 1917	Ecuador
20	loriculata Meyrick, 1932	Bolivia

*21	metallicella Busck, 1909	USA (Az)
*22	perichalca Meyrick, 1922	USA (NM)
23	quinqueguttata Walsingham, 1897	Jamaica
23.1	rubella (Blanchard, 1852) (Elachista), n. comb. [44]	Chile
24	schulzella (Fabricius, 1794) (Tinea)	"West Indies"
*25	sexpunctella Walsingham, 1892	USA (Az)
*26	tripunctella Walsingham, 1892	USA (Tx)
27	urichi Busck, 1910	Trinidad

LAMPROLOPHUS Busck, 1900
  Embola Walsingham, 1909

28	dentifera (Walsingham, 1909) (Embola)	Mexico (Mor)
29	marginata (Walsingham, 1892) (Heliodines)	St. Vincent
30	obolarcha Meyrick, 1909	Bolivia
31	xanthocephala (Walsingham, 1909) (Embola)	Mexico (Tab)

LITHARIAPTERYX Chambers, 1876

*32	abroniaeella Chambers, 1876	USA (Co)
	abroniella Meyrick, 1913, emend.	
*33	jubarella Comstock, 1940	USA (Ca)
*34	mirabilinella Comstock, 1940	USA (Ca)

PSEUDASTASIA Walsingham, 1909

35	opulenta Walsingham, 1909	Panama

SCELORTHUS Busck, 1900

36	calcifera Walsingham, 1909	Mexico (Mor)
*37	pisoniella Busck, 1900	USA (Fl)

THRASYDOXA Meyrick, 1912

38	tyrocopa Meyrick, 1912	Colombia

Immoidea

41. IMMIDAE

by J. B. Heppner

MOCA Walker, 1863
  Adricara Walker, 1863
  Alicadra Walker, [1866]
  Jobula Walker, 1866
  Callartona Hampson, [1893]

1	albodiscata (Walker, 1863) (Adricara)	Brazil (Am)
2	antiquata (Meyrick, 1913) (Imma)	Guyana
3	aphrodora (Meyrick, 1922) (Imma)	Brazil (Pa)
4	ethirastis (Meyrick, 1922) (Imma)	Peru
5	mitrodeta (Meyrick, 1922) (Imma)	Peru
6	mniographa (Meyrick, 1931) (Imma)	Peru
7	nephallactis (Meyrick, 1906) (Imma)	Venezuela
8	nipharcha (Meyrick, 1931) (Imma)	Brazil (Am)
9	niphostoma (Meyrick, 1922) (Imma)	Brazil (Am)
10	paratma (Meyrick, 1912) (Imma)	Guyana
11	pelomacta (Meyrick, 1922) (Imma)	Brazil (Pa)
12	phthorosema (Meyrick, 1912) (Imma)	Colombia
13	roscida (Meyrick, 1922) (Imma)	Brazil (Am)
14	rugosella (Busck, 1913) (Imma)	Guyana
15	vexatalis (Walker, [1866]) (Alicadra)	Brazil
16	zophodes (Meyrick, 1909) (Imma)	Bolivia

LOXOTROCHIS Meyrick, 1906

17	sepias Meyrick, 1906	Brazil (ES)

IMMA Walker, [1859]
  Pingrassa Walker, [1859]
  Tortricomorpha C. Felder, 1861
  Topaza Walker, 1864
  Vinzela Walker, [1866]
  Thylacopleura Meyrick, 1886
  Davendra Moore, 1887
  Pseudotortrix Turner, 1900

18	atialis (Walker, 1859) (Pyralis)	Brazil
19	boeta (Druce, 1898) (Thalpochares)	Panama
20	cancanopis Meyrick, 1906	Colombia
21	chloromelalis (Walker, [1866]) (Aglossa)	Brazil (Pa)
22	cincta (Druce, 1898) (Eustrotia)	Guatemala
23	ciniata (Druce, 1898) (Thalpochares)	Panama
24	confluens Meyrick, 1931	Brazil (Pa)
25	cuneata Meyrick, 1906	Brazil
26	cyanospora Meyrick, 1926	Colombia
27	eriospila Meyrick, 1922	Brazil (Pa)
28	euglypta Meyrick, 1931	Colombia
29	leniflua Meyrick, 1931	Colombia
30	metachlora Meyrick, 1906	Brazil
31	prasinospora Meyrick, 1915	Ecuador
32	protocrossa Meyrick, 1909	Bolivia

33	**quadrivittana** (Walker, 1863) (Gauris)	Brazil (RJ)
34	**sciophanes** Walsingham, 1914	Guatemala
35	**thymora** Meyrick, 1906	Brazil (Am)
36	**varipes** (Walker, 1862) (Pyralis)	Brazil

Addendum

22. ELACHISTIDAE

ARISTOPTILA Meyrick, 1932

0.1   **smaragdophanes** Meyrick, 1932   Ecuador

ATMOZOSTIS Meyrick, 1932

0.2   **hilda** Meyrick, 1932   Colombia

ELACHISTA Treitschke, 1833

3.1   **petalistis** Meyrick, 1932   Guyana

23. BLASTOBASIDAE

AUXIMOBASIS Walsingham, 1892

53.1   **invigorata** Meyrick, 1932   Virgin Is. (St. Thomas)

BLASTOBASIS Zeller, 1855

63.1   **athymopa** Meyrick, 1932   Virgin Is. (St. Thomas)

26. AGONOXENIDAE

DIACHOLOTIS Meyrick, 1937

0.1   **iopyrrha** Meyrick, 1937   Costa Rica

HELCANTHICA Meyrick, 1932

0.2   **spermotoca** Meyrick, 1932   Virgin Is. (St. Thomas)

PALAEOMYSTIS Meyrick, 1931

9.1   **chalcopeda** Meyrick, 1931   Brazil (RJ)

PROCHOLA Meyrick, 1915

22.1   **chalcothorax** Meyrick, 1932   Brazil (ES)
34.1   **sancticola** Meyrick, 1932   Virgin Is. (St. John)

ZARATHA Walker, 1864

39.1   **cervinella** (Walsingham, 1897) (Syntomactis), Virgin Is. (St. Croix)
         n. comb.

27. COSMOPTERIGIDAE

COSMOPTERIX Hübner, [1825]

6.1   **albicaudis** Meyrick, 1932 (Cosmopteryx [sic]) Virgin Is. (St. Thomas)

29. GELECHIIDAE

RECURVARIA Haworth, 1828

204.1 **hippurista** Meyrick, 1932   Virgin Is. (St. Thomas)

UNTOMIA Busck, 1906

494.1 **formularis** Meyrick, 1929   Brazil (Am)/Colombia

HYPATIMA Hübner, [1825]

503.1 **disposita** (Meyrick, 1931) (Chelaria), **n. comb.**   Brazil (ES)

MYROPHILA Meyrick, 1923

617.1 **caryoplecta** Meyrick, 1930   Brazil (Pa)

NOTES

1. "Dalaca" species of authors that are not placed under **Dalaca** Walker, 1856, belong to other genera presently undefined, or the species need further study. [p. 17]

2. This is the first time Palaeosetidae have been represented in the Neotropical fauna (Nielsen and Robinson, 1983). [p. 17]

3. Nepticulinae is used in place of Stigmellinae, as proposed by Scoble (1983), due to the need for nominate subordinate taxa (ICZN Art. 37). [p. 17]

4. Nepticulini is used, as in footnote 3, in place of Stigmellini, as proposed by Scoble (1983) (ICZN Art. 37). [p. 17]

5. **Prodoxoides** is included, although described in 1984, due to the significance of including the first discovery of Prodoxidae from temperate South America. [p. 18]

6. Tineidae genera are listed alphabetically pending a generic revision, since the author considers the present subfamily limits too uncertain and ill defined. [p. 19]

7. The named subspecies of **Acrolophus macrogaster** (Walsingham) are listed as synonyms since they are not geographically isolated populations but apparently only forms of the same species. [p. 19]

8. **Choropleca terpsichorella** (Busck) was described from Hawaii from specimens introduced from Central America. [p. 21]

9. Many species remaining in **Tinea** Linnaeus may be found to belong to other genera following further study. [p. 23]

10. **Tinea galeatella** Mabille, 1888, has been listed as a synonym of **Plutella xylostella** (Linnaeus, 1758) but has been found to be a synonym of **Tinea pallescentella** Stainton, 1851 (teste E. S. Nielsen). [p. 23]

11. **Harmaclona** is retained in Arrhenophanidae pending further study to determine its possible placement in Tineidae. [p. 25]

12. **Dasycarea** is considered a valid representative of the Old World family Amphitheridae, the only known representative of this family in the New World. [p. 25]

13. The Madinier name, "noctuella," was originally used in the sense of "a small moth," and not as a binomial name, as subsequently cited. Article 11 (ICZN) indicates that the name is unavailable and, in not being binomial or scientific, a vernacular name. [p. 25]

14. **Cremastobombycia lantanella** Busck was described from Hawaii from specimens introduced from Mexico. [p. 26]

15. The placement of "Elachista" **rubella** Blanchard is still uncertain. It was thought to belong to **Phyllocnistis** but now appears best placed in Heliodinidae, tentatively in **Heliodines** (teste D. R. Davis). [p. 27]

16. Most Neotropical Oecophoridae generic type species have been examined by the author for assignment to subfamily. [p. 27]

17. **Exaeretia** Stainton, 1849, is considered by the author to be the senior synonym of **Depressariodes** Turati, 1924, and **Martyrhilda** Clarke, 1941, but some authors restrict **Exaeretia** to a few Palearctic species only. [p. 27]

18. Oecophoridae genera described by Clarke (1978) were not assigned to subfamilies by Clarke. They are placed in the checklist according to their apparently valid subfamily and tribal assignments but in alphabetical order under each higher category. [p. 27]

19. **Eumimographe** Dognin, 1905, is considered a separate genus by some researchers. [p. 28]

20. **Atteria lydia** Druce was transferred from Tortricidae to Oecophoridae by Obraztsov (1966. Proc. US NM., 118:620). The species is herein assigned to the **Eumimographe** group of species, now in **Hypercallia**. [p. 28]

21. **Erysiptila** Meyrick was not included in the New World monograph of Ethmiidae (Powell, 1973) but is closely related to **Ethmia** (teste V. O. Becker). [p. 29]

22. **Macrocirca** Meyrick is another genus not treated by Powell (1973), but apparently is closely related to **Ethmia**; this genus may even be a synonym of **Ethmia** (teste V. O. Becker). [p. 30]

23. The new name, **Rectiostoma** Becker (1982. J. N. Y. Ent. Soc., 89:270), was named to replace the preoccupied **Setiostoma** Zeller, 1875. The Zeller description inadvertently appeared after the publication of Felder and Rogenhofer (1875. "Reise Novara...") wherein the name "Setiostoma" was used for species now assigned to Glyphipterigidae. [p. 35]

24. Numerous species of **Stenoma** will in the future be assigned to new genera but are retained in **Stenoma** pending further study. [p. 35]

25. The generic assignment of **Dasycera auricollis** Walsingham is uncertain and is based on Walsingham's congeneric inclusion of the species with "Dasycera" **newmanella** Clemens, now assigned to **Mathildana**. [p. 40]

26. **Eupolella** Fletcher appears to be related to **Glyphidocera** and is, thus, transferred to Symmocinae (Blastobasidae) (teste V. O. Becker). [p. 41]

27. Syntypes of **Coleophora picticornis** Walsingham are from Haiti, St. Thomas, and Grenada. [p. 42]

28. The Neotropical genera of Gelechiidae are arranged by subfamily except for a large group of genera too little known to make any reasonable assessment of their subfamily position; these latter genera are arranged alphabetically as unplaced genera. [p. 44]

29. **Crambodoxa** Meyrick may belong to Oecophoridae (teste V. O. Becker). [p. 52]

30. **Iphimachaera** Meyrick may be related to **Tinaegeria** in Oecophoridae (teste V. O. Becker). [p. 52]

31. **Leistogenes** Meyrick, 1927, may be a junior synonym of **Logisis** Walsingham, 1909 (teste V. O. Becker). [p. 52]

32. **Ptilostonychia** Walsingham may be related to **Glyphidocera,** in which case it would belong in Symmocinae (Blastobasidae) (teste V. O. Becker). [p. 53]

33. "Gaea" **lilloi** Köhler is transferred from Sesiidae to Gelechiidae (teste T. D. Eichlin) but is as yet unassigned to a gelechiid genus. [p. 53]

34. The superfamily name Copromorphoidea is used, since higher category names need not be based on priority of authorship. Thus, Carposinoidea, Alucitoidea and Glyphipterygoidea are not used. [p. 53]

35. **Sericostola** Meyrick is here transferred from Yponomeutidae to Glyphipterigidae following examination of the type-species (teste J. B. Heppner). [p. 54]

36. Attevinae, **new status,** is here assigned as a new subfamily of Yponomeutidae for **Atteva, Lactura, Pygmocrates,** and the Old World **Anticrates.** These genera all differ from other yponomeutids in possessing chaetosemata on the head vertex. [p. 55]

37. **Syncerastis** Meyrick is here transferred from Tineidae to Yponomeutidae following examination of the type-species (teste D. R. Davis). [p. 56]

38. The subfamily assignment of **Urodus** requires further study. [p. 56]

39. **Pexicnemidia** Möschler, 1890, is transferred from Tineidae to synonymy of **Urodus** Herrich-Schäffer, 1854, following examination of the type-species (teste D. R. Davis). [p. 56]

40. There are undoubtedly a number of Nearctic species of **Argyresthia** that also occur in the oak zone of northern Mexico. [p. 56]

41. **Adela selectella** Walker was transferred from Adelidae to Acrolepiidae as a species of **Antispastis** by Davis (1980. J. Lepid. Soc., 32:187). [p. 57]

42. Schreckensteiniinae is retained as a subfamily of Heliodinidae. Having spines on the hind tibiae does not in itself signify a separate family of Lepidoptera, especially while other characters show relationships to other Heliodinidae. The apparent similarities to Epermeniidae are unconfirmed. [p. 57]

43. New World **Heliodines** require study to determine if they are actually congeneric with the European type-species of the genus. [p. 57]

44. "Elachista" **rubella** Blanchard, as noted in note 15, is only tentatively assigned to **Heliodines** (teste D. R. Davis). [p. 57]

45. **Mnesichara** Walsingham, 1912, is now thought to be a distinct genus in Oecophorinae, including the species **M. dyctiota** Walsingham, 1912, and **M.ithymeta** (Meyrick, 1926). [p. 28]

INDEX TO SPECIES

abatiae E. M. Hering, Phyllocnistis, 20-137
abboti Holland, Oiketicus, 16-56
abbotii Grote, Oiketicus, 16-56
abdita Walsingham, Acrolophus, 15-1
abdominella Busck, Ethmia, 21-329
abiasta Meyrick, Glyphidocera, 23-2
abligatella (Walker), Niditinea, 15-306
abnormalis Walsingham, Cosmopterix, 27-6
abolitella (Walker), Compsolechia, 29-365
abortiva (Walsingham), Cymotricha, 29-528
abraxasella (Walker), Ethmia, 21-309
abraxasella (Walker), Ethmia, 21-309a
abraxella Meyrick, Ethmia, 21-309
abroniaeella Chambers, Lithariapteryx, 40-32
abroniella Meyrick, Lithariapteryx, 40-32
abrupta Walsingham, Opostega, 8-1
abruptella (Walker), Compsolechia, 29-366
abscensella (Walker), Amydria, 15-201
abscessella (Walker), Vazugada, 29-680
absconditella (Walker), Antaeotricha, 21-450
absconditella (Walker), Monochroa, 29-3
absoluta (Meyrick), Scrobipalpula, 29-231
absolutrix Meyrick, Tinea, 15-368
absyrtus Schaus, Phassus, 5-52
acaciella (Busck), Friseria, 29-105
acajuti Becker, Cerconota, 21-842
acalles (Walsingham), Alucita, 31-2
accessoriella Frey & Boll, Opostega, 8-2
accincta Meyrick, Nealyda, 29-4
accinctella (Walker), Compsolechia, 29-367
accurata (Meyrick), Anadasmus, 21-421
acertella (Busck), Antequera, 27-1
achatina (Zeller), Cerconota, 21-843
achnodes Meyrick, Acrocercops, 20-45
achroea Walsingham, Logisis, 29-779
acicularis Meyrick, Untomia, 29-493
acidata Meyrick, Opostega, 8-3
acmaea Clarke, Lelita, 21-1683
acompsa Walsingham, Durrantia, 21-407
aconitis Clarke, Stenoma, 21-1131
acontiella (Walker), Stenoma, 21-1115
acornus Hasbrouck, Acrolophus, 15-2
acosmeta (Walsingham), Compsolechia, 29-368
acratodes Meyrick, Stenoma, 21-1116
acribota Meyrick, Stenoma, 21-1453
acridula (Meyrick), Orphnolechia, 21-1077
acritomorpha Clarke, Palinorsa, 21-260
acrobapta Meyrick, Antaeotricha, 21-751
acrobatica Meyrick, Struthoscelis, 21-1709
acrocosma (Meyrick), Gonionota, 21-112
acrodelta Meyrick, Plutella, 35-11
acrodisca Meyrick, Choropleca, 15-220
acroglypta Meyrick, Battaristis, 29-341
acrograpta (Meyrick), Antaeotricha, 21-445
acrolychna Meyrick, Dichomeris, 29-564
acronephela Meyrick, Antaeotricha, 21-446
acronitis (Busck), Chlamydastis, 21-910
acropelta Meyrick, Commatica, 29-727
acrosticta Walsingham, Stenoma, 21-1319
actiella Barnes & Busck, Isophrictis, 29-1
actinota (Walsingham), Neomachlotica, 34-17
actista (Meyrick), Antaeotricha, 21-447
acuminata (Staudinger), Dichomeris, 29-565
acuminata (Walsingham), Copticostola, 29-751
adaequata (Meyrick), Onebala, 29-619

additella (Walker), Antaeotricha, 21-450
addon (Busck), Antaeotricha, 21-448
adductella (Walker), Antaeotricha, 21-449
adequata Clarke, Onebala, 29-619
adjunctella (Walker), Antaeotricha, 21-450
adjutrix Meyrick, Holcocera, 23-82
adminiculata Meyrick, Stenoma, 21-1117
administra Meyrick, Auximobasis, 23-44
admixta (Walsingham), Antaeotricha, 21-451
adoratrix Meyrick, Stenoma, 21-1118
adornata (Meyrick), Antaeotricha, 21-452
adulans Meyrick, Stenoma, 21-1119
adusta Walsingham, Opostega, 8-4
adustella (Walker), Stenoma, 21-1120
adustella (Zeller), Ethmia, 21-369
adustipennis (Walsingham), Brachyacma, 29-524
advocata (Meyrick), Gonioterma, 21-1001
adytodes Meyrick, Stenoma, 21-1121
aedificata Meyrick, Anacampsis, 29-306
aedilis Meyrick, Prochola, 26-18
aenconivella (Walker), Tiquadra, 15-420
aeneiceps (R. Felder & Rogenhofer), Tinaegeria, 21-1730
aeneoalbida (Walsingham), Pompostolella, 15-339
aeneocapitella (Walsingham), Caloptilia, 20-1
aeneonivella (Walker), Tiquadra, 15-420
aenigmatica Clarke, Trichotaphe, 29-649
aeolastis (Meyrick), Caloptilia, 20-2
aeolella (Walsingham), Proboloptilia, 15-342
aequabilis (Meyrick), Antaeotricha, 21-453
aequanima (Meyrick), Plumana, 16-4
aequata Meyrick, Dichomeris, 29-566
aequatoriella Kristensen & Nielsen, Squamicornia, 1-2
aequilibris Meyrick, Compsolechia, 29-369
aequivoca Meyrick, Blastobasis, 23-60
aequorea (Meyrick), Evippe, 29-95
aerifica (Meyrick), Stigmella, 7-1
aerinotata (Butler), Antaeotricha, 21-454
aesiocopia (Walsingham), Gonioterma, 21-1002
aethiops R. Felder & Rogenhofer, Adela, 12-14
aethographa Clarke, Gonionota, 21-113
aethoptera Clarke, Gonionota, 21-114
aethostola Meyrick, Machimia, 21-231
affinis R. Felder & Rogenhofer, Antaeotricha, 21-455
affirmata (Meyrick), Marmara, 20-116
affirmatella Busck, Stenoma, 21-1122
agathelpis Meyrick, Stenoma, 21-1134
aggerata Meyrick, Stenoma, 21-1249
aggravata (Meyrick), Antaeotricha, 21-456
aggregata Meyrick, Stenoma, 21-1123
aglaogramma Meyrick, Triclonella, 27-45
aglypta Meyrick, Antaeotricha, 21-457
agonistes (Walsingham), Mompha, 25-2
agramma Becker, Timocratica, 21-1473
agraria (Meyrick), Cerconota, 21-844
agrestis Meyrick, Auximobasis, 23-45
agrifoliella (Braun), Cameraria, 20-134
agrionides Walker, Phassus, 5-48
agrotipennella (Grote), Acrolophus, 15-138
agypsota Meyrick, Prochola, 26-19
agyrtodes (Meyrick), Pyramidobela, 21-1701
ahenea Walsingham, Heliozela, 14-10
alba Zeller, Tegeticula, 13-2
albaciliella Busck, Heliodines, 40-13
albastrigulella (Kearfott), Pleurota, 21-1732

albella Amsel, Timocratica, 21-1476
albella (Blanchard), Aliciana, 21-1518
albella (Zeller), Timocratica, 21-1474
albicella Busck, Antaeotricha, 21-458
albicellata Meyrick, "Scaeosopha", 27-40
albiciliella Meyrick, Heliodines, 40-13
albicilla (Zeller), Antaeotricha, 21-458
albicornella Busck, Walshia, 27-106
albicostella (Beutenmüller), Ethmia, 21-310
albida (Walker), Stenoma, 21-1124
albifrons (Walsingham), Trycherodes, 21-304
albifrons Zeller, Antaeotricha, 21-459
albilimbella (R. Felder & Rogenhofer), Antaeotricha, 21-460
albilingua Walsingham, Arogalea, 29-53
albimacula (Walsingham), Phyllonorycter, 20-122
albinervella (Kieffer & Jörgensen), Tecia, 29-820
albipectus (Walsingham), Fascista, 29-103
albipennis Meyrick, Acrolophus, 15-3
albipes (R. Felder & Rogenhofer), Zetesima, 21-1512
albisquamella Zeller, Elachista, 22-2
albistrigella (Möschler), Pyroderces, 27-38
albitincta (Meyrick), Antaeotricha, 21-461
albitogata Becker, Timocratica, 21-1475
albivitella Zeller, Pammeces, 26-10
albodiscata (Walker), Moca, 41-1
albomarginata (Walsingham), Acrocercops, 20-46
albovenosa Walsingham, Plutella, 35-18
albovenosa Zeller, Antaeotricha, 21-462
alcyonis Meyrick, Stilbosis, 27-91
aletha Duckworth, Lethata, 21-1035
aleuropis Meyrick, Setiarcha, 15-357
alexandra (Meyrick), Hypercallia, 21-199
algidella (Walker), Antaeotricha, 21-650
algosa (Meyrick), Gonioterma, 21-1025
allardi Clarke, Coptotelia, 21-30
alligans (Butler), Stenoma, 21-1125
alloea Walsingham, Oestomorpha, 29-786
allura Viette, Pfitzneriana, 5-61
allutella (Rebel), Phereoeca, 15-335
alluvialis Meyrick, Stenoma, 21-1126
alsiosum Walsingham, Gonioterma, 21-1003
alsocoma Meyrick, Stilbosis, 27-92
alticolans Meyrick, Untomia, 29-494
alticolens Walsingham, Untomia, 29-494
altisona (Meyrick), Symmetrischema, 29-259
altivola (Meyrick), Taygete, 29-269
amabilis Walsingham, Durrantia, 21-408
amauropis Meyrick, Cymotricha, 29-529
amauroptera Clarke, Gonionota, 21-115
amaurota (Meyrick), Compsolechia, 29-370
amazona Duckworth, Lethata, 21-1036
amazonensis Viette, 1950, Druceiella, 5-31
amazonica Meyrick, Compsolechia, 29-371
ambiens Meyrick, Stenoma, 21-1127
ambigua (R. Felder & Rogenhofer), Psittacastis, 21-294
amblystoma (Meyrick), Chlamydastis, 21-971
ambusta (Walsingham), Compsolechia, 29-372
amentata Meyrick, Opsodoca, 15-325
amicula Zeller, Antaeotricha, 21-463
ammitis (Meyrick), Exaeretia, 21-1
ammodes (Walsingham), Antaeotricha, 21-464
ammoxanthus (Meyrick), Dichomeris, 29-565
ampherista Meyrick, Antaeotricha, 21-821
amphibola Walsingham, Stilbosis, 27-93

amphichroma Clarke, Corita, 21-1545
amphicoma (Meyrick), Vazugada, 29-681
amphicosma Meyrick, Cymotricha, 29-530
amphicrena (Meyrick), Gonionota, 21-116
amphilocha Meyrick, Urodus, 36-34
amphilychna Meyrick, Sematoptis, 27-42
amphilyta Meyrick, Antaeotricha, 21-465
amphiptera Meyrick, Stenoma, 21-1128
amphiscolia Meyrick, Battaristis, 29-342
amphizyga Meyrick, Antaeotricha, 21-466
amseli Duckworth, Timocratica, 21-1476
anachasta Meyrick, Blastobasis, 23-61
anachoreta Bradley, Ecpathophanes, 17-5
anaclintris Meyrick, Antaeotricha, 21-467
analis (Busck), Inga, 21-1583
analytica Meyrick, Tinea, 15-369
anamochla (Meyrick), Chlamydastis, 21-911
anaphanta (Meyrick), Orphnolechia, 21-1078
anaphorella (Walsingham), Acrolophus, 15-4
anaphrictis Meyrick, Homosetia, 15-265
anathyrsa Meyrick, Acrolophus, 15-5
anaxesta Meyrick, Stenoma, 21-1129
ancalopa (Meyrick), Alucita, 31-3
ancalota (Meyrick), Chlamydastis, 21-912
anceps (Butler), Anadasmus, 21-422
anceps Walsingham, Amydria, 15-202
ancillaris Meyrick, Stenoma, 21-1130
ancobathra Meyrick, Ussara, 34-22
anconitis Meyrick, Stenoma, 21-1131
ancora Clarke, Nanodacna, 26-5
ancorata (Walsingham), Inga, 21-1584
ancylacma Meyrick, Stenoma, 21-1132
ancyristis Meyrick, Ussara, 34-23
ancyropis Meyrick, Homosetia, 15-266
andesae Davis & Nielsen, 1980, Synempora, 3-1
andina (Meyrick), Stigmella, 7-2
anelaea (Meyrick), Timocratica, 21-1477
anelicta (Meyrick), Gonionota, 21-117
anetodes Meyrick, Stenoma, 21-1133
angulatella (Walsingham), Acrolophus, 15-6
angulifera Walsingham, Acompsia, 29-515
angusta Duckworth, Lethata, 21-1037
angusta Meyrick, Auximobasis, 23-46
angustalatella Powell, Ethmia, 21-311
angustella (Walsingham), Dyotopasta, 15-251
anisectis (Meyrick), Faculta, 29-102
anisodes (Meyrick), Gonionota, 21-118
anisopa (Meyrick), Onebala, 29-620
anisota Meyrick, Coleophora, 24-1
anita Busck, Gonioterma, 21-1022
anna Busck, Gonioterma, 21-1004
annixa Meyrick, Antaeotricha, 21-484
annosa (Butler), Stenoma, 21-1134
annulata Busck, Perimede, 27-86
annulata Clarke, Doina, 21-69
annulatellus Wood, Plutella, 35-18
annulicornis (Walsingham), Recurvaria, 29-200
anoma Walsingham, Holophysis, 29-479
anomala Clarke, Talitha, 21-9
anonella (Sepp), Cerconota, 21-845
anophthalma (Meyrick), Lethata, 21-1038
antarctica Staudinger, Endrosis, 21-1563
antarcticus (Staudinger), Callipielus, 5-11
anthestias Meyrick, Syncraternis, 15-360
anthogramma Meyrick, Acrocercops, 20-47

anthracella Forbes, Antipolistes, 15-212
anthraceuta (Meyrick), Ithome, 27-74
anthracospora Meyrick, Machimia, 21-232
anthracura (Meyrick), Compsolechia, 29-373
anticharis (Meyrick), Antaeotricha, 21-455
anticosma Meyrick, Hybroma, 15-273
anticrates Meyrick, Trichotaphe, 29-650
antidectis (Meyrick), Triclonella, 27-46
antillia Forbes, Cosmopterix, 27-9
antilyra Meyrick, Stenoma, 21-1169
antiquata (Meyrick), Moca, 41-2
antipathetica (Forbes), Xystrologa, 15-466
antiplaca Meyrick, Compsolechia, 29-374
antischema Meyrick, Stathmopoda, 21-1725
antisticha Meyrick, Dichomeris, 29-567
antitacta Meyrick, Stenoma, 21-1135
antitoxa (Meyrick), Phyllonorycter, 20-123
antitypa Meyrick, Falculina, 21-995
antonellus (Barnes & McDunnough), Acrolophus, 15-63
apachella Dietz, Amydria, 15-203
apertella Busck, Acrolophus, 15-7
aphaidropa (Dyar), Naevipenna, 16-16
aphanes (Walsingham), Cerconota, 21-846
aphanes (Zeller), Holcocera, 23-83
aphanodesma Meyrick, Stenoma, 21-1444
aphanoptis Meyrick, Urodus, 36-35
aphilodes Meyrick, Blastobasis, 23-62
aphiltra Meyrick, Aristotelia, 29-9
aphroditeella Chambers, Strobisia, 29-488
aphroditella Frey, Strobisia, 29-488
aphrodora (Meyrick), Moca, 41-3
aphrogenes (Meyrick), Chlamydastis, 21-986
aphrogramma (Meyrick), Gonioterma, 21-1002
aphrophanes Meyrick, Stenoma, 21-1136
apicalis (Busck), Chlamydastis, 21-958
apicalis (Hampson), Biopsyche, 16-47
apicella Walker, Taruda, 21-298
apicepuncta (Busck), Semophylax, 29-512
apicepunctella (Walsingham), Acrocercops, 20-48
apicilinella (Clemens), Battaristis, 29-352
apicipuncta (Meyrick), Semophylax, 29-512
apicistrigella (Chambers), Battaristis, 29-352
apiculata Meyrick, Cosmopterix, 27-7
aplytopis Meyrick, Stenoma, 21-1137
apoclina (Meyrick), Chlamydastis, 21-913
aporodes Meyrick, Antaeotricha, 21-486
apsorrhoa Meyrick, Stenoma, 21-1138
aptila Meyrick, Stenoma, 21-1139
aquilina (Meyrick), Symmetrischema, 29-260
arachnia (Meyrick), Antaeotricha, 21-468
arachniotis Meyrick, Antaeotricha, 21-546
aratella (Walker), Antaeotricha, 21-469
araucana (Bartlett-Calvert), Gonionota, 21-166
arcanella (Busck), Durrantia, 21-409
arcasalis (Walker), Acrolophus, 15-8
arcei (Druce), Acrolophus, 15-158
archaea Walsingham, Arogalea, 29-54
archigrapha Meyrick, Onebala, 29-621
arctella (Walker), Sitotroga, 29-513
arctostaphylella (Walsingham), Ethmia, 21-312
arcturella (Walker), Acrolophus, 15-9
arda Clarke, Perzelia, 21-264
ardeola (Meyrick), Symmetrischema, 29-261
ardesiella (Walsingham), Trichotaphe, 29-651
ardiophora Meyrick, Battaristis, 29-343

arenaria (Walsingham), Chlamydastis, 21-914
arenosa (Meyrick), Anadasmus, 21-423
arenosus Butler, Callipielus, 5-11
argentata Meyrick, Psittacastis, 21-281
argentata Ureta, Callipielus, 5-17
argentea (Busck), Euceratia, 35-3
argentea Busck, Pavolechia, 29-799
argentidisca Dognin), Eomichla, 21-1564
argentidorsella (Busck), Hastamea, 21-197
argentiferus Walker, Phassus, 5-36
argentifrons Walsingham, Strobisia, 29-487
argentifrontella (Walsingham), Phyllonorycter, 20-124
argentiliniella Busck, Euprora, 19-1
argentinana Brèthes, Oliera, 11-3
argentinella (Berg), Dichomeris, 29-568
argentinus Walsingham, Acrolophus, 15-10
argicerauna (Meyrick), Gonioterma, 21-1005
argigastra (Walsingham), Trichotaphe, 29-652
argilla Clarke, Osmarina, 21-259
argillacea Walsingham, Blastobasis, 23-63
argillacea (Zeller), Stenoma, 21-1140
argo Meyrick, Hexeretmis, 31-22
argocorys (Meyrick), Antaeotricha, 21-470
argocosma Meyrick, Acrocercops, 20-49
argocymba (Meyrick), Chlamydastis, 21-915
argogastra Meyrick, Trichotaphe, 29-652
argonais (Meyrick), Timocratica, 21-1478
argonias Clarke, Timocratica, 21-1478
argophthalma Meyrick, Syncamaris, 30-12
argopleura Clarke, Gonionota, 21-119
argosema (Meyrick), Chionodes, 29-70
argospila Meyrick, Scoliographa, 21-297
argospora Meyrick, Stenoma, 21-1141
argotoma Meyrick, Stenoma, 21-1404
argoxantha Meyrick, Thiotricha, 29-299
argyracma Meyrick, Compsolechia, 29-375
argyracta Meyrick, Hapalonoma, 29-609
argyractis Meyrick, Aristotelia, 29-10
argyritis Meyrick, Stilbosis, 27-94
argyrobasis (Duckworth), Rectiostoma, 21-1098
argyropa Meyrick, Hypercallia, 21-200
argyropasta Walsingham, Psilocorsis, 21-273
argyrophaea Forbes, Protodarcia, 15-344
argyrophorum E. M. Hering, Theama, 21-1715
arida Walsingham, Acrolophus, 15-11
arimasalis (Walker), Acrolophus, 15-12
arista Walsingham, Hypercallia, 21-201
aristula Walsingham, Homodoxus, 15-264
arizonella Dietz, Amydria, 15-204
arizonellus Walsingham, Acrolophus, 15-13
armata (Zeller), Stenoma, 21-1142
armiferella (Walker), Cerconota, 21-847
aromatica (Meyrick), Lethata, 21-1039
arotrosema (Walsingham), Trichotaphe, 29-653
arquata Meyrick, Ussara, 34-24
arridens Meyrick, Stenoma, 21-1348
arriguttii (Pastrana), Alucita, 31-4
artifex (Kieffer & Jörgensen), Cecidoses, 11-1
arystis Meyrick, Antaeotricha, 21-471
asaphogramma Meyrick, Acrocercops, 20-50
ascalopa (Meyrick), Plumana, 16-5
ascetica (Meyrick), Exaeretia, 21-2
ascodes Meyrick, Stenoma, 21-1143
asemanta Dognin, Cryptolechia, 21-49
aspera (Zeller), Tiquadra, 15-421

asperula Clarke, Doina, 21-70
asphalopis (Meyrick), Antaeotricha, 21-817
asphaltitis Meyrick, Dissoptila, 29-755
assa (Druce), Aepytus, 5-65
assecta Zeller, Antaeotricha, 21-472
assignata Meyrick, Stenoma, 21-1144
assimilis Vázquez, Oiketicus, 16-53
associata Meyrick, Stenoma, 21-1380
assymmetra Nielsen & Davis, Prodoxoides, 13-6
astacopis (Meyrick), Cerconota, 21-900
asthenopa (Meyrick), Lethata, 21-1040
astragalopa Meyrick, Hormantris, 15-272
astrapias Walsingham, Cosmopterix, 27-8
astrella Walsingham, Adela, 12-15
astrocharis Meyrick, Panthytarcha, 15-331
astroconis (Meyrick), Battaristis, 29-349
astronoma (Meyrick), Menesta, 21-1069
astuta Meyrick, Pigritia, 23-109
astynoma (Meyrick), Antaeotricha, 21-473
atalantis Meyrick, Acrocercops, 20-51
atelesta Meyrick, Battaristis, 29-344
atelura Meyrick, Glyphipterix, 34-32
aterpes Walsingham, Stenoma, 21-1145
aterrima (Trelease), Tegeticula, 13-1
atialis (Walker), Imma, 41-18
atmastra Meyrick, Compsolechia, 29-376
atmodes Meyrick, Stenoma, 21-1345
atmosema Meyrick, Blastobasis, 23-64
atmospora (Meyrick), Antaeotricha, 21-474
atmozona Meyrick, Blastobasis, 23-65
atomosella (Walker), Praeacedes, 15-341
atractelia (Meyrick), Neomachlotica, 34-18
atricassis (Meyrick), Cerconota, 21-848
atrifascis (Meyrick), Symmetrischema, 29-262
atriguttata Meyrick, Trichotaphe, 29-654
atriplicella Kieffer & Jörgensen, Gnorimoschema, 29-156
atrivallata Meyrick, Halimarmara, 21-1577
atrolinea (Barnes & McDunnough), Lactura, 36-10
atropicta (Zeller), Inga, 21-1666
attenuata (Walsingham), Acrocercops, 20-52
attenuatella (Walker), Cosmopterix, 27-9
atteria Busck, Ancipita, 21-11
attonita Meyrick, Oxycryptis, 29-787
auchmera Walsingham, Urodus, 36-36
aucupatrix Meyrick, Thiotricha, 29-300
audax Walsingham, Glyphidocera, 23-3
augescens Meyrick, Stenoma, 21-1146
aulonota (Meyrick), Evippe, 29-96
aulorrhoa (Meyrick), Scrobipalpula, 29-230
aurea Busck, Hamadera, 21-196
aurea (Fitch), Atteva, 36-8
aureella (Blanchard), Acrolepiopsis, 39-1
aureoapicella (Möschler), Ethmia, 21-309
aureoflamma Walsingham, Heliodines, 40-14
aurera (Stretch), Atteva, 36-8
auriciliella Busck, Fortinea, 29-772
auricollis (Walsingham), Mathildana, 21-1687
auricoma Meyrick, Stenoma, 21-1147
aurigenus Pfitzner, Phassus, 5-42
auriinea Zeller, Phyllocnistis, 20-138
auripennis Boisduval, Myrsila, 34-3
aurisulcata Meyrick, Trichotaphe, 29-655
auritogata (Walsingham), Plocamosaris, 29-640
auromaculata Walsingham, Tinea, 15-370

australis (Walsingham), Acrolophus, 15-14
autampyx Meyrick, Promenesta, 21-1087
autocrena (Meyrick), Gonionota, 21-120
autodesma (Meyrick), Holophysis, 29-480
autoplecta (Meyrick), Plumana, 16-6
auxoleuca (Meyrick), Timocratica, 21-1501
auxiliaris (Meyrick), Holophysis, 29-481
auxoptila Meyrick, Telphusa, 29-281
avida Meyrick, Stenoma, 21-1405
avitella Walker, Tiquadra, 15-421
axena Meyrick, Antaeotricha, 21-475
aztecana Walsingham, Stenoma, 21-1148
baccharidis E. M. Hering, Phyllocnistis, 20-139
baccata Meyrick, Dichomeris, 29-569
bactra Busck, Acrolophus, 15-15
badiella (Amsel), Lethata, 21-1038
bahiensis (Perty), Antaeotricha, 21-750
baja Powell, Ethmia, 21-313
balanitis Meyrick, Idiocrates, 21-229
balanocentra (Meyrick), Antaeotricha, 21-476
balanoptis Meyrick, Stenoma, 21-1149
baldufi Hasbrouck, Acrolophus, 15-16
baleni (Zeller), Exaeretia, 21-3
balia (Walsingham), Compsolechia, 29-377
baliandra Meyrick, Stenoma, 21-1150
baliopa Meyrick, Lindera, 15-291
baliostola Walsingham, Ethmia, 21-314
baliostoma Busck, Ethmia, 21-314
ballista (Meyrick), Antaeotricha, 21-477
balsamopa (Meyrick), Taygete, 29-270
barbaropis Busck, Tinea, 15-373
barberella (Busck), Eudactylota, 29-93
barbipalpus Busck, Acrolophus, 15-17
barema Durrant, Acrolophus, 15-18
barnesii (Dyar), Acrolophus, 15-178
barydelta (Meyrick), Taygete, 29-271
barydesma (Meyrick), Holophysis, 29-482
baryspila Meyrick, Acrolophus, 15-19
basalis Walker, Arauzona, 21-1721
basalis Zeller, Antaeotricha, 21-478
basichlora Meyrick, Prochola, 26-20
basiferella (Walker), Antaeotricha, 21-479
basilaris (Busck), Antaeotricha, 21-480
basilica Hodges, Neodactylota, 29-179
basimacula Möschler, Antaeotricha, 21-602
basinigra R. Felder, "Psychoglene", 16-61
basiplagata Walsingham, Cycloplasis, 40-10
basiplagata Walsingham, Holcocera, 23-84
basiplagiata Forbes, Cycloplasis, 40-10
basisignella (Zeller), Mompha, 25-3
basirei Schaus, Phassus, 5-37
basirosella Busck, Costoma, 21-46
basirubra (Schaus), Druceiella, 5-30
basirubrella (Walker), Antaeotricha, 21-481
basistriga (Barnes & McDunnough), Lactura, 36-11
basqueella (Chambers), Stegasta, 29-250
batesella (Walker), Antaeotricha, 21-765
bathrocentra Meyrick, Stenoma, 21-1151
bathrochlora Meyrick, Gelechia
bathrogramma (Meyrick), Stenoma, 21-1152
bathrotoma (Meyrick), Antaeotricha, 21-482
bathyntis Meyrick, Stenoma, 21-1153
bathyphaea (Meyrick), Cerconota, 21-849
batrachopis (Meyrick), Chlamydastis, 21-916
battis Hodges, Perimede, 27-87

belella (Walker), Aristotelia, 29-41
bella (Chambers), Heliodines, 40-15
bella Duckworth, Falculina, 21-996
benigna Meyrick, Stenoma, 21-1154
bergii (Weyenbergh), Oiketicus, 16-48
bermudensis Hodges, Batrachedra, 24-21a
betulinella (Hübner), Endrosis, 21-1563
biannulata Meyrick, Stenoma, 21-1155
biarcuata Meyrick, Antaeotricha, 21-483
biatomella Walsingham, Pigritia, 23-110
bibula Meyrick, Exosphrantis, 21-1576
bicensa Meyrick, Stenoma, 21-1156
bicolor (Walsingham), Nealyda, 29-5
bicolor (Zeller), Antaeotricha, 21-484
bicolorella Forbes, Protodarcia, 15-345
bicoloripennis Hodges, Triclonella, 27-47
bicornuta Becker, Timocratica, 21-1479
bidens Walsingham, Acrolophus, 15-20
biedermanni (Viette), Aepytus, 5-73
bifida (Meyrick), Chlamydastis, 21-917
bifissa Meyrick, Stibarenches, 29-815
bifurcata Busck, Acrolophus, 15-21
bifuscella (Forbes), Commatica, 29-728
bilinguis (Meyrick), Antaeotricha, 21-485
bimarginata Meyrick, Doliotechna, 21-1552
bimarginellum Walsingham, Compsolechia, 19-9
bimendella (Zeller), Scardia, 15-349
bimetallica Walsingham, Ecballogonia, 27-33
biniveipunctata (Walsingham), Stegasta, 29-249
binocularis Meyrick, Anthistarcha, 29-701
binodis Meyrick, Stenoma, 21-1360
binotatella (Walker), Compsolechia, 29-378
binubila Zeller, Antaeotricha, 21-486
bipectinicornis Hasbrouck, Acrolophus, 15-98
bipunctalis (Warren), Coptotelia, 21-31
bipunctella (Ragonot), Tinea, 15-398
bipupillata Meyrick, Antaeotricha, 21-487
biruptella Zeller, Argyresthia, 37-1
biseriata (Zeller), Stenoma, 21-1157
bisignata Meyrick, Stenoma, 21-1158
bistrigella (Busck), Battaristis, 29-345
bittenella (Busck), Ethmia, 21-315
bivirgella R. Felder & Rogenhofer, Tinea, 15-371
blandula Meyrick, Stenoma, 21-1159
blepharopa (Meyrick), Compsolechia, 29-379
boeta (Druce), Imma, 41-19
bogotatella Alpheraky, Setomorpha, 15-359
bogotatella (Walker), Lindera, 15-294
bogotensis (Walsingham), Acrolophus, 15-22
bolistis (Meyrick), Gonioterma, 21-1006
boliviana Busck, Tinea, 15-372
boliviensis Viette, Pfitzneriana, 5-59b
bombaulia Meyrick, Acrolophus, 15-23
bonniwelli Barnes & Benjamin, Oiketicus, 16-55
borboropis Meyrick, Tinea, 15-373
borboropis Meyrick, Phereoeca, 15-337
borboropis Meyrick, Tinea, 15-373
borquiniella (Köhler), Gonionota, 21-121
borsaniella Köhler, Gnorimoschema, 29-157
borsanii Köhler, Oiketicus, 16-49
bosqueella (Chambers), Stegasta, 29-250
bosquella (Chambers), Stegasta, 29-250
boucardi Druce, Acrolophus, 15-24
bougainvilleae E. M. Hering, Nealyda, 29-6
bourgognei Viette, Roseala, 5-114

bourquini Clarke, Gonionota, 21-122
bourquini Pastrana, Phyllocnistis, 20-140
boviceps Walsingham, Choropleca, 15-221
bovinella (Busck), Lethata, 21-1041
bracatingae (Köhler), Antaeotricha, 21-488
brachyanches Meyrick, Urodus, 36-37
brachycasis Meyrick, Amiantastis, 16-2
brachymetra Meyrick, Dichomeris, 29-570
brachyplaca (Meyrick), Cerconota, 21-850
brachysaris Meyrick, Antaeotricha, 21-489
brachyxista Meyrick, Philomusaea, 21-1693
bradleyi (Viette), Aepytus, 5-112
brasiliensis (Heylaerts), Lumacra, 16-27
brasiliensis (Viette), Aepytus, 5-109
brasiliensis (Zagulajev), Scardia, 15-350
brassicella Fitch, Plutella, 35-18
brevipalpella Walsingham, Auximobasis, 23-47
brevipalpis (Walsingham), Psittacastis, 21-293
breviramis (Meyrick), Triclonella, 27-59
brevisella (Walker), Inga, 21-1585
brevistrigata (Walsingham), Lepyrotica, 15-283
breyeri Bourquin, Acrocercops, 20-53
breyeri Pastrana, Coleophora, 24-2
bridarollii (Pastrana), Alucita, 31-5
brochospila (Meyrick), Compsolechia, 29-372
brochota Meyrick, Antaeotricha, 21-490
bromeliae Walsingham, Valentinia, 23-106
bruneri Busck, Hypercallia, 21-202
brunnea (Fletcher), Alucita, 31-6
brunnea Povolný, Keiferia, 29-168
brunnea (Schaus), Aepytus, 5-112
brunnescens Robinson, Callipielus, 5-12
brunneus Busck, Prostomeus, 29-511
brunniceps (R. Felder & Rogenhofer), Filinota, 21-93
bryochlora Meyrick, Sorotacta, 29-809
bryocosma Meyrick, Stenoma, 21-1160
bryophanes (Meyrick), Gonioterma, 21-1007
bryotrophoides (Zeller), Borkhausenia, 21-1532
bryoxyla Meyrick, Stenoma, 21-1161
bufo Walsingham, Gelechia, 29-114
bugabae Walsingham, Acrolophus, 15-25
bullata Meyrick, Carposina, 32-6
burmanniana (Stoll), Gonioterma, 21-1008
burmeisteri Weyenbergh, "Psyche", 16-62
burrowsi Jones, Dendropsyche, 16-26
burserella (Busck), Caloptilia, 20-3
busckellum Walsingham, Synallagma, 25-29
buscki Duckworth, Lethata, 21-1042
busckiella Engel, Synallagma, 25-29
butyranthes Meyrick, Tiquadra, 15-422
butyrota (Meyrick), Timocratica, 21-1480
byrsinites (Meyrick), Anadasmus, 21-424
byrsocyma (Meyrick), Coptotelia, 21-31
byssina (Zeller), Stenoma, 21-1162
byssophanes (Meyrick), Chlamydastis, 21-918
bythitis Meyrick, Stenoma, 21-1163
bythochroa (Meyrick), Cerconota, 21-881
bythodes (Meyrick), Acrolepiopsis, 39-2
cabima Busck, Gonada, 21-106
cacahuamilpensis (Herrera), Monopis, 15-299
cachrydias Meyrick, Dichomeris, 29-571
cacoderma Walsingham, Gelechia, 29-115
cacoeciella (Amsel), Gonioterma, 21-1012
caduca (Walsingham), Machimia, 21-233
caducella Zeller, Tinea, 15-374

caecata (Meyrick), Chlamydastis, 21-919
caeligena (Meyrick), Charistica, 29-715
caementosa Meyrick, Acrocercops, 20-54
caenochytis (Meyrick), Antaeotricha, 21-491
caerula Meyrick, Tinea, 15-375
caerulipalpis Meyrick, Compsistis, 21-25
caesarea Meyrick, Stenoma, 21-1164
caesia Meyrick, Stenoma, 21-1165
caespitella Zeller, Gelechia, 29-116
caieta Hodges, Friseria, 29-106
calator Hodges, Batrachedra, 24-15
calcarata (Meyrick), Teinoptila, 36-32
calcarata Walsingham, Walshia, 27-107
calcifera Walsingham, Scelorthus, 40-36
calculatrix Meyrick, Aristotelia, 29-11
calidaria (Meyrick), Coptotelia, 21-32
caliginea (Meyrick), Anadasmus, 21-425
callichalca Meyrick, Cosmopterix, 27-10
callichlora Meyrick, Promenesta, 21-1088
callichora (Meyrick), Caloptilia, 20-4
callichroma (Meyrick), Charistica, 29-716
callicoma Meyrick, Stenoma, 21-1166
callidelta Meyrick, Glyphipterix, 34-33
callidora (Meyrick), Rectiostoma, 21-1099
callierastis (Meyrick), Inga, 21-1586
calligera (Zeller), Urodus, 36-38
callimnestra Meyrick, Rhodanassa, 21-1114
calliscelis Meyrick, Copocentra, 40-4
callitechna Meyrick, Telphusa, 29-282
calumnians Meyrick, Molopostola, 29-784
calumniella Powell, Ethmia, 21-316
calycocentra (Meyrick), Inga, 21-1587
calyptrodes Meyrick, Triclonella, 27-48
camarina Meyrick, Antaeotricha, 21-492
camarodes Meyrick, Stenoma, 21-1167
camaronae (Zeller), Caloptilia, 20-5
camelopis (Meyrick), Inga, 21-1588
campalea (Walsingham), Compsolechia, 29-380
camptochrysa Meyrick, Acrocercops, 20-55
camptospila Meyrick, Stenoma, 21-1168
campylodes Meyrick, Antaeotricha, 21-493
cana (R. Felder & Rogenhofer), Stenoma, 21-1169
cancanodes (Meyrick), Inga, 21-1589
cancanopis Meyrick, Imma, 41-20
cancellata (Meyrick), Pessograptis, 29-507
candidata Meyrick, Blastobasis, 23-66
candidella (Blanchard), Eucalliathla, 35-2
caneodes Meyrick, Anacampsis, 29-336
canescens Butler, Thanatopsyche, 16-41
canescens Walsingham, Ectaga, 21-85
cannescens Clarke, Pseuderotis, 21-418
canofusella (Walker), Compsolechia, 29-381
canonias Meyrick, Stenoma, 21-1170
cantatrix Meyrick, Stenoma, 21-1205
cantharitis (Meyrick), Antaeotricha, 21-494
capax Meyrick, Acrolophus, 15-26
capillata Walsingham, Anorthosia, 29-518
capitatus (Fabricius), Stegasta, 29-251
capitella (Fabricius), Stegasta, 29-251
capnobola Meyrick, Stenoma, 21-1171
capnocoma (Meyrick), Promenesta, 21-1089
capnocrossa (Meyrick), Anadasmus, 21-426
capnodes Walsingham, Anchimacheta, 36-24
capnosphaera (Meyrick), Cerconota, 21-851
capnota Meyrick, Tischeria, 9-1

capraria Meyrick, Glyphidocera, 23-4
caprimulga (Walsingham), Antaeotricha, 21-495
caprimulgus Walsingham, Acrolophus, 15-158
capsaria (Meyrick), Inga, 21-1584
capsica (Bradley & Povolný), Symmetrischema, 29-263
capsiformis (Meyrick), Antaeotricha, 21-496
capsivorum Povolný, Symmetrischema, 29-264
capsulata Meyrick, Antaeotricha, 21-497
capsulifex Kieffer & Jörgensen, Dicranoses, 11-5
captans (Meyrick), Gonionota, 21-123
capyrodes Meyrick, Anacampsis, 29-307
carabodes (Meyrick), Antaeotricha, 21-498
carabopa Meyrick, Urodus, 36-39
carabophanes Meyrick, Antaeotricha, 21-499
carbasea (Meyrick), Antaeotricha, 21-500
carbonifer (Busck), Cerconota, 21-852
carcinomatella Zeller, Argyresthia, 37-2
cardinata (Meyrick), Carposina, 32-3
caribbea Meyrick, Glyphidocera, 23-5
carinella (Walsingham), Trichotaphe, 29-656
cariosella (Dietz), Setomorpha, 15-359
carmodiella Busck, Leicophasma, 15-289
carphitis Meyrick, Antaeotricha, 21-821
carpocapsella (Walker), Psilocorsis, 21-274
carribea Busck, Glyphidocera, 23-5
caryaefoliella (Chambers), Dichomeris, 29-572
carycastis (Meyrick), Onebala, 29-622
carycina (Meyrick), Myrophila, 29-616
caryifoliella Meyrick, Dichomeris, 29-572
caryodesma Meyrick, Stenoma, 21-1172
caryograpta (Meyrick), Antaeotricha, 21-501
caryophragma Meyrick, Myrophila, 29-617
caryoterma Meyrick, Compsolechia, 29-382
cassiae (Weyenbergh), Curtorama, 16-33
cassiaella Jörgensen, Bruchiana, 29-710
cassicordis Dyar, Acrolophus, 15-51
cassidata (Meyrick), Compsolechia, 29-383
cassigera Meyrick, Stenoma, 21-1173
cassiteranthes Meyrick, Filinota, 21-94
castellana Meyrick, Stenoma, 21-1174
castigata (Meyrick), Inga, 21-1666
castilloi Robinson, Callipielus, 5-12
catacentra Meyrick, Prochola, 26-21
catagnampta Meyrick, Acrolophus, 15-27
catalinella (Busck), Epilechia, 29-502
catalytica Meyrick, Tinea, 15-376
catapasta Walsingham, Perimede, 27-88
catapeltica Meyrick, Ethmia, 21-317
catapsecta (Meyrick), Anadasmus, 21-441
catasticta (Meyrick), Inga, 21-1590
catectis Meyrick, Glyphidocera, 23-6
catenella (Zeller), Hypercallia, 21-203
catenifer Walsingham, Stenoma, 21-1175
cathagnista Meyrick, Antaeotricha, 21-502
catharactis Meyrick, Antaeotricha, 21-503
catharinae (Viette), Aepytus, 5-81
catharmosta Meyrick, Stenoma, 21-1176
catharosema Meyrick, Stagmaturgis, 29-812
cathecta Walsingham, Acrolophus, 15-140
cathidrota (Meyrick), Ithome, 27-75
catholica Meyrick, Prochola, 26-22
cathostiota Meyrick, Stenoma, 21-1177
catorthota Meyrick, Choropleca, 15-222
caudatella Walsingham, Glyphipterix, 34-34
caumatias (Meyrick), Inga, 21-1591

caustonota (Meyrick), Cotyloscia, 29-525
caustopis Meyrick, Falculina, 21-997
caymana Sattler, Deltophora, 29-84
cecropia Meyrick, Stenoma, 21-1292
cedroxyla Meyrick, Antaeotricha, 21-504
celidotis Meyrick, Antaeotricha, 21-505
cellicoma Meyrick, Ethmia, 21-318
celophora (Meyrick), Taeniostolella, 34-13
cenelpis (Walsingham), Anacampsis, 29-308
cenotes (Walsingham), Antaeotricha, 21-738
censoria (Meyrick), Cerconota, 21-853
centrocrossa Meyrick, Calliprora, 29-60
centrodina Meyrick, Stenoma, 21-1439
ceramocha (Meyrick), Acrolophus, 15-28
ceramoxantha Meyrick, Thyrsomnestis, 29-825
ceratistes (Walsingham), Antaeotricha, 21-506
cerealella (Olivier), Sitotroga, 29-513
cerinura (Meyrick), Onebala, 29-623
cerochra Meyrick, Glyphidocera, 23-7
cerophaea (Meyrick), Inga, 21-1592
certiorata (Meyrick), Cerconota, 21-854
cerussata Walsingham, Gelechia, 29-117
cervicolor Meyrick, Acrolophus, 15-29
cestivora Hayward, Gnorimoschema, 29-158
cestrella (Busck), Acrolepiopsis, 39-3
cestrivora Clarke, Gnorimoschema, 29-158
cestrota Meyrick, Glyphipterix, 34-35
chalastis (Meyrick), Antaeotricha, 21-507
chalcobaphes Walsingham, Phyllonorycter, 20-125
chalcochra Meyrick, Mythoplastis, 15-303
chalcochtha Meyrick, Habrophylax, 21-195
chalcodesma Meyrick, Ussara, 34-25
chalcodora Meyrick, Ethmia, 21-319
chalcogramma Powell, Ethmia, 21-320
chalcolampra (Meyrick), Acrolepiopsis, 39-4
chalcopera Walsingham, Perioristica, 29-801
chalcoptila (Meyrick), Leptozestis, 27-84
chaldaica Meyrick, Coptotelia, 21-33
chalepa Walsingham, Stenoma, 21-1178
chalinopa Meyrick, Acrocercops, 20-56
chalinophanes (Meyrick), Antaeotricha, 21-508
chalinopis Meyrick, Trichotaphe, 29-657
chalybaeella (Walker), Stenoma, 21-1179
chalybea (R. Felder & Rogenhofer), Parelectroides, 29-794
chalybeichroa (Walsingham), Aristotelia, 29-12
chalybochroa Meyrick, Aristotelia, 29-12
chalyburga (Meyrick), Onebala, 29-624
chambersella Dyar, Lita, 29-176.1
chambersella (Murtfeldt), Brachmia, 29-520
championella (Walsingham), Psittacastis, 21-282
championi Druce, Phassus, 5-43
chaquensis Köhler, "Platoeceticus", 16-63
charagma Clarke, Gonionota, 21-190
charigramma Meyrick, Polyhymno, 29-187
charipepla (Meyrick), Pompostolella, 15-340
chariphanes (Meyrick), Acrolepiopsis, 39-5
charitarcha Meyrick, Stenoma, 21-1180
charitopis Meyrick, Acrocercops, 20-57
chelidonia Meyrick, Compsolechia, 29-384
chelobathra Meyrick, Antaeotricha, 21-800
chemsaki Powell, Ethmia, 21-321
chersadacta Meyrick, Homostinea, 15-271
chersopa Meyrick, Syrmologa, 15-362
chersopsamma Meyrick, Anapatris, 21-444

chersota (Meyrick), Ithome, 27-76
chiarelliae Pastrana, Coleophora, 24-3
chilensis Davis, Apoplania, 3-4
chilensis (Philippi), Thanatopsyche, 18-40
chilensis (Ureta), Andeabatis, 5-29
chilensis (Viette), Dalaca, 5-2
chili (Povolný), Ptycerata, 29-197
chilibrella (Busck), Trycherodes, 21-305
chiliensis Viette, Callipielus, 5-11
chiliensis (Viette), Dalaca, 5-2
chiliensis Viette, Puermytrans, 5-24
chiloides Kristensen, Neotheora, 4-1
chilosema (Meyrick), Antaeotricha, 21-509
chionastra Meyrick, Hypercallia, 21-204
chionocrossa Meyrick, Ceromitia, 12-1
chionodora Meyrick, Stenoma, 21-1415
chionogramma (Meyrick), Stenoma, 21-1181
chionopis Meyrick, Hypercallia, 21-205
chionoptila (Meyrick), Chlamydastis, 21-920
chionosphena (Meyrick), Chlamydastis, 21-921
chionostigma (Walsingham), Parastega, 29-790
chionozona Meyrick, Triclonella, 27-49
chionura Meyrick, Commatica, 29-729
chiquita (Busck), Urodus, 36-40
chiquita (Busck), Ecpathophanes, 17-6
chiquitella (Busck), Scrobipalpula, 29-232
chiriquensis Pfitzner, "Dalaca", 5-115
chlorina (Kearfott), Gonioterma, 21-1009
chlorobasis (Zeller), Rectiostoma, 21-1107
chlorocephala Meyrick, Gelechia, 29-118
chloroceros Meyrick, Tinea, 15-377
chlorochroa (Meyrick), Inga, 21-1593
chloroloba Meyrick, Stenoma, 21-1182
chloromelalis (Walker), Imma, 41-21
chloromis (Meyrick), Antaeotricha, 21-774
chloroneura Meyrick, Aritstotelia, 29-13
chloronympha Meyrick, Acrocercops, 20-58
chloropeda Meyrick, Holcocera, 23-85
chloropelta Meyrick, Acrolophus, 15-30
chlorophthalma Meyrick, Metabolaea, 29-782
chlorophylla Walsingham, Phytomimia, 21-267
chloropis Meyrick, Prochola, 26-23
chloroplaca (Meyrick), Stenoma, 21-1183
chloroptilia (Meyrick), Caloptilia, 20-6
chlorosema Meyrick, Antispila, 14-1
chlorosticta (Meyrick), Chlamydastis, 21-922
chlorothrota (Meyrick), Anadasmus, 21-427
chloroxantha Meyrick, Stenoma, 21-1184
cholerocrossa Meyrick, Stenoma, 21-1185
choleroptila (Meyrick), Gonioterma, 21-1010
chonactis Meyrick, Acrolophus, 15-31
choneuta Meyrick, Heliodines, 40-16
chordostoma (Meyrick), Anoncia, 27-3
choreutidea Butler, Hyperskeles, 21-1581
choritica Meyrick, Cnismorectis, 15-236
chorrera (Busck), Machimia, 21-234
christocoma Meyrick, Antaeotricha, 21-510
chromatopa Meyrick, Stenoma, 21-1186
chromolitha (Meyrick), Gonioterma, 21-1011
chromotechna Meyrick, Stenoma, 21-1187
chrysallacta Meyrick, Glyphipterix, 34-36
chrysampyx Meyrick, Promenesta, 21-1090
chrysangela Meyrick, Ussara, 34-26
chrysibasis (Duckworth), Rectiostoma, 21-1100
chrysippa (Druce), Lactura, 36-12

chrysoconis Meyrick, Urodus, 36-41
chrysocosma Meyrick, Acrocercops, 20-59
chrysodeta Meyrick, Machlotica, 34-15
chrysodidyma Dyar, Phassus, 5-54
chrysogastra Meyrick, Stenoma, 21-1263
chrysoplaca (Meyrick), Compsolechia, 29-385
chrysorma Meyrick, Phalerarcha, 34-4
chrysorrhabda Meyrick, Stilbosis, 27-95
chrysostoma (Meyrick), Pavolechia, 29-799
cicadella(Sepp), Antaeotricha, 21-511
cicadella (Sepp), Antaeotricha, 21-448
ciccella Barnes & Busck, Heliodines, 40-17
cicerella (Rondani), Plutella, 35-18
cinclidias (Meyrick), Eunebristis, 29-604
cincta (Druce), Imma, 41-22
cinctella (Walker), Cymotricha, 29-531
cinerea Butler, Toecorhychia, 36-33
cinerea (Geoffroy), Plutella, 35-18
cinereocervina (Walsingham), Menesta, 21-1070
ciniata (Druce), Imma, 41-23
cinnamicostella (Zeller), Cymotricha, 29-532
cinnamomea Clarke, Alynda, 21-1521
circographa Meyrick, Paraspastis, 21-1082
circumdata (Zeller), Tiquadra, 15-423
cirrhantha Meyrick, Acrocercops, 20-60
cirrhobasis (Duckworth), Rectiostoma, 21-1101
cirrhogramma Meyrick, Stenoma, 21-1134
cirrhophaea (Meyrick), Costoma, 21-47
cirrhoscia Meyrick, Zelleria, 36-88
cirrhoxantha (Meyrick), Antaeotricha, 21-512
cissiella Busck, Acrocercops, 20-61
cistulata Meyrick, Anacampsis, 29-309
citranthes (Meyrick), Taygete, 29-272
citraula Meyrick, Pammeces, 26-11
citraulax Meyrick, Nematochares, 21-258
citrina (Busck), Lactura, 36-13
citrinopa Meyrick, Cosmopterix, 27-11
citrocarpa Meyrick, "Scaeosopha", 27-41
citroclista Meyrick, Hypercallia, 21-206
citrodeta Meyrick, Cryptolechia, 21-50
citronota (Meyrick), Gonionota, 21-124
citrophaea (Meyrick), Antaeotricha, 21-525
citroptila Meyrick, Apotactis, 29-703
citroscia Meyrick, Promenesta, 21-1087
citroxantha Meyrick, Stenoma, 21-1188
claripennis Busck, Stenoma, 21-1189
clarissa Busck, Ethmia, 21-309
clarkei Amsel, Stenoma, 21-1421
clarkei Pastrana, Antispastis, 39-19
clarkei Powell, Ethmia, 21-322
claudescens Meyrick, Timocratica, 21-1499
clava Powell, Ethmia, 21-323
clavifera Meyrick, Stenoma, 21-1400
clemensella Chambers, Xylesthia, 15-445
cleodorella (Zeller), Thiotricha, 29-301
cleopatra Meyrick, Antaeotricha, 21-513
cleptica Walsingham, Acrolophus, 15-32
clerotoma (Meyrick), Phyllonorycter, 20-126
clevelandi (Busck), Erysiptila, 21-308
clistogramma Meyrick, Calliprora, 29-61
clistrodoma (Meyrick), Faculta, 29-102
clitellaria Meyrick, Pachygeneia, 29-789
clitoriella Busck, Acrocercops, 20-62
clitozona Meyrick, Pachydyta, 15-329
clivosa Meyrick, Antaeotricha, 21-791

clopica Meyrick, Gelechia, 29-119
closterias Meyrick, Chariphylla, 21-18
clotho (Meyrick), Eomichla, 21-1565
clysmographa Meyrick, Stenoma, 21-1190
clytandra Meyrick, Orthenches, 35-6
clytosema Meyrick, Acrocercops, 20-63
cnecobasis (Duckworth), Rectiostoma, 21-1102
cnecodes (Meyrick), Inga, 21-1594
cnemosaris (Meyrick), Antaeotricha, 21-514
cnephaea (Walsingham), Hypercallia, 21-207
coarctella Zeller, Sitotroga, 29-514
cocae (Busck), Psittacastis, 21-283
cocama Pfitzner, "Dalaca", 5-116
cockerelli (Busck), Friseria, 29-107
cockerelli (Dyar), Acrolophus, 15-33
codicata Meyrick, Stenoma, 21-1191
coffeaella Busck, 1925, Auximobasis, 23-48
coffeella (Guérin-Méneville), Perileucoptera, 19-8
cognatella (Walker), Compsolechia, 29-386
colleta Walsingham, Polyhymno, 29-188
colligata Meyrick, Stenoma, 21-1192
collina Meyrick, Pachysaris, 29-635
collocatella (Walker), Compsolechia, 29-466
colluta Meyrick, Thrypsigenes, 29-823
collybista (Meyrick), Stenoma, 21-1193
cologramma Clarke, Gonionota, 21-125
colombiana Povolný, Keiferia, 29-169
colorata Meyrick, Glyphipterix, 34-37
colpodes (Walsingham), Coptotelia, 21-34
colposaris (Meyrick), Antaeotricha, 21-515
columbaris Meyrick, Stenoma, 21-1194
columnaris Meyrick, Glyphipterix, 34-38
comastis Meyrick, Gonionota, 21-126
comissata Meyrick, Stegasta, 29-252
comma Busck, Stenoma, 21-1195
commendata Meyrick, Blastobasis, 23-67
commixta Meyrick, Borkhausenia, 21-1533
commutata (Meyrick), Stenoma, 21-1196
comosa (Walsingham), Antaeotricha, 21-516
comosae Hodges, Batrachedra, 24-16
complanata (Meyrick), Deoclona, 29-687
completella (Walker), Stenoma, 21-1197
complexa (Meyrick), Chlamydastis, 21-923
complicata Clarke, Coptotelia, 21-35
compressa (Walsingham), Gonioterma, 21-1012
compsacma Meyrick, Diataga, 15-246
compsocharis Meyrick, Stenoma, 21-1198
compsocoma Meyrick, Stenoma, 21-1199
compsographa Meyrick, Antaeotricha, 21-517
compsoneura (Meyrick), Antaeotricha, 21-518
compta (Clemens), Atteva, 36-8
compulsa Meyrick, Pyramidobela, 21-1702
conchita Busck, Gonioterma, 21-1013
conchylitis Meyrick, Tinea, 15-378
concinna (Meyrick), Inga, 21-1595
concinna Walsingham, Gelechia, 29-120
concisa Meyrick, Battaristis, 29-346
concitata Meyrick, Batrachedra, 24-17
concolorella (Chambers), Ithome, 27-77
condaliavorella Busck, Trichotaphe, 29-658
condemnatrix Meyrick, Stenoma, 21-1200
condita Durrant, Acrolophus, 15-34
condylota Meyrick, Stilbosis, 27-96
confarreatella (Zeller), Borkhausenia, 21-1534
confederata (Grote & Robinson), Astala, 16-34

confinella (R. Felder & Rogenhofer), Gonionota, 21-127
confirmata Meyrick, Oxylechia, 29-788
confixella (Walker), Antaeotricha, 21-519
conflicta Meyrick, Polyhymno, 29-189
confluens Meyrick, Imma, 41-24
confusella (Walker), Ethmia, 21-324
confusellastra Powell, Ethmia, 21-325
confusellus J. B. Smith, Acrolophus, 15-138
congelata Meyrick, Antaeotricha, 21-520
congeminatella Zeller, Xylesthia, 15-445
conglobata Meyrick, Ethmia, 21-326
congregata (Jones), Cryptothelea, 16-22
congregatella Brèthes, Ridiaschinia, 11-4
congressella (Walker), Cerconota, 21-855
congrua Meyrick, Stenoma, 21-1201
congruens Walsingham, Opostega, 8-5
conifera (Meyrick), Scrobipalpula, 29-233
coniferarum (Packard), Thyridopteryx, 16-60
coniomicta Meyrick, Episyrta, 15-256
coniopa (Meyrick), Antaeotricha, 21-521
coniophaea Meyrick, Stenoma, 21-1202
coniosema Meyrick, Battaristis, 29-347
conisticta Walsingham, Anacampsis, 29-310
conosema Meyrick, Glyphipterix, 34-39
conserva (Meyrick), Inga, 21-1596
considerata Meyrick, Anacampsis, 29-311
consobrina (Meyrick), Cerconota, 21-856
consociata (Meyrick), Taygete, 29-273
consociella (Walker), Stenoma, 21-1203
consociella (Walker), Antaeotricha, 21-765
consona (Meyrick), Chionodes, 29-71
consonella (Busck), Antaeotricha, 21-765
conspersa Butler, Argyresthia, 37-3
conspersa Meyrick, Batrachedra, 24-18
conspersa (Walsingham), Mompha, 25-4
constans Walsingham, Auximobasis, 23-49
constellaris Meyrick, Mompha, 25-5
constellata (Meyrick), Gonionota, 21-128
constituta (Meyrick), Antaeotricha, 21-522
constricta (Meyrick), Antaeotricha, 21-523
constrictivalva Becker, Timocratica, 21-1481
contophora Meyrick, Stenoma, 21-1416
contorta Meyrick, Acrocercops, 20-64
contortella (Walker), Antaeotricha, 21-677
contrariella (Walker), Inga, 21-1666
contrariella Zeller, Atemelia, 35-26
contrasta Clarke, Gonionota, 21-129
contrita Meyrick, Pachysaris, 29-636
controversella (Zeller), Holcocera, 23-86
contubernalis Meyrick, Acrolophus, 15-35
contubernatella (Fitch), Dichomeris, 29-586
contumax Meyrick, Stenoma, 21-1436
conturbatella (Walker), Antaeotricha, 21-524
conveniens Meyrick, Stenoma, 21-1204
convergens Walsingham, Polyhymno, 29-190
conversa Meyrick, Phelotropa, 21-1085
convexicostata (Zeller), Stenoma, 21-1205
conviva Meyrick, Mompha, 25-6
convoluta Duckworth, Peleopoda, 21-413
convolvuli (Walsingham), Brachmia, 29-521
cophodes Meyrick, Blastobasis, 23-68
copia(Clarke), Batrachedra, 24-19
copromima Meyrick, Antaeotricha, 21-525
coquillettella Busck, Ethmia, 21-327
cora (Busck), Cerconota, 21-887

coracophila Meyrick, Tiquadra, 15-440
coracopis Meyrick, Tinea, 15-379
corallina (Meyrick), Inga, 21-1674
corallina Walsingham, Aristotelia, 29-14
corculata Meyrick, Comotechna, 21-19
cordia Powell, Ethmia, 21-328
cordicaria Meyrick, Spanioptila, 20-111
cordiella Busck, Acrocercops, 20-65
cordillerella (Kieffer & Jörgensen), Ypsolopha, 35-20
cordillerella (Strand), Ypsolopha, 35-20
coriodes Meyrick, Antaeotricha, 21-526
cornifer Walsingham, Anacampsis, 29-312
cornifera Meyrick, Anacampsis, 29-312
cornigerella Zeller, Glyphipterix, 34-40
cornuta (Busck), Teuchophanes, 29-646
coronata Walsingham, Ethmia, 21-329
coronta Druce, Osrhoes, 6-1
corrientis (Walsingham), Acrolophus, 15-36
corticinella Snellen, Setomorpha, 15-359
corticinicolor Strand, Acrolophus, 15-37
corvigera Meyrick, Antaeotricha, 21-527
corvula (Meyrick), Stenoma, 21-1206
corvula Walsingham, Acrolophus, 15-38
corymba Durrant, Acrolophus, 15-39
corymbas Meyrick, Compsolechia, 29-387
corymbota Meyrick, Myrmecozela, 15-301
corystes (Meyrick), Inga, 21-1597
coscinophora (Pfitzner), Aepytus, 5-71
cosmeta Walsingham, Acrolophus, 15-40
cosmocrossa Meyrick, Acrophiletes, 29-690
cosmodoxa Meyrick, Psittacastis, 21-284
cosmogona Meyrick, Atteva, 36-1
cosmographa Meyrick, Aristotelia, 29-15
cosmorapa Meyrick, Homosetia, 15-267
cosmophragma Meyrick, Athrinacia, 21-12
cosmoterma Meyrick, Antaeotricha, 21-528
cossidella Busck, Harmaclona, 17-7
cossoides R. Felder & Rogenhofer, Acrolophus, 15-41
costalimai (Bourquin), Stigmella, 7-3
costalis (Busck), Vazugada, 29-682
costaricae (Busck), Urodus, 36-42
costaricensis Druce, Phassus, 5-51
costaricensis (Schaus), Lumacra, 16-27
costatella (Walker), Antaeotricha, 21-529
costipunctella (Möschler), Stegasta, 29-250
costistrigella (Dietz), Nemapogon, 15-305
costotristrigella (Chambers), Nemapogon, 15-305
crambina Busck, Stenoma, 12-1207
crambinella (Zeller), Polyhymno, 29-191
crassa Meyrick, Timocratica, 21-1499
crassicornis Walsingham, Aristotelia, 29-16
crassinodis (Zeller), Mompha, 25-7
craterias Meyrick, Philomusaea, 21-1694
crateroptila (Meyrick), Chlamydastis, 21-924
creberrima Walsingham, Gelechia, 29-121
cremastis (Meyrick), Antaeotricha, 21-530
crepitana Busck, Stenoma, 21-1208
crepitans Meyrick, Stenoma, 21-1208
cretata Davis, Carposina, 32-4
cretella Walsingham, Tinea, 15-380
cretifera (R. Felder & Rogenhofer), Stenoma, 21-1314
crimnodes Meyrick, Borkhausenia, 21-1535
crinifrons Walsingham, Acrolophus, 15-42
crinita Meyrick, Glyphipterix, 34-41
criodes Meyrick, Eripnura, 29-764

crispulella (Berg), Ypsolopha, 35-21
cristata Walsingham, Gonionota, 21-130
critica (Walsingham), Taygete, 29-274
crocatella Zeller, Hypercallia, 21-208
crocatus (Ureta), Dalaca, 5-1
crocidura Meyrick, Tiquadra, 15-424
crocipuntella (Walsingham), Arogalea, 29-55
crocodeta Meyrick, Tinea, 15-381
crocodilopa Meyrick, Compsolechia, 29-388
crocodora Meyrick, Dissoptila, 29-756
crocogramma Meyrick, Glyphidocera, 23-8
croconympha Meyrick, Tinaegeria, 21-1727
crocoptila (Meyrick), Gonioterma, 21-1014
crocorrhoa Meyrick, Hybroma, 15-274
crocosticta Meyrick, Stenoma, 21-1209
crocoxysta Meyrick, Pammeces, 26-12
crocuta (R. Felder & Rogenhofer), Antaeotricha, 21-589
crossospila Meyrick, Dichomeris, 29-573
crossota (Walsingham), Inga, 21-1598
crossotorna Meyrick, Commatica, 29-730
crotalariella (Busck), Brachyacma, 29-524
crotalistis Meyrick, Acrocercops, 20-66
crotospila Meyrick, Blastobasis, 23-69
crucifera (Busck), Inga, 21-1599
cruciferarum Zeller, Plutella, 35-18
cruciferella (Dietz), Lindera, 15-294
cruciformis Meyrick, Triclonella, 27-50
cruda Meyrick, Machimia, 21-235
crustaria (Meyrick), Scrobipalpula, 29-234
cruttwellae Davis, Naevipenna, 16-17
cruttwelli Davis, Naevipenna, 16-17
cryeropis Meyrick, Antaeotricha, 21-531
cryphiodes (Meyrick), Glyphidocera, 23-9
crypsangela Meyrick, Stenoma, 21-1210
crypsastra Meyrick, Stenoma, 21-1211
crypsetaera Meyrick, Stenoma, 21-1212
crypsilychna Meyrick, Brachmia, 29-521
crypsiphaea (Meyrick), Antaeotricha, 21-532
crypsiphragma Meyrick, Orphnolechia, 21-1079
crypsithias (Meyrick), Antaeotricha, 21-668
crypticopa (Meyrick), Anacampsis, 29-313
cryptina (Walsingham), Commatica, 29-731
cubana Duckworth, Mothonica, 21-1072
cubensis Busck, Ethmia, 21-330
culminata Meyrick, Plutella, 35-12
culminicola (Staudinger), Tinea, 15-382
cumulata Walsingham, Urodus, 36-43
cumulatella Zeller, Tinea, 15-383
cuneata Meyrick, Imma, 41-25
cuneatella Walker, Taruda, 21-299
cuneifera Walsingham, Gelechia, 29-122
cupidinea (Meyrick), Inga, 21-1600
cuprata (Meyrick), Stigmella, 7-4
cuprea (Dognin), Arctopoda, 21-1526
cuprea Walsingham, Heliozela, 14-11
cupreata (Dognin), Hypercallia, 21-209
cupreella Walsingham, Eucosmophora, 20-18
cupreonivella (Walsingham), Ethmia, 21-331
cuprescens Walsingham, Philonome, 19-2
cuprifera Pfitzner, "Dalaca", 5-117
curcassi Busck, Neurobathra, 20-37
curiata (Meyrick), Lethata, 21-1041
curtella (Busck), Battaristis, 29-348
curtipennis (Butler), Stenoma, 21-1213
curvatella Busck, Parastega, 29-792

curviliniella (Busck), Chlamydastis, 21-925
curvipunctella (Walsingham), Ithome, 27-78
curvula Clarke, Irenia, 21-1681
custodita (Meyrick), Inga, 21-1601
cyanactis Meyrick, Pessograptis, 29-508
cyanarcha Meyrick, Stenoma, 21-1214
cyanaspis (Meyrick), Gonionota, 21-131
cyanastra (Meyrick), Lygronoma, 21-1684
cyanea Walsingham, Ethmia, 21-332
cyanochlora Meyrick, Enteucha, 7-13
cyanocoma Meyrick, Labdia, 27-35
cyanolampra Meyrick, Parascaeas, 21-1081
cyanombra (Meyrick), Urodus, 36-44
cyanoneura Meyrick, Trichotaphe, 29-659
cyanoplecta Meyrick, Erithyma, 21-89
cyanorrhoa Meyrick, Commatica, 29-732
cyanospora Meyrick, Imma, 41-26
cyathopa Meyrick, Coptotelia, 21-36
cyathopoides Clarke, Coptotelia, 21-37
cycladopa (Meyrick), Setomorpha, 15-358
cyclobasis Meyrick, Antaeotricha, 21-533
cyclogramma Meyrick, Acrocercops, 20-67
cyclophora Meyrick, Acrolophus, 15-43
cyclophthalama Hodges, Inga, 21-1602
cyclophthalma (Meyrick), Inga, 21-1602
cyclopica Meyrick, Urodus, 36-45
cycloptila (Meyrick), Cerconota, 21-855
cyclosema Meyrick, Antispila, 14-2
cyclospila (Meyrick), Cymotricha, 29-533
cycnographa Meyrick, Stenoma, 21-1215
cycnolopha (Meyrick), Antaeotricha, 21-534
cycnomorpha Meyrick, Antaeotricha, 21-535
cylindrota Meyrick, Holcocera, 23-87
cymbalista Meyrick, Stenoma, 21-1216
cymella Forbes, Acrocercops, 20-68
cymogramma (Meyrick), Antaeotricha, 21-536
cynegetis Meyrick, Phytomimia, 21-268
cynopis (Meyrick), Antaeotricha, 21-795
cynthia Meyrick, Aristotelia, 29-17
cyphoxantha Meyrick, Stenoma, 21-1217
cypraeella (Zeller), Ethmia, 21-333
cypraspis Meyrick, Ethmia, 21-334
cyprodeta Meyrick, Antaeotricha, 21-537
cystiodes (Meyrick), Chlamydastis, 21-926
cytheraea Meyrick, Aristotelia, 29-18
dactylota Meyrick, Parectopa, 20-23
daduchus Hodges, Batrachedra, 24-20
daedalea (Walsingham), Onebala, 29-625
daguerrei Köhler, "Chalia", 16-64
damina Walsingham, Acrolophus, 15-44
danieli Viette, Aepytus, 5-80
darwini Robinson, Erechthias, 15-258
dasyleuca Meyrick, Ordrupia, 30-7
dasyneura Meyrick, Stenoma, 21-1218
dasypoda Walsingham, Aristotelia, 29-19
dasytricha Meyrick, Limnaecia, 27-36
daturae (Zeller), Scrobipalpula, 29-235
davisella Powell, Ethmia, 21-335
decapitata Meyrick, Iphimachaera, 29-776
decedens Walsingham, Sceptea, 23-42
decens Meyrick, Urodus, 35-46
decipiens Riley, Prodoxus, 13-4
decipiens Walsingham, Coleophora, 24-4
decoctor Hodges, Batrachedra, 24-21
decora (Zeller), Stenoma, 21-1219

decoratella Walker, Ussara, 34-27
deflexa (Meyrick), Chlamydastis, 21-927
deflua (Meyrick), Chlamydastis, 21-928
delapsa Meyrick, Gelechia, 29-123
delatrix Meyrick, Telphusa, 29-283
delector Hodges, Chedra, 24-32
delenita Meyrick, Stenoma, 21-1295
deligata (Meyrick), Inga, 21-1603
deliquescens Meyrick, Tischeria, 9-2
delliella (Fernald), Ethmia, 21-336
delotoma (Meyrick), Lepyrotica, 15-284
delphinodes Meyrick, Stenoma, 21-1220
deltochlora Meyrick, Alsodryas, 29-695
deltodes (Walsingham), Cronicombra, 34-6
deltodoma Meyrick, Plutella, 35-13
deltomis Meyrick, Stenoma, 21-1221
deltopis Meyrick, Antaeotricha, 21-538
deluccae Amsel, Praeacedes, 15-341
demarcha Meyrick, Heliodines, 40-18
demas (Busck), Antaeotricha, 21-539
demotes Walsingham, Acrocercops, 20-69
demotica (Walsingham), Antaeotricha, 21-540
dendrokomos Jones, Oiketicus, 16-55
densata (Meyrick), Scrobipalpula, 29-236
dentifera Meyrick, Comotechna, 21-20
dentifera (Walsingham), Lamprolophus, 40-28
dentiger Walsingham, Acrolophus, 15-102
depressa Meyrick, Scythris, 28-1
depressariella (Walker), Atelosticha, 21-1527
derelicta Meyrick, Spiladarcha, 36-30
deridens Meyrick, Antaeotricha, 21-541
descitum Walsingham, Gonioterma, 21-1015
desecta (Meyrick), Antaeotricha, 21-542
desectella (Zeller), Compsolechia, 29-389
desertorum (Berg), Machimia, 21-236
desidiosa (Meyrick), Gonioterma, 21-1013
designata Meyrick, Doliotechna, 21-1553
designatella (Walker), Gymotricha, 29-534
desmochares Meyrick, Acrocercops, 20-70
desmodiella (Clemens), Porphyrosela, 20-135
despecta (Meyrick), Praeacedes, 15-341
destillata (Zeller), Antaeotricha, 21-543
determinata Clarke, Gonionota, 21-132
detracta Walsingham, Walshia, 27-108
deuteropa Meyrick, Stenoma, 21-1222
devoluta Meyrick, Stilbosis, 27-97
diachelota Meyrick, Acrolophus, 15-45
diacmota Meyrick, Gelechia, 29-124
diacnista Meyrick, Myrophila, 29-618
diacta (Meyrick), Antaeotricha, 21-544
diagrapha Meyrick, Machimia, 21-237
dialis Hodges, Lita, 29-172
diametrica Meyrick, Stenoma, 21-1223
diaphora Walsingham, Cosmopterix, 27-12
diarthra Meyrick, Leptochersa, 15-282
diasticta Meyrick, Isembola, 29-777
diatriba (Walsingham), Gonioterma, 21-1016
diazeucta Meyrick, Compsolechia, 29-472
dicax (Meyrick), Compsolechia, 29-390
dichroa Herrich-Schäffer, Animula, 16-44
dichroanthes Meyrick, Brithyceros, 15-218
dicommatias (Meyrick), Simacauda, 10-2
dictyogramma (Meyrick), Anadasmus, 21-443
dictyopsamma Meyrick, Acrolophus, 15-46
dietzi (Duckworth), Rectiostoma, 21-1107

diffinis (R. Felder & Rogenhofer), Teresita, 21-1711
diffracta Meyrick, Antaeotricha, 21-545
diffractella Zeller, Argyresthia, 37-4
diffusa Kristensen & Nielsen, Heterobathmia, 2-1
digesta Meyrick, Holcocera, 23-88
digitata Robinson, Callipielus, 5-12
diglypta Meyrick, Triclonella, 27-51
dignella Walsingham, Dichomeris, 29-574
dilecta (Meyrick), Inga, 21-1604
dilinopa (Meyrick), Stenoma, 21-1224
diluta Meyrick, Plutella, 35-14
diluticornis (Walsingham), Lepyrotica, 15-285
dimetropis (Meyrick), Gonioterma, 21-1017
dimidiatus (Berg), Dalaca, 5-3
dimidiella (Walsingham), Acrolophus, 15-47
dimorpha Duckworth, Cerconota, 21-857
dimota Meyrick, Scythris, 28-2
diolcella Forbes, Aristotelia, 29-20
diorista (Meyrick), Stenoma, 21-1225
diortha (Meyrick), Compsolechia, 29-391
diplaca Meyrick, Perilicmetis, 15-333
diplarcha Meyrick, Antaeotricha, 21-546
diplodelta Meyrick, Anacampsis, 29-314
diplolychna Meyrick, Compsolechia, 29-392
diplophaea Meyrick, Antaeotricha, 21-547
diplosaris (Meyrick), Antaeotricha, 21-548
diplosticha Meyrick, Cryptolechia, 21-51
diplotrocha Meyrick, Hypercallia, 21-210
diptila Meyrick, Colpocrita, 15-237
directa (Meyrick), Cymotricha, 29-535
directrix Meyrick, Thioscelis, 21-1468
directus Busck, Acrolophus, 15-48
dirempta (Zeller), Antaeotricha, 21-549
discalis (Busck), Antaeotricha, 21-550
discatella Walker, Choropleca, 15-223
discipunctella Rebel, Setomorpha, 15-359
discolor (Walsingham), Antaeotricha, 21-551
discors (Meyrick), Chlamydastis, 21-929
discostrigella (Chambers), Ethmia, 21-337
discostrigella (Chambers), Ethmia, 21-337a
discrepana Busck, Stenoma, 21-1226
discrepans Meyrick, Stenoma, 21-1226
disjecta (Zeller), Antaeotricha, 21-552
dispar Köhler, "Chalia", 16-65
dispersa Duckworth, Lethata, 21-1043
dispilella (Walker), Stenoma, 21-1227
displicitella (Walker), Compsolechia, 29-419
disrupta Meyrick, Dissoptila, 29-757
dissimilis Kearfott, Antaeotricha, 21-484
dissimilis Walsingham, Tinea, 15-384
dissimulatrix Meyrick, Orthenches, 35-7
dissita Clarke, Gonionota, 21-133
dissona (Meyrick), Antaeotricha, 21-553
distans (Gozmány), Niditinea, 15-306
disticha (Meyrick), Chlamydastis, 21-930
distictella Forbes, Telphusa, 29-284
distincta (Strand), Urodus, 36-47
distorta (Meyrick), Inga, 21-1605
diveni (Heinrich), Anoncia, 27-4
diversella (Busck), Arla, 29-51
diversus Busck, Acrolophus, 15-88
dives Walsingham, Eucosmophora, 20-19
dividua Meyrick, Scythris, 28-3
dividuella Zeller, Tinea, 15-371
divisa Walsingham, Tinea, 15-401

doeri (Walsingham), Acrolophus, 15-49
dolbyi (Walsingham), Gelechia, 29-125
doleropis (Meyrick), Antaeotricha, 21-554
dolichoscia Meyrick, Batrachedra, 24-22
dolopis (Walsingham), Machimia, 21-238
dominicae Davis, Carposina, 32-5
dominicae Duckworth, Chlamydastis, 21-931
dominicella Walsingham, Glyphidocera, 23-10
donatella (Walker), Stegasta, 29-253
dorcadopa Meyrick, Stenoma, 21-1228
dorcas Meyrick, Phyllocnistis, 20-141
dorita (Schaus), Aepytus, 5-92
dormita (Schaus), Aepytus, 5-98
dorsalis (Busck), Battaristis, 29-349
dorsella (Fabricius, 1787), Antaeotricha, 21-821
dorsella (Fabricius, 1794), Antaeotricha, 21-821
dorsivittella (Zeller), Coleotechnites, 29-83
drachmaea Meyrick, Compsolechia, 29-393
drapetica Meyrick, Tiquadra, 15-425
dromica (Meyrick), Antaeotricha, 21-555
drosocycla Meyrick, Bythocrates, 15-219
drosophaea Meyrick, Glyphidocera, 23-11
dryadopa Meyrick, Brachmia, 29-521
dryas (Butler), Setomorpha, 15-359
dryaula Meyrick, Stenoma, 21-1229
dryobathra (Meyrick), Chionodes, 29-72
dryochrossa Meyrick, Compsolechia, 29-394
dryoconis Meyrick, Stenoma, 21-1230
dryocosma Meyrick, Stenoma, 21-1231
dryocrypta (Meyrick), Gonionota, 21-134
dryodesma (Meyrick), Gonionota, 21-134
dryograpta Meyrick, Pedaliotis, 15-332
dryoscia (Meyrick), Cerconota, 21-858
dryosphaera (Meyrick), Chlamydastis, 21-932
dryotechna (Meyrick), Antaeotricha, 21-738
dubiosella (Beutenmüller), Plutella, 35-18
dubitatrix (Meyrick), Phereoeca, 15-336
duckworthi Powell, Ethmia, 21-338
ductifera Meyrick, Acrolophus, 15-50
dudiella Busck, Gnorimoschema, 29-159
dudiosella (Moriuti), Plutella, 35-18
dulica Walsingham, Hybroma, 15-275
duplicata Sattler, Deltophora, 29-85
durangensis Deschka, Phyllonorycter, 20-126.1
dyctiota (Walsingham), Filinota, 21-95
dynastis Meyrick, Antaeotricha, 21-821
earobasis (Duckworth), Rectiostoma, 21-1103
ebenocosta (Meyrick), Cerconota, 21-859
ebria Meyrick, Stenoma, 21-1365
eburata (Meyrick), Chionodes, 29-73
echinon (Druce), Acrolophus, 15-51
echinura Meyrick, Acrolophus, 15-52
eclampsis Durrant, Acrocercops, 20-71
ectenes Walsingham, Acrolophus, 15-53
edax Hodges, Ithome, 27-79
edithella Busck, Atteva, 36-8
edmondsii (Butler), Doina, 21-71
effera (Meyrick), Brachmia, 29-521
effluxa (Meyrick), Timocratica, 21-1482
ehrhornella (Dietz), Lindera, 15-294
eidmannella Busck, Atticonviva, 15-213
ejiciens Meyrick, Scythris, 28-4
elachista Walsingham, Opostega, 8-6
elachistella (Zeller), Aristotelia, 29-21
elaeodes (Walsingham), Antaeotricha, 21-556
elaeostola (Meyrick), Chlamydastis, 21-933

elaeurga Meyrick, Stenoma, 21-1232
elaphodes (Meyrick), Inga, 21-1606
elaphrodes (Meyrick), Acrolepiopsis, 39-6
elatior (R. Felder & Rogenhofer), Antaeotricha, 21-557
eldorado Pfitzner, Phassus, 5-46
elegans Köhler, "Oiketicus", 16-66
elegans Zeller, Loxotoma, 21-1067
elena Clarke, Coptotelia, 31-38
elephantopis Meyrick, Gelechia, 29-126
elephas (Walsingham), Compsolechia, 29-395
elissa Meyrick, Philomusaea, 21-1695
ellipsias (Meyrick), Cymotricha, 29-536
elliptica (Forbes), Onebala, 29-626
elliptica Meyrick, Triclonella, 27-52
elmorei (Keifer), Keiferia, 29-170
eloisella (Clemens), Mompha, 25-8
elongata Nielsen & Davis, Basileura, 10-1
elongata Walsingham, Tischeria, 9-3
elpista Walsingham, Glyphidocera, 23-12
elucidella Barnes & Busck), Coleotechnites, 29-82
elutella Busck, Ethmia, 21-339
emancipata Meyrick, Adullamitis, 29-691
embythia Meyrick, Stenoma, 21-1233
emicans Meyrick, Locharca, 29-178
emigrans (Meyrick), Brachmia, 29-521
eminens Meyrick, Stenoma, 21-1234
eminula Meyrick, Stenoma, 21-1409
emissurella (Walker), Battaristis, 29-349
emma (Busck), Cerconota, 21-860
emollita Meyrick, Antaeotricha, 21-652
empedocles Meyrick, Acrolophus, 15-54
emphatica Meyrick, Stenoma, 21-1235
emphytopa Meyrick, Acrolophus, 15-55
emplasta Meyrick, Commatica, 29-733
empyrea (Meyrick), Inga, 21-1607
empyrota Meyrick, Stenoma, 21-1236
encamina (Meyrick), Inga, 21-1608
encentris Meyrick, Acrocercops, 20-72
enchytopa Meyrick, Tinea, 15-385
encyclia Meyrick, Antaeotricha, 21-558
endochra (Meyrick), Anadasmus, 21-438
engalactis Meyrick, Carposina, 32-1
enodata Meyrick, Antaeotricha, 21-559
enormis Meyrick, Batrachedra, 24-23
entaphrota (Meyrick), Inga, 21-1609
entemopa Walsingham, Leucopera, 19-6
entephras Meyrick, Stenoma, 21-1421
entherastis Meyrick, Barymochtha, 15-216
entryphopa Meyrick, Coleostoma, 29-725
enumerata Meyrick, Stenoma, 21-1237
eolampis (Meyrick), Caloptilia, 20-7
eothina Meyrick, Machimia, 21-239
epastra Meyrick, Glyphipterix, 34-42
ephaptis Meyrick, Tischeria, 9-4
ephemeraeformis (Haworth), Thyridopteryx, 16-60
ephialtes Walsingham, Tabernillaia, 29-818
ephoria Meyrick, Aristotelia, 29-22
ephorista Meyrick, Thomictis, 19-12
epibola (Walsingham), Compsolechia, 29-396
epibryas Meyrick, Pyramidobela, 21-1703
epicentra Meyrick, Phthorimaea, 29-185
epichorda Turner, Brachyacma, 29-524
epicnesta Meyrick, Stenoma, 21-1238
epicosma (Meyrick), Stigmella, 7-5
epicrossa (Meyrick), Antaeotricha, 21-560

epicta Walsingham, Stenoma, 21-1239
epidesma Walsingham, Cryptolechia, 21-52
epignampta Meyrick, Antaeotricha, 21-561
epigramma (Herrich-Schäffer), Aepytus, 5-82
epilygella Powell, Ethmia, 21-340
epipacta Meyrick, Stenoma, 21-1240
epiphantha Meyrick, Promolopica, 29-803
episema (Walsingham), Compsolechia, 29-444
episimbla (Meyrick), Antaeotricha, 21-562
epispila (Meyrick), Phyllonorycter, 20-127
epitricha (Meyrick), Barticeja, 29-59
epophrysta (Meyrick), Chlamydastis, 21-934
equatorialis Viette, Aepytus, 5-63
erasella Zeller, Tinea, 15-386
erasicosma (Meyrick), Inga, 21-1610
erasmia Meyrick, Cosmopterix, 27-13
erebantha (Meyrick), Lipomerinx, 15-295
erebodelta Meyrick, Compsolechia, 29-397
eremarcha Meyrick, Ceromitia, 12-2
eremia Clarke, Gonionota, 21-135
eremita Curtis, Cecidoses, 11-1
eremna Meyrick, Commatica, 29-734
eremnogramma Clarke, Doina, 21-72
erethismia Meyrick, Acrolophus, 15-56
erethistis Meyrick, Calliprora, 29-62
ergastulella Zeller, Blastobasis, 23-70
ergates (Walsingham), Antaeotricha, 21-563
ergatica Walsingham, Atteva, 36-2
eriacma (Meyrick), Cerconota, 21-861
eriobotryae Busck, Deoclona, 29-688
eriochrysa Meyrick, Tinea, 15-387
eriocnista (Meyrick), Inga, 21-1611
eriospila Meyrick, Imma, 41-27
eromene (Walsingham), Recurvaria, 29-201
erotarcha (Meyrick), Cerconota, 21-893
erotias (Meyrick), Inga, 21-1612
erotica (Meyrick), Antaeotricha, 21-564
erotopis (Meyrick), Gonionota, 21-136
errantella (Walsingham), Parornix, 20-39
erschoffii (Zeller), Antaeotricha, 21-773
erycina Meyrick, Aristotelia, 29-23
erythema (Walsingham), Inga, 21-1613
erythroleuca (Meyrick), Gonionota, 21-137
erythropennis (Dognin), Stenoma, 21-1458
erythrorma Meyrick, Crembalastis, 40-9
eschara Clarke, Revonda, 21-1707
escharitis Meyrick, Phycomorpha, 30-9
essedaria Meyrick, Cronicombra, 34-7
etearcha Meyrick, Triclonella, 27-53
ethirastis (Meyrick), Moca, 41-4
eucentra Meyrick, Doliotechna, 21-1554
eucharacta (Meyrick), Harpograptis, 27-34
eucharistis Meyrick, Cryptolechia, 21-53
euchorda (Meyrick), Hypatima, 29-504
euchthonia (Meyrick), Gnorimoschema, 29-160
euclina Meyrick, Prochola, 26-24
eucnemis Walsingham, Spanioptila, 20-112
eucoma Meyrick, Antaeotricha, 21-565
eudactyla R. Felder & Rogenhofer, Alucita, 31-7
euglypta Meyrick, Imma, 41-28
eumenodora Meyrick, Stenoma, 21-1241
eumitrella (Busck), Phalerarcha, 34-5
eumolybda Meyrick, Psittacastis, 21-285
euparypha Meyrick, Trichotaphe, 29-660
eupatoriella Busck, Aristotelia, 29-24

eupecta (Meyrick), Compsolechia, 29-398
euphanes Meyrick, Baeonoma, 21-829
euphracta (Meyrick), Charistica, 29-723
euphrantis Meyrick, Oenoe, 15-308
euporia Walsingham, Acrolophus, 15-57
eurinella (Zagulajev), Niditinea, 15-306
eurrhipis Meyrick, Glyphidocera, 23-13
eurychalca Meyrick, Acrocercops, 20-73
eurychrysa Meyrick, Psittacastis, 21-286
eurydelta Meyrick, Calliprora, 29-63
eurydesma (Meyrick), Stigmella, 7-6
eurydryas (Meyrick), Gonionota, 21-138
eurygypsa Meyrick, Compsolechia, 29-399
eurymolybda Meyrick, Machlotica, 34-16
eurypepla (Meyrick), Ypsolopha, 35-22
euryspoda (Lower), Setomorpha, 15-359
eurythmiella Busck, Ussara, 34-28
eusaris Meyrick, Holcocera, 23-89
eusema (Walsingham), Rectiostoma, 21-1104
eusoma Walsingham, Yponomeuta, 36-86
eusticta Meyrick, Stenoma, 21-1242
euteles Walsingham, Acrolophus, 15-58
euthoracica (Schaus), Lactura, 36-14
euthrinca Meyrick, Antaeotricha, 21-566
euthyrsa (Meyrick), Gonionota, 21-139
euzosta Walsingham, Triclonella, 27-54
eva Meyrick, Stenoma, 21-1243
evanescens (Butler), Stenoma, 21-1244
evippella (Forbes), Evippe, 29-97
evitata (Walsingham), Trichotaphe, 29-661
exacta (Meyrick), Gnorimoschema, 29-161
exagitata (Meyrick), Psilocorsis, 21-275
exalbata Meyrick, Ceromitia, 12-3
exallacta Meyrick, Cymotricha, 29-537
exanthes Meyrick, Mythoplastis, 15-304
exarata (Zeller), Stenoma, 21-1245
exarga Meyrick, Coleophora, 24-5
exasperata (Meyrick), Antaeotricha, 21-567
excavata (Busck), Trichotaphe, 29-662
excavata Clarke, Gonionota, 21-140
excisa Meyrick, Antaeotricha, 21-568
excisorella (Walker), Dichomeris, 29-578
excitata Meyrick, Dromiaulis, 27-32
exclamans (Herrich-Schäffer), Aepytus, 5-74
exclamationis Pfitzner, Phassus, 5-41
exclarella Möschler, Gelechia, 29-127
exempta Meyrick, Stenoma, 21-1246
exercitata Meyrick, Tiquadra, 15-426
exhalata Meyrick, Stenoma, 21-1247
exigua Meyrick, Acrolophus, 15-59
exilis Walsingham, Megacraspedus, 29-2
exodias Meyrick, Mompha, 25-9
exornata (Zeller), Ethmia, 21-341
exornatella Busck, Ethmia, 21-341
exorycha Meyrick, Parectopa, 20-24
expansa (Meyrick), Gonioterma, 21-1018
expilata Meyrick, Stenoma, 21-1426
explicita Meyrick, Stenoma, 21-1248
expurgata Meyrick, Glyphipterix, 34-43
exquisita Busck, Atteva, 36-8
exquisita Duckworth, Gonioterma, 21-1019
exsiccata Meyrick, Glyphidocera, 23-14
exstimulata Meyrick, Antequera, 27-1
exsultans (Zeller), Mompha, 25-10
extenta (Busck), Antaeotricha, 21-569

exteriorella (Walker), Charistica, 29-717
externella (Walker), Stenoma, 21-1249
extima Clarke, Gonionota, 21-141
extracta Meyrick, Tinea, 15-388
extranea (H. Edwards), Tegeticula, 13-1
extranea (Walsingham), Telphusa, 29-285
extremella (Walker), Commatica, 29-735
exusta Meyrick, Antaeotricha, 21-570
exuta Clarke, Eudolichura, 35-4
fabricata Meyrick, Ordrupia, 30-7
facunda Hodges, Cosmopterix, 27-14
faecosa (R. Felder & Rogenhofer), Stenoma, 21-1260
falcata Clarke, Utilia, 21-1716
falcatella (Walker), Commatica, 29-736
falcigera Meyrick, Glyphipterix, 34-44
falculinella Busck, Gonada, 21-107
fallax (Butler), Stenoma, 21-1250
falsidica (Meyrick), Antaeotricha, 21-571
falsus Butler, Palaephatus, 15-330
familiaris Zeller, Tinea, 15-389
famulata Meyrick, Dichomeris, 29-575
fanniella Busck, Ordrupia, 30-6
farinacea (Walsingham), Mompha, 25-11
farracea Meyrick, Acrolophus, 15-60
farraria Meyrick, Stenoma, 21-1251
farrella Powell, Ethmia, 21-342
fasciata (Busck), Antaeotricha, 21-573
fasciata Walker, Tinaegeria, 21-1728
fasciatipedella (Zeller), Dita, 21-1550
fascicularis Zeller, Antaeotricha, 21-572
fasciculata Meyrick, Acrocercops, 20-74
fasciculatus Vázquez, Oiketicus, 16-57
fasciella (R. Felder & Rogenhofer), Compsolechia, 29-400
fasciolata (Butler), Ceromitia, 12-4
fascomaculella (Dietz), Nemapogon, 15-305
fassliana (Dognin), Stenoma, 21-1252
fasslii (Pfitzner), Aepytus, 5-111
fastigata (Meyrick), Gonioterma, 21-1020
fastuosa (Zeller), Atteva, 36-8
favigera (Meyrick), Urodus, 36-48
favillata (Meyrick), Baeonoma, 21-830
febriculella (Zeller), Recurvaria, 29-202
felisae Clarke, Homoeoprepes, 26-1
felix Busck, Stenoma, 21-1404
fenestella (Zeller), Lucyna, 21-230
fenestra Busck, Stenoma, 21-1253
fenestrella (Scopoli), Endrosis, 21-1563
fenestrella Zeller, Coptotelia, 21-39
ferculata Meyrick, Stenoma, 21-1254
ferella (Berg), Phthorimaea, 29-181
fermentata (Meyrick), Cerconota, 21-862
fernaldella (Riley), Rectiostoma, 21-1105
fernandezyepezi Duckworth, Lethata, 21-1044
ferrarenella (Walker), Acrolophus, 15-61
ferreata (Meyrick), Compsolechia, 29-401
ferricanella Walsingham, Stenoma, 21-1255
ferrocanella (Walker), Stenoma, 21-1255
ferruginea (Kirby), Aepytus, 5-98
ferruginea (Walsingham), Acrolophus, 15-62
ferruginosa (Walker), Aepytus, 5-98
fervescens Meyrick, Amphiclada, 40-3
fervida (Zeller), Inga, 21-1614
fervidus Busck, Acrolophus, 15-63
festicola (Meyrick), Gonionota, 21-142

festiva Busck, Ethmia, 21-343
fida Meyrick, Dichomeris, 29-576
fiebrigi (Köhler), Epichnopterix, 16-11
figularis (Meyrick), Cerconota, 21-863
filicicornis (Walsingham), Acrolophus, 15-64
filicornis (Dyar), Acrolophus, 15-64
filicornis (Zeller), Recurvaria, 29-203
filicula Clarke, Stathmopoda, 21-1726
filiferella (Walker), Antaeotricha, 21-574
fimbriata Clarke, Gonionota, 21-143
finitrix Meyrick, Stenoma, 21-1256
flaccescens Meyrick, Durrantia, 21-410
flagellifer Walsingham, Recurvaria, 29-204
flagellifera Meyrick, Recurvaria, 29-204
flagelliformis Meyrick, Pholcobates, 21-266
flammulella Walsingham, Gelechia, 29-128
flava (Zeller), Inga, 21-1615
flavicaudata Walsingham, Ethmia, 21-344
flaviceps (R. Felder & Rogenhofer), Ussara, 34-29
flavicillata Walsingham, Auximobasis, 23-50
flavicincta (Walsingham), Alucita, 31-8
flavicosta (R. Felder & Rogenhofer), Costoma, 21-48
flavida Meyrick, Auximobasis, 23-51
flavidella (Busck), Compsosaris, 29-749
flavidorsis Meyrick, Gonada, 21-108
flavipennis (R. Felder & Rogenhofer), Snellenia, 21-1723
flavivitta (Walker), Atteva, 36-3
flavivittella (Clemens), Dichomeris, 29-586
flavocincta Sattler, Deltophora, 29-86
flavofasciata Wollaston, Cosmopterix, 27-9
flectella Walker, Tinea, 15-390
flexibilis (Meyrick), Cerconota, 21-864
flexiloqua Meyrick, Anacampsis, 29-315
flexuosa Walsingham, Bucculatrix, 19-13
flinti Clarke, Doina, 21-73
flinti (Duckworth), Rectiostoma, 21-1106
flocculosa (Meyrick), Antaeotricha, 21-575
florens Meyrick, Compsocrita, 15-239
floridana (Neumoegen), Atteva, 36-8
florinda Clarke, Utilia, 21-1717
fluctuans Meyrick, Cymotricha, 29-538
fluminata (Meyrick), Mothonica, 21-1073
fluvialis Meyrick, Scythris, 28-5
foetterlei (Viette), Aepytus, 5-84
fonteboae (Strand), Urodus, 36-49
forbesi Bourquin, Leucanthiza, 20-115
forcipata (Meyrick), Chlamydastis, 21-935
forficulella (Walsingham), Urodus, 36-50
formosus Butler, Ithutomus, 36-29
formulata (Meyrick), Dichomeris, 29-577
forreri Walsingham, Acrolophus, 15-65
forreri (Walsingham), Antaeotricha, 21-576
forsteri Amsel, Antaeotricha, 21-821
forsteri Viette, Aepytus, 5-75
fortificata Gozmány, Tinea, 15-416
fractilinea (Walsingham), Antaeotricha, 21-577
fractiliniella (Dietz), Setomorpha, 15-359
fractinubes (Walsingham), Antaeotricha, 21-578
fragilella (Walsingham), Lepyrotica, 15-286
fragmentella (Dognin), Chlamydastis, 21-936
fraterna (R. Felder & Rogenhofer), Antaeotricha, 21-579
fraternella (Busck), Timocratica, 21-1483
frigidella (Packard), Niditinea, 15-306
friserella Busck, Ordrupia, 30-7

fritillella Powell, Ethmia, 21-345
frondifer Busck, Stenoma, 21-1257
frontalis (Zeller), Antaeotricha, 21-580
frontella (Walsingham), Proboloptilia, 15-343
frontestrigata Walsingham, Tinea, 15-391
fuegensis Nielsen & Robinson, Calada, 5-22
fulcrata Meyrick, Stenoma, 21-1258
fulgerator Grote, Oiketicus, 16-57
fulgurator Herrich-Schäffer, Oiketicus, 16-57
fuliginosa (R. Felder & Rogenhofer), Battaristis, 29-349
fulminalis Meyrick, Urodus, 36-51
fulminata (Meyrick), Cerconota, 21-865
fulta Meyrick, Antaeotricha, 21-581
fulvicilia Meyrick, Trichotaphe, 29-663
fulvicolor Meyrick, Xystrologa, 15-447
fulvidella (Walsingham), Schistonoea, 21-420
fulviguttata (Zeller), Atteva, 36-4
fumida Walsingham, Acrolophus, 15-66
fumifica (Walsingham), Antaeotricha, 21-582
fumipennis (Busck), Antaeotricha, 21-827
fumosa Nielsen & Robinson, Callipielus, 5-16
fumosa (Zeller), Urodus, 36-52
fundigera (Meyrick), Inga, 21-1616
funerana (Sepp), Stenoma, 21-1259
funeratella (Zeller), Scardia, 15-351
funicularis (Meyrick), Chlamydastis, 21-937
furcata (Walsingham), Antaeotricha, 21-583
furva (Meyrick), Inga, 21-1617
furvescens Meyrick, Thrypsigenes, 29-824
furvida (Walker), Inga, 21-1614
fusca Duckworth, Lethata, 21-1045
fuscata Duckworth, Thioscelis, 21-1469
fuscella Forbes, Schistophila, 29-228
fuscella (Linnaeus), Niditinea, 15-306
fuscipalpalis Becker, Timocratica, 21-1484
fuscipunctella Haworth, Niditinea, 15-306
fuscomaculella (Chambers), Nemapogon, 15-305
fuscoochrella (Chambers), Chionodes, 29-77
fuscorectangulata Duckworth, Antaeotricha, 21-584
fuscostrigella Chambers, Polyhymno, 29-194
fuscula Forbes, Prochola, 26-25
fuscus (Mabille), Dalaca, 5-8
fusigera (Meyrick), Cerconota, 21-866
fusistrigella (Walker), Stenoma, 21-1260
futura Meyrick, Stenoma, 21-1261
gaiophanes Clarke, Gonionota, 21-144
galactura Meyrick, Tiquadra, 15-427
galeata (Walker), Acrolophus, 15-67
galeatella Mabille, Tinea, 15-403
galeomorpha (Meyrick), Chlamydastis, 21-938
gallegoi Clarke, Maesara, 21-254
garleppi (Druce), Acrolophus, 15-68
gaulica Meyrick, Psittacastis, 21-287
gelidella (Walker), Ethmia, 21-346
gemellata Meyrick, Stenoma, 21-1262
gemina (Zeller), Chlamydastis, 21-939
geminata Clarke, Aliciana, 21-1519
gemistis (Meyrick), Exoncotis, 15-261
gemmans Walsingham, Acrocercops, 20-75
gemmata (Grote), Atteva, 36-8
gemmula Walsingham, Trapeziophora, 34-20
generatrix Meyrick, Antaeotricha, 21-585
genetta (R. Felder & Rogenhofer), Stenoma, 21-1352
geniatella (Busck), Anthistarcha, 29-702
gentilii Nielsen & Robinson, Callipielus, 5-15

genuina (Meyrick), Inga, 21-1618
geranomorpha Meyrick, Thioscelis, 21-1470
gerda (Busck), Anadasmus, 21-440
germana Walsingham, Choropleca, 15-224
germinans (Meyrick), Anadasmus, 21-429
geyeri Berg, Oiketicus, 16-50
gigantea Busck, Ethmia, 21-347
gigantea (Zeller), Oiketicus, 16-57
giganteus (Druce), Acrolophus, 15-69
giganteus (Herrich-Schäffer), Trichophassus, 5-34
ginocchionus Köhler, "Oiketicus", 16-67
gioia Clarke, Coptotelia, 21-40
glaciata Meyrick, Antaeotricha, 21-821
gladiata Meyrick, Polyhymno, 29-192
glandiferella (Zeller), Deltophora, 29-87
glaphyra (Walsingham), Compsolechia, 29-402
glaphyrodes (Meyrick), Antaeotricha, 21-586
glaucescens (Meyrick), Antaeotricha, 21-587
glaucopa (Meyrick), Lethata, 21-1046
glaucophanes (Meyrick), Diploschizia, 34-73
glaucopidella (Guenée), Atteva, 35-4
glebula Clarke, Doina, 21-74
glischrodes Meyrick, Cryptolechia, 21-54
gloverii (Packard), Cryptolechia, 16-24
glycerostoma Meyrick, Antaeotricha, 21-588
glycinopis Meyrick, Eunomarcha, 29-770
gnathodes Meyrick, Cycloplasis, 40-11
gnathodoxa Meyrick, Gelechia, 29-129
gnomonica Meyrick, Stilbosis, 27-98
gnorisma (Walsingham), Hypercallia, 21-211
godmani (Walsingham), Thiotricha, 29-302
goniocentra Meyrick, Acrolophus, 15-70
goniospila Meyrick, Gelechia, 29-130
gossypiella Morrill, Bucculatrix, 19-14
gossypiella (Saunders), Pectinophora, 29-506
gossypii (Forbes), Stigmella, 7-7
gracilella (Chambers), Anorthosia, 29-519
gracilis Walsingham, Blastobasis, 23-71
grandaeva (Zeller), Stenoma, 21-1263
grandis (Perty), Timocractica, 21-1485
granella (Linnaeus), Nemapogon, 15-305
granivora Meyrick, Dichomeris, 29-575
granulata Meyrick, Cronicombra, 34-8
granulatella (Walker), Acrolophus, 15-140
graphica Busck, Stenoma, 21-1264
graphiphorella (Walker), Stenoma, 21-1410
graphopterella (Walker), Antaeotricha, 21-589
gratiosa (R. Felder & Rogenhofer), Filinota, 21-96
gravescens Meyrick, Antaeotricha, 21-590
gregalis (Meyrick), Scrobipalpula, 29-237
gregariella (Murtfeldt), Porphyrosela, 20-135
gregariella (Zeller), Scrobipalpula, 29-238
grenadella (Walsingham), Xystrologa, 15-448
grenadensis Walsingham, Blastobasis, 23-72
griseana (Fabricius), Antaeotricha, 21-821
griseana (Sepp), Antaeotricha, 21-591
griseanomima Busck, Antaeotricha, 21-591
griseella (Chambers), Niditinea, 15-306
griseus (Walsingham), Acrolophus, 15-71
griseus (Walsingham), Acrolophus, 15-71a
guarani Becker, Timocratica, 21-1486
guarani Pfitzner, "Dalaca", 5-118
gubernata (Meyrick), Gonioterma, 21-1021
gubernatrix Meyrick, Antaeotricha, 21-592
gudmanella (Walsingham), Tildenia, 29-305

gugelmanni Viette, Aepytus, 5-67
guianensis Schaus, Phassus, 5-53
guilandinae Busck, Blastobasis, 23-73
guittonae (Bourquin), Stigmella, 7-8
gunni (Busck), Antaeotricha, 21-572
gunniella (Busck), Neurostrota, 20-21
guttata Busck, Eritarbes, 27-72
guyanensis Viette, Aepytus, 5-105
gymnastis Meyrick, Stenoma, 21-1265
gymnolopha Meyrick, Antaeotricha, 21-593
gypsolitha (Meyrick), Lethata, 21-1047
gypsomicta Meyrick, Tinea, 15-392
gypsopa Meyrick, Otocharea, 15-327
gypsoterma (Meyrick), Antaeotricha, 21-594
gyralea (Meyrick), Eunebristis, 29-605
habilis (Meyrick), Antaeotricha, 21-595
habrarcha Meyrick, Cycloplasis, 40-12
habristis (Meyrick), Gonionota, 21-145
habrochitona (Walsingham), Trichotaphe, 29-664
haemataula (Meyrick), Inga, 21-1619
haematula Hodges, Inga, 21-1619
haemitheia (R. Felder & Rogenhofer), Rectiostoma, 21-1107
haemoplecta (Meyrick), Taruda, 21-300
haemotheia (Meyrick), Rectiostoma, 21-1107
haesitans (Walsingham), Antaeotricha, 21-596
hagenella (Chambers), Ethmia, 21-348
hagenella (Chambers), Ethmia, 21-348a
haitiensis Davis, Lumacra, 16-29
halidora Meyrick, Acrolophus, 15-72
haliplancta Meyrick, Demobrotis, 15-245
hallwachsae Clarke, Eomichla, 21-1565.1
halmas Meyrick, Stenoma, 21-1266
halmeuta Meyrick, Polypsecta, 15-338
halmyra (Meyrick), Compsolechia, 29-403
halobapta Meyrick, Hypercallia, 21-212
halosema (Meyrick), Acrolepiopsis, 39-7
halosphora (Meyrick), Inga, 21-1620
hamifera Walsingham, Thaumatolita, 21-1713
hamiferella (Hübner), Acrolophus, 15-73
hammella Busck, Ethmia, 21-349
hamon (Busck), Cerconota, 21-845
haplocentra Meyrick, Antaeotricha, 21-597
haplodoxa Meyrick, Promenesta, 21-1091
haploxyla Meyrick, Stenoma, 21-1267
hapsicora Meyrick, Antaeotricha, 21-598
hapsidota Meyrick, Acrocercops, 20-76
harmoniella Busck, Acrolophus, 15-74
harparsen Forbes, Acrolophus, 15-75
harpella Walsingham, Acrolophus, 15-87
harpobathra Meyrick, Antaeotricha, 21-478
harpoceros Meyrick, Stenoma, 21-1268
hastigera Meyrick, Acrocercops, 20-77
hayecki (Foetterle), Trichophassus, 5-34
haywardi Busck, Timocratica, 21-1501
haywardi Köhler, "Zamopsyche", 16-68
haywardi Pastrana, Coleophora, 24-6
hebes (Dognin), Anadasmus, 21-440
hecate Butler, Melaneulia, 21-255
hectorea (Meyrick), Stenoma, 21-1269
hedemanni (Walsingham), Acrolophus, 15-76
heimlichi (Ureta), Parapielus, 5-27
helga Schaus, Aepytus, 5-78
heliaca Meyrick, Melochrysis, 21-1688
helicias Meyrick, Antaeotricha, 21-599

helicomitra Meyrick, Acrocercops, 20-78
helicopis (Meyrick), Parelectroides, 29-795
helicosema Meyrick, Neophylarcha, 30-5
heliocephala (Meyrick), Simacauda, 10-3
heliodepta Meyrick, Hypercallia, 21-213
heliomima Meyrick, Hypercallia, 21-214
helobia (Meyrick), Inga, 21-1621
helotypa Meyrick, Baeonoma, 21-831
hemeropa Meyrick, Ilingiotis, 29-611
hemibathra Meyrick, Antaeotricha, 21-600
hemichlora (Meyrick), Chlamydastis, 21-940
hemichrysea (Pfitzner), Aepytus, 5-107
hemichrysella (Walker), Dichomeris, 29-578
hemicycla Meyrick, Telphusa, 29-286
hemidryas Meyrick, Opogona, 15-313
hemilampra Meyrick, Stenoma, 21-1270
hemileuca Butler, Dalaca, 5-3
hemileucas Meyrick, Compsolechia, 29-404
hemilitha (Clarke), Scrobipalpula, 29-239
hemiommata (Zeller), Taruda, 21-301
hemiphanta Meyrick, Stenoma, 21-1271
hemiscia (Walsingham), Antaeotricha, 21-601
hemisigna Clarke, Parastega, 29-791
hemiteles Walsingham, Holcocera, 23-90
hemitephras Meyrick, Antaeotricha, 21-763
henshawiella (Busck), Scrobipalpula, 29-240
hephaestiella (Zeller), Urodus, 36-53
heptametra Meyrick, Parectopa, 20-25
heptastica Walsingham, Ethmia, 21-350
herbacea (Meyrick), Lethata, 21-1048
herifuga Meyrick, Stenoma, 21-1272
herilis (R. Felder & Rogenhofer), Antaeotricha, 21-602
hermosella Busck, Filinota, 21-97
hesitans Busck, Antaeotricha, 21-596
hesmarcha (Meyrick), Stenoma, 21-1273
hesperidella (Hübner), Plutella, 35-16
hessitans (Heinrich), Antaeotricha, 21-596
hetaera (Meyrick), Symphanactis, 29-816
hetaeria Walsingham, Gelechia, 29-131
heteracma Meyrick, Dichomeris, 29-579
heteractis (Meyrick), Beltheca, 29-707
heterochroma Clarke, Hypercallia, 21-215
heterolychna Meyrick, Mompha, 25-12
heteropa (Meyrick), Antaeotricha, 21-685
heterosaris (Meyrick), Antaeotricha, 21-603
heterosema Meyrick, Stenoma, 21-1416
heteroxantha Meyrick, Stenoma, 21-1274
hexacentra Meyrick, Commatica, 29-737
hexacentris Meyrick, Urodus, 36-54
hexaleuca Meyrick, Clepticodes, 15-234
hexameris (Meyrick), Caloptilia, 20-8
hexascia (Meyrick), Cerconota, 21-867
hexasticta Walsingham, Dichomeris, 29-580
hieroglyphica Powell, Ethmia, 21-351
hieroglyphica Walsingham, Aristotelia, 29-25
himaea Meyrick, Antaeotricha, 21-604
himerodes Meyrick, Stenoma, 21-1275
hippuris Meyrick, Acrocercops, 20-79
hiramella Busck, Ethmia, 21-352
hirsutevestita (Walsingham), Acrolophus, 15-77
hirsutus Busck, Acrolophus, 15-144
hirsutus (Walsingham), Acrolophus, 15-78
hispaniolae Davis, Paucivena, 16-13
hodgesella Powell, Ethmia, 21-353
hoffmanni (Köhler), Cryptothelea, 16-21

hoffmanni (Vázquez), Astala, 16-35
holarga Meyrick, Baeonoma, 21-832
holcadica Meyrick, Stenoma, 21-1276
holocapna Meyrick, Tinea, 15-393
hologramma Meyrick, Glyphipterix, 34-45
holomorpha Meyrick, Prochola, 26-26
holophaea (Meyrick), Stenoma, 21-1277
holopyrrha Meyrick, Cryptolechia, 21-55
homala Walsingham, Stenoma, 21-1278
homochorda Meyrick, Compsistis, 21-26
homochromatica Walsingham, Holcocera, 23-91
homologa (Meyrick), Antaeotricha, 21-605
hopfferi (Zeller), Stenoma, 21-1279
hoplitica Meyrick, Stenoma, 21-1280
hoplodoxa Meyrick, Chalcomima, 29-714
hora (Busck), Hypatima, 29-505
hordei (Kirby), Sitotroga, 29-513
horiarcha Meyrick, Perisceptis, 16-9
horiodes Meyrick, Dichomeris, 29-581
horista Walsingham, Untomia, 29-495
horizontias (Meyrick), Antaeotricha, 21-606
horni Köhler, "Oiketicus", 16-69
horocharis Meyrick, Stenoma, 21-1281
horocompsa Meyrick, Dichomeris, 29-582
horocyma Meyrick, Stenoma, 21-1282
horometra (Meyrick), Cerconota, 21-868
horosema Meyrick, Tinea, 15-403
horridalis (Walker), Acrolophus, 15-79
horridula (Zeller), Xylesthia, 15-443
hospitalis Meyrick, Stenoma, 21-1283
howardi Walsingham, Aristotelia, 29-26
howdeni Powell, Ethmia, 21-354
huebneri (Geyer), Phassus, 5-36
hulstellus Beutenmüller, Acrolophus, 15-178
humerella (Walker), Antaeotricha, 21-607
humeriferella (Walker), Antaeotricha, 21-773
humilis Hodges, Anacampsis, 29-316
humilis Powell, Ethmia, 21-355
hyacinthitis Meyrick, Stenoma, 21-1284
hyalinacra Davis, Lumacra, 16-31
hyalocryptis Meyrick, Stenoma, 21-1285
hyalophaea Meyrick, Doliotechna, 21-1555
hyalophanta (Meyrick), Antaeotricha, 21-608
hydara Walsingham, Cryptolechia, 21-56
hydraea (Meyrick), Ypsolopha, 35-23
hydraena Meyrick, Stenoma, 21-1286
hydrelaeas (Meyrick), Cerconota, 21-869
hydrochroa Meyrick, Pycnotarsa, 21-1699
hydrogramma (Meyrick), Gonionota, 21-146
hydrophora Meyrick, Antaeotricha, 21-609
hylomaga (Meyrick), Stigmella, 7-9
hypanthes Meyrick, Stilbosis, 27-99
hyperbolica (Meyrick), Inga, 21-1622
hypochloa (Walsingham), Glyphidocera, 23-15
hypocirrha Meyrick, Stenoma, 21-1287
hypoleuca Clarke, Gonionota, 21-147
hypophaea Meyrick, Acrolophus, 15-80
hypsicrates Meyrick, Urodus, 36-55
hyptiotes Clarke, Gonionota, 21-148
hysginiella (Wallengren), Atteva, 36-5
iambica Meyrick, Homosetia, 15-268
ianthes (Meyrick), Dichomeris, 29-565
ianthina (Walsingham), Antaeotricha, 21-610
iatma Meyrick, Stenoma, 21-1288
icarus Busck, Acrolophus, 15-81

ichnitis Meyrick, Cryptolechia, 21-57
ichnota Meyrick, Battaristis, 29-350
ichthyodes (Meyrick), Chlamydastis, 21-941
icriodes (Meyrick), Chionodes, 29-74
icteropis Meyrick, Stenoma, 21-1289
icterota (Meyrick), Inga, 21-1623
idiarcha Meyrick, Trichembola, 29-827
idiocentra Meyrick, Anacampsis, 29-317
ignavella (Zeller), Taygete, 29-275
ignicolor (Busck), Machimia, 21-240
ignita (Busck), Filinota, 21-98
ignobilis (Zeller), Gonioterma, 21-1022
ignotella (Walker), Niditinea, 15-306
iinvicta Clarke, Glyphipterix, 34-47
illepida Meyrick, Antaeotricha, 21-611
illita (Meyrick), Chlamydastis, 21-942
illiterata Meyrick, Glyphidocera, 23-16
illucidella (Walker), Antaeotricha, 21-524
illudens Meyrick, Acrolophus, 15-82
illuminella (Busck), Machimia, 21-241
illustra Duckworth, Lethata, 21-1049
ilyella (Zeller), Scrobipalpula, 20-241
ilyodes Meyrick, Ceromitia, 12-5
ilyopis Meyrick, Saridacma, 30-11
imitans (R. Felder & Rogenhofer), Urodus, 36-56
imitata Druce, Urodus, 36-57
immersa Walsingham, Stenoma, 21-1290
imminens (Meyrick), Antaeotricha, 21-612
immota Meyrick, Antaeotricha, 21-613
immunda (Zeller), Stenoma, 21-1291
immunis Meyrick, Scythris, 28-6
immuricata (Meyrick), Caloptilia, 20-9
impactella (Walker), Antaeotricha, 21-614
impedita (Meyrick), Antaeotricha, 21-615
imperiella (Busck), Eomichla, 21-1566
impressella (Busck), Stenoma, 21-1292
impressella (Walker), Cerconota, 21-870
impressipenella (Bilimek), Monopis, 15-299
impudica Walsingham, Phthorimaea, 29-182
impurata Meyrick, Stenoma, 21-1293
impurgata Walsingham, Gelechia, 29-132
inaequalis (Busck), Faculta, 29-102
inaequalis (Walsingham), Faculta, 29-102
inaequepulvella (Chambers), Brachmia, 29-520
inamoenella Zeller, Setomorpha, 15-359
inanima Meyrick, Scythris, 28-7
inardescens Meyrick, Stenoma, 21-1294
inaugurata Meyrick, Cosmopterix, 27-15
inca Meyrick, Arrhenophanes, 17-1
incalescens (Meyrick), Gonionota, 21-149
incensatella (Walker), Inga, 21-1624
incensella (Zeller), Hypercallia, 21-216
incincta Walsingham, Stilbosis, 27-100
incisa Meyrick, Gonionota, 21-150
incisa (Walsingham), Psittacastis, 21-288
incisoria Meyrick, Ephedroxena, 15-255
incisurella (Walker), Antaeotricha, 21-616
incitatrix (Meyrick), Anadasmus, 21-430
incommoda Meyrick, Philomusaea, 21-1696
incompleta Meyrick, Antaeotricha, 21-617
incongrua Meyrick, Antaeotricha, 21-618
inconspicua Clarke, Doina, 21-75
inconspicua Forbes, Acrocercops, 20-80
incontigua Clarke, Gonionota, 21-151
incrassata Meyrick, Antaeotricha, 21-619

increpans Meyrick, Exoncotis, 15-263
increta (Butler), Doina, 21-76
incretata Meyrick, Auximobasis, 23-52
incudella Forbes, 15-297
incurva (Meyrick), Compsolechia, 29-405
indalma Walsingham, Psilocorsis, 21-276
indecora (Walker), Acrolophus, 15-107
indecora (Zeller), Gonioterma, 21-1023
indicata (Strand), Aepytus, 5-108
indicatella (Walker), Antaeotricha, 21-620
indigna (Walsingham), Dichomeris, 29-583
indiscriminata Clarke, Nanodacna, 26-6
indistincta (Amsel), Lethata, 21-1041
indocilis Meyrick, Glyphidocera, 23-17
indomita Meyrick, Glyphipterix, 34-46
infamis Meyrick, Baeonoma, 21-833
infecta (Meyrick), Antaeotricha, 21-621
inferiorella Zeller, Schreckensteinia, 40-1
infida Meyrick, Acrolophus, 15-83
inflammata (Meyrick), Inga, 21-1625
inflata (Butler), Stenoma, 21-1295
infracta (Walsingham), Friseria, 29-108
infrenata (Meyrick), Antaeotricha, 21-622
infusa Meyrick, Stenoma, 21-1296
infuscata Hodges, Crasimorpha, 29-500
inga Busck, Gonioterma, 21-1024
ingloria Meyrick, Dichomeris, 29-584
ingricella (Möschler), Ethmia, 21-324
injucunda Meyrick, Stenoma, 21-1297
innexa (Meyrick), Antaeotricha, 21-623
inopina Walsingham, Energia, 21-993
inquieta (Meyrick), Anacampsis, 29-318
inquinata Meyrick, Lophaeola, 29-780
inquinula Zeller, Antaeotricha, 21-624
inscitella (Walker), Inga, 21-1666
inscitella Walker, Tiquadra, 15-428
inscitum (Busck), Chlamydastis, 21-943
insectella auct., Setomorpha, 15-359
insequens Meyrick, Recurvaria, 29-205
insidiata (Meyrick), Antaeotricha, 21-625
insignata Clarke, Gonionota, 21-152
insimulata Meyrick, Antaeotricha, 21-626
inspectrix (Meyrick), Chlamydastis, 21-944
inspiciens Meyrick, Cymotricha, 29-539
instans Meyrick, Dichomeris, 29-585
insulana Clarke, Gonionota, 21-153
insularis Davis, Pterogyne, 16-10
insularis Walsingham, Anacampsis, 29-319
insularis Walsingham, Auximobasis, 23-53
insulella (Walsingham), Acrocercops, 20-81
intaminata Meyrick, Machimia, 21-241.1
integra Meyrick, Doliotechna, 21-1556
intentella (Walker), Cymotricha, 29-540
interfracta Meyrick, Cosmopterix, 27-16
interfusa Meyrick, Acrolophus, 15-84
interjuncta Meyrick, Phthorimaea, 29-183
intermedia (R. Felder & Rogenhofer), Antaeotricha, 21-607
intermediella (Chambers), Aristotelia, 29-38
intermedius (Riley), Tegeticula, 13-2
intermissella (Zeller), Recurvaria, 29-206
interpolata Meyrick, Ethirostoma, 29-768
intersecta (Meyrick), Antaeotricha, 21-627
intexta Meyrick, Coleophora, 24-7
intonans (Meyrick), Gonionota, 21-154
intricatus Riley, Prodoxus, 13-5

inturbatella (Walker), Cerconota, 21-871
inurbana Meyrick, Glyphidocera, 23-18
inusta (Meyrick), Compsolechia, 29-406
invicta Meyrick, Glyphipterix, 34-47
invida Durrant, Acrolophus, 15-85
invidiosa Meyrick, Xystrologa, 15-449
invigilans (Meyrick), Lethata, 21-1060
involucralis Meyrick, Stenoma, 21-1306
involuta Meyrick, Gnorimoschema, 29-162
invulgata Meyrick, Stenoma, 21-1298
io (Busck), Rhodanassa, 21-1114
iobapta (Meyrick), Eudactylota, 29-94
ioclista Meyrick, Glyphipterix, 34-48
iocoma Meyrick, Stenoma, 21-1299
iodes Walsingham, Anchimacheta, 36-25
ioleuca (Meyrick), Gonionota, 21-155
ioloncha (Meyrick), Brachyacma, 29-524
iopercna Meyrick, Stenoma, 21-1300
iopetra (Meyrick), Antaeotricha, 21-628
iophlebia (Zeller), Urodus, 36-58
ioploca (Meyrick), Charistica, 29-718
ioplocama Meyrick, Hapalothyma, 19-10
ioptila (Meyrick), Antaeotricha, 21-629
iostalacta Meyrick, Stenoma, 21-1301
iphicleia Meyrick, Triclonella, 27-55
ipomoeae Busck, Acrocercops, 20-82
iracunda (Meyrick), Inga, 21-1626
iras Meyrick, Antaeotricha, 21-630
irascens Meyrick, Stenoma, 21-1302
irene (Barnes & Busck), Antaeotricha, 21-631
irenella (Busck), Eomichla, 21-1567
irenias (Meyrick), Antaeotricha, 21-632
iresiarcha (Meyrick), Ithome, 27-80
iriantha (Meyrick), Charistica, 29-719
iridea Forbes, Glaucacna, 29-774
iridella Powell, Ethmia, 21-356
iridipennella Clemens, Strobisia, 29-488
iriphanes (Meyrick), Phyllonorycter, 20-128
irresoluta Duckworth, Lethata, 21-1050
irridipennella Busck, Strobisia, 29-488
irrisoria Meyrick, Acrolophus, 15-86
irrorata (Busck), Lactura, 36-15
irrubricata Walsingham, Cosmopterix, 27-17
isabella (R. Felder & Rogenhofer), Stenoma, 21-1162
isarga (Meyrick), Timocratica, 21-1487
isastra (Meyrick), Gonionota, 21-156
isaura Clarke, Teresita, 21-1712
ischioptila (Meyrick), Anadasmus, 21-431
ischnoptera Meyrick, Compsolechia, 29-407
ischnoscia (Meyrick), Cerconota, 21-872
ischnotoma (Meyrick), Marmara, 20-17
isochlora (Meyrick), Scrobipalpula, 29-242
isochyta (Meyrick), Antaeotricha, 21-633
isodisca Meyrick, Choropleca, 15-225
isodonta Meyrick, Tinea, 15-394
isodryas (Meyrick), Gonionota, 21-157
isographa Meyrick, Timocratica, 21-1499
isoleura Meyrick, Heliodines, 40-19
isomeris (Meyrick), Antaeotricha, 21-634
isophylla Meyrick, Gonionota, 21-158
isoplintha (Meyrick), Antaeotricha, 21-635
isoporphyra (Meyrick), Antaeotricha, 21-636
isortha (Meyrick), Marmara, 20-118
isosticta (Meyrick), Antaeotricha, 21-637
isotoma Meyrick, Cosmopterix, 27-18

isotona Meyrick, Antaeotricha, 21-638
isotrocha Meyrick, Promenesta, 21-1092
isoxesta Meyrick, Urodus, 36-59
isthmiella Busck, Scardia, 15-352
isthmiella (Busck), Urodus, 36-60
ithycosma (Meyrick), Exoteleia, 29-101
ithymetra Meyrick, Filinota, 21-99
ithytona Meyrick, Antaeotricha, 21-639
iulina (Walsingham), Scardia, 15-353
izquierdoi (Ureta), Callipielus, 5-19
jalapae Walsingham, Acrolophus, 15-87
jamaicensis (Walsingham), Phthorimaea, 29-184
janzeni Powell, Ethmia, 21-357
jaspidata (Meyrick), Acrolepiopsis, 39-8
javarica (Butler), Stenoma, 21-1314
jeanneli (Viette), Aepytus, 5-72
jocularis Walsingham, Schreckensteinia, 40-2
johannis (Zeller), Stigmella, 7-10
jonesi (Barnes & Benjamin), Cryptothelea, 16-24
jonesi Schaus, Oiketicus, 16-50
jordani (Viette), Aepytus, 5-87
josephinella Dyar, Ethmia, 21-348
joviella Walsingham, Ethmia, 21-358
jubarella Comstock, Lithariapteryx, 40-33
jucunda Meyrick, Stenoma, 21-1303
jugata (Walsingham), Cymotricha, 29-541
jujuyensis (Pastrana), Alucita, 31-9
julia Powell, Ethmia, 21-359
juniperi Hodges, Periploca, 27-90
juvenalis (Meyrick), Antaeotricha, 21-640
juvenis Walsingham, Stilbosis, 27-101
juventella (Walsingham), Untomia, 29-496
kasyi Duckworth, Falculina, 21-998
katharinae Pfitzner, "Dalaca", 5-119
kearfottella Dietz, Xylesthia, 15-445
kearfotti (Dyar), Acrolophus, 15-88
kennicottella Clemens, Endrosis, 21-1563
kerbyi Lahille, Oiketicus, 16-58
keifferi Kieffer & Jörgensen, Tecia, 29-821
kincaidiella (Busck), Lotisma, 30-3
kirbii Walker, Oiketicus, 16-57
kirbyi Guilding, Oiketicus, 16-57
kirbyi (Möschler), Ethmia, 21-360
kitella (Walsingham), Recurvaria, 29-207
klemaniana (Stoll), Stenoma, 21-1304
klotsi Hasbrouck, Acrolophus, 15-88.1
knabi (Walsingham), Batrachedra, 24-24
koehleri Bourquin, Tischeria, 9-5
krahmeri Nielsen & Robinson, Callipielus, 5-18
kuenckelii (Heylaerts), Lumacra, 16-28
küenkeli (Köhler), Lumacra, 16-28
künckeli (Köhler), Lumacra, 16-28
künckelii (Heylaerts), Lumacra, 16-28
labyrinthias Meyrick, Compsistis, 21-27
lacera (Zeller), Antaeotricha, 21-641
lacertosa Meyrick, Antaeotricha, 21-813
laciniosa (Meyrick), Scrobipalpula, 29-236
lactaria Meyrick, Alsodryas, 29-696
lactella (Denis & Schiffermüller), Endrosis, 21-1563
lacticaput (Walsingham), Friseria, 29-109
lacticaudellum Walsingham, Drepanoterma, 29-760
lacticeps (Meyrick), Friseria, 29-109
lacticoma (Meyrick), Chionodes, 29-75
lactirivis (Meyrick), Xystrologa, 15-450
lactis (Busck), Chlamydastis, 21-945

lacunata (Meyrick), Inga, 21-1627
laetifica (Busck), Chlamydastis, 21-948
laetifica Durrant, Acrolophus, 15-89
laevinella (Zeller), Mompha, 25-13
laeviuscula Zeller, Stenoma, 21-1404
lagneia Clarke, Doina, 21-77
lagopus (Möschler), Aepytus, 5-106
lagunculariella (Busck), Compsolechia, 29-408
lamella (Busck), Cronicombra, 34-9
laminata Nielsen & Robinson, Dalaca, 5-7
laminensis Pastrana, Ceromitia, 12-6
laminicornis Hasbrouck, Acrolophus, 15-98
lamprocosma Meyrick, Filinota, 21-100
lampyridella (Busck), Antaeotricha, 21-642
lanceella Sattler, Deltophora, 29-88
lanceolata (Walsingham), Erechthias, 15-260
languens Meyrick, Anacampsis, 29-330
languescens (Meyrick), Cerconota, 21-873
languida (Meyrick), Inga, 21-1628
languidalis (Walker), Acrolophus, 15-107
lanosa Duckworth, Lethata, 21-1051
lantanella Busck, Cremastobombycia, 20-121
lapidea Meyrick, Stenoma, 21-1305
lapidella Walsingham, Anacampsis, 29-321
lapidescens Meyrick, Gelechia, 29-133
lapilella (Busck), Stenoma, 21-1306
lasciva (Walsingham), Taygete, 29-276
lasia Walsingham, Zetesima, 21-1513
latebricola Meyrick, Telphusa, 29-287
latebrivora (Meyrick), Amydria, 15-212
lateralis (Dyar), Lactura, 36-16
laterestriata (Walsingham), Thiotricha, 29-303
latescens Walsingham, Paranoea, 29-639
lathiptila (Meyrick), Antaeotricha, 21-643
latiberbis Meyrick, Acrolophus, 15-90
latipennella Zeller, Tinea, 15-395
latipennis (Zeller), Gonioterma, 21-1025
latipes (Walker), Snellenia, 21-1724
latistriga Walsingham, Untomia, 29-497
latitans (Dognin), Stenoma, 21-1307
lativittella (Walker), Antaeotricha, 21-821
laudata Meyrick, Antaeotricha, 21-644
lavata Walsingham, Stenoma, 21-1308
laxa (Meyrick), Antaeotricha, 21-645
lebetias (Meyrick), Antaeotricha, 21-646
lecaniella Busck, Blastobasis, 23-74
lecithaula Meyrick, Antaeotricha, 21-647
lecithitis (Meyrick), Gonionota, 21-159
leioptera Clarke, Leuroperna, 35-5
lembifera (Meyrick), Cerconota, 21-843
leniflua Meyrick, Imma, 41-29
lenta Clarke, Ectaga, 21-86
lenta (Meyrick), Keiferia, 29-170
lentiginosa (Zeller), Tiquadra, 15-429
leonhardi Petersen, Tinea, 15-416
leontodes (Meyrick), Anadasmus, 21-432
lepidocarpa (Meyrick), Antaeotricha, 21-648
lepidocyma Meyrick, Glyphidocera, 23-19
lepidopteris (Dyar), Astala, 16-34
lepidota Meyrick, Falculina, 21-999
lepisma Walsingham, Catarata, 21-839
leprosa (R. Felder & Rogenhofer), Antaeotricha, 21-572
leptobelisca (Meyrick), Chlamydastis, 21-946
leptocona Meyrick, Glyphipterix, 34-49
leptogma Meyrick, Stenoma, 21-1309

leptogramma (Meyrick), Antaeotricha, 21-649
leptophragma (Meyrick), Inga, 21-1629
leptosceles Walsingham, Diataga, 15-247
leptospora Meyrick, Gelechia, 29-134
leptynta Meyrick, Opogona, 15-314
lepyropis Meyrick, Coleophora, 24-8
lerodes Durrant, Acrolophus, 15-91
leucactis Meyrick, Petasanthes, 21-1084
leucallactis Meyrick, Acrolophus, 15-71
leucana (Sepp), Stenoma, 21-1310
leucaniella (Walker), Stenoma, 21-1311
leucillana (Zeller), Antaeotricha, 21-650
leucobasilaris Davis, Lumacra, 16-32
leucocapna (Meyrick), Timocratica, 21-1488
leucocephala Becker, Timocratica, 21-1510b
leucocephala (Walsingham), Chionodes, 29-80
leucochna Meyrick, Taruda, 21-302
leucochrysis Meyrick, Mompha, 25-14
leucoclista Meyrick, Eomichla, 21-1568
leucoclistra Meyrick, Syrmologa, 15-363
leucocras Walsingham, Polyhymno, 29-193
leucocryptis (Meyrick), Antaeotricha, 21-651
leucodelta (Meyrick), Baeonoma, 21-834
leucodocis (Zeller), Acrolophus, 15-92
leucogona Zeller, Blastobasis, 23-75
leucogramma Meyrick, Antaeotricha, 21-572
leucographa Clarke, Acrocercops, 20-83
leucographa Walsingham, Athrinacia, 21-13
leucomias Meyrick, Promenesta, 21-1093
leuconota (Zeller), Acrocercops, 20-84
leuconota (Zeller), Evippe, 29-98
leuconympha (Meyrick), Rectiostoma, 21-1108
leucophaeella (Walker), Baeonoma, 21-835
leucoplasta (Meyrick), Chlamydastis, 21-947
leucopleura Meyrick, Teuchophanes, 29-647
leucopogon Walsingham, Acrolophus, 15-93
leucoporpa (Meyrick), Gonionota, 21-160
leucoptila (Meyrick), Chlamydastis, 21-948
leucorectis (Meyrick), Timocratica, 21-1489
leucorrhapta (Meyrick), Compsolechia, 29-409
leucosaris (Meyrick), Cerconota, 21-897
leucoschista Meyrick, Zelleria, 36-89
leucostena (Walsingham), Vazugada, 29-683
leucothea (Busck), Lethata, 21-1052
leucotricha Meyrick, Acrolophus, 15-94
leucoxantha Clarke, Irenia, 21-1682
leucozostra Meyrick, Nealyda, 29-7
leucozyga Meyrick, Blastobasis, 23-76
leukogramma Bryk, Callipielus, 5-11
lianthes (Meyrick), Anadasmus, 21-425
liberatella (Walker), Auximobasis, 23-54
leucura (Walsingham), Strobisia, 29-489
libertina Meyrick, Stenoma, 21-1312
libidinosa (Meyrick), Inga, 21-1630
libitina Druce, Acrolophus, 15-95
lichenias (Meyrick), Chlamydastis, 21-949
licheniphilus (Köhler), Prochalia, 16-14
lichenista (Meyrick), Gonionota, 21-161
lichyi Powell, Ethmia, 21-361
licmaea Meyrick, Stenoma, 21-1452
lictor Walsingham, Ectaga, 21-87
lignicolor Zeller, Antaeotricha, 21-652
lignicolora Forbes, Aristotelia, 29-27
ligulella Hübner, Dichomeris, 29-586
lilloi Köhler, "Gaea", 29-830

limbata Walsingham, Isocorypha, 15-281
limbipennella Clemens, Plutella, 35-18
limicola Meyrick, Holcocera, 23-92
limpia Dognin, Animula, 16-42
linaria Clarke, Batrachedra, 24-25
linda Busck, Ethmia, 21-362
lindenella (Busck), Friseria, 29-107
lindseyi (Barnes & Busck), Antaeotricha, 21-653
lineola Clarke, Deia, 21-1547
lingulata Meyrick, Compsolechia, 29-410
liniella Busck, Stenoma, 21-1205
linsdalei Powell, Ethmia, 21-363
linteata (Meyrick), Gonioterma, 21-1026
linus Druce, Acrolophus, 15-96
lipara Duckworth, Thioscelis, 21-1471
liquescens Meyrick, Mysaromima, 21-1076
lissopeda Meyrick, Pygmocrates, 36-23
literatella (Busck), Palinorsa, 21-261
lithochroma Busck, Promenesta, 21-1094
lithochroma Walsingham, Pammeces, 26-13
lithocolletina (Zeller), Parectopa, 20-26
lithodelta Meyrick, Anacampsis, 29-322
lithodes Walsingham, Gelechia, 29-133
lithograpta (Meyrick), Chlamydastis, 21-950
lithogypsa (Meyrick), Anadasmus, 21-433
lithomacha Meyrick, Parectopa, 16-27
lithomorpha (Meyrick), Compsolechia, 29-411
lithopa Durrant, Acrolophus, 15-97
lithophaea (Meyrick), Urodus, 36-61
lithopola Walsingham, Adela, 12-16
lithoxesta Meyrick, Stenoma, 21-1253
litigiosa (Meyrick), Chionodes, 29-76
litura (R. Felder & Rogenhofer), Taeniostolella, 34-14
litura Zeller, Stenoma, 21-1313
liturosella (Zeller), Chionodes, 29-77
lizeri Köhler, "Oiketicus", 16-70
lizeri Pastrana, Ceromitia, 12-7
lobitarsis Zeller, Peleopoda, 21-414
lobitarsus Hodges, Peleopoda, 21-414
logistica (Meyrick), Nanodacna, 26-7
lonchocarpella Busck, Stilbosis, 27-102
longicilia Becker, Timocratica, 21-1490
longimaculata (Dognin), Hypercallia, 21-217
longipalpis Meyrick, Borkhausenia, 21-1536
longitudinella Busck, Baeciva, 29-708
lophandra Meyrick, Glyphidocera, 23-20
lophoptycha (Meyrick), Antaeotricha, 21-654
lophosaris (Meyrick), Antaeotricha, 21-655
loquax (Meyrick), Symmetrischema, 29-265
loriculata Meyrick, Heliodines, 40-20
lotoxantha Meyrick, Opogona, 15-315
loxobathra (Meyrick), Inga, 21-1631
loxochorda Meyrick, Hypercallia, 21-218
loxogramma Meyrick, Compsolechia, 29-412
loxogrammos (Zeller), Antaeotricha, 21-656
loxotoma (Busck), Timocratica, 21-1491
lucicola (Maassen), Pfitzneriella, 5-128
lucidiorella (Walker), Stenoma, 21-1314
luciliella Zeller, Elachista, 22-3
lucrifuga Meyrick, Dichomeris, 29-587
lucrosa (Meyrick), Antaeotricha, 21-657
luctifica Zeller, Stenoma, 21-1315
luctuosa Meyrick, Acrocercops, 20-85
luctuosa (Walsingham), Scardia, 15-354
ludicra Meyrick, Comotechna, 21-21

luminosa (Busck), Teuchophanes, 29-647
lunaris Gaede, Chezala, 21-1544
lunatica Meyrick, Amblytenes, 24-14
lunimaculata (Dognin), Antaeotricha, 21-658
lunularis Meyrick, Chezala, 21-1544
lunuligera Walker, Paelia, 31-24
lupata Meyrick, Commatica, 29-738
luridella (Zeller), Utilia, 21-1718
luscina (Zeller), Antaeotricha, 21-607
lusciosa (Meyrick), Exaeretia, 21-4
luteela Forbes, Infurcitinea, 15-279
luteicornis (Berg), Parapielus, 5-25
luteola (R. Felder & Rogenhofer), Gonionota, 21-162
luteostrigella Chambers, Polyhymno, 29-194
lutulenta (Zeller), Cerconota, 21-874
lycopersicella (Busck), Keiferia, 29-170
lycopersicella (Walsingham), Keiferia, 29-170
lydia (Druce), Hypercallia, 21-219
lyonetiella (Chambers), Mompha, 25-8
lypetica Walsingham, Dichomeris, 29-588
lyrella (Walsingham), Brachmia, 29-522
lysalges (Walsingham), Cerconota, 21-875
lysimeris Meyrick, Antaeotricha, 21-659
machetes (Walsingham), Antaeotricha, 21-660
machinatrix (Meyrick), Cerconota, 21-876
macleaii (Westwood), Cryptothelea, 16-20
macleayi (Guilding), Cryptothelea, 16-20
macleayii (Meyrick), Cryptothelea, 16-20
macraulax Meyrick, Stenoma, 21-1316
macrocera R. Felder & Rogenhofer, Zaratha, 26-40
macrocera Walsingham, Iconisma, 23-104
macrochorda Meyrick, Compsistis, 21-28
macrotenis Meyrick, Ordrupia, 30-8
macrogaster (Walsingham), Acrolophus, 15-98
macroleuca (Meyrick), Timocratica, 21-1492
macronota (Meyrick), Antaeotricha, 21-661
macrophallus Hasbrouck, Acrolophus, 15-99
macroptera Meyrick, Dichomeris, 29-578
macroptycha Meyrick, Stenoma, 21-1317
macrosphena (Meyrick), Vazugada, 29-684
macrozancla Meyrick, Acrolophus, 15-100
maculata Clarke, Altiura, 21-1520
maculata Duckworth, Lethata, 21-1038
maculata (Riley), Tegeticula, 13-1
maculata Walsingham, Acrolophus, 15-101
maculata Walsingham, Oecia, 23-41
maculella (Blanchard), Acrolepiopsis, 39-9
maculicostella Kieffer & Jörgensen, Cecidolechia, 21-1543
maculicostella Strand, Cecidolechia, 21-1543
maculipennis (Curtis), Plutella, 35-18
maculisecta Busck, Acrolophus, 15-102
maculosa Butler, Arctopoda, 21-1526
magnatella (Zeller), Mompha, 25-8
mahagoniatus (Pfitzner), Aepytus, 5-101
major (Busck), Timocratica, 21-1493
majorella Dietz, Setomorpha, 15-359
malachita Meyrick, Antaeotricha, 21-662
malacodoxa (Meyrick), Ypsolopha, 35-24
malacoscia Meyrick, Compsistis, 21-29
malacozesta Meyrick, Stenoma, 21-1119
malifoliella (Fitch), Dichomeris, 29-586
malindella (Busck), Friseria, 29-107
mallodeta Meyrick, Tiquadra, 15-430
malthacopa Meyrick, Taphrosaris, 29-819

manceps Meyrick, Antaeotricha, 21-663
manella (Möschler), Trichotaphe, 29-665
mangelivora (Walsingham), Compsolechia, 29-413
manicola Meyrick, Amiantastis, 16-1
manoa Pfitzner, "Dalaca", 5-120
mansita Busck, Ethmia, 21-364
manticodes Meyrick, Acrolophus, 15-103
maranthaceae Busck, Acrocercops, 20-86
marantica Walsingham, Urodus, 36-62
marcella (Busck), Inga, 21-1613
marcida (Butler), Stenoma, 21-1255
marcida Walsingham, Acrolophus, 15-104
marcius Druce, Phassus, 5-40
margalaestriata Keuchenius, Setomorpha, 15-359
margaritacea Gozmány, Tinea, 15-416
margaritacea (Meyrick), Coptotelia, 21-41
marginata Busck, Stenoma, 21-1345
marginata (Walsingham), Lamprolophus, 40-29
marginella Busck, Promenesta, 21-1095
margorieella Dietz, Amydria, 15-205
margoriella Dietz, Amydria, 15-205
marjorieella Dietz, Amydria, 15-205
marjoriella Busck, Amydria, 15-205
marmaritis Walsingham, Acrocercops, 20-87
marmarocyma Meyrick, Endothamna, 30-2
marmaropis (Meyrick), Acrolepiopsis, 39-10
marmorata Butler, Dalaca, 5-3
marmorea R. Felder & Rogenhofer, Antaeotricha, 21-664
marmorea (Walsingham), Ethmia, 21-365
marmoreipennis Walsingham, Ditrigonophora, 36-27
marmorella (Chambers), Nemapogon, 15-305
marona (Schaus), Lumacra, 16-27
maroni (Busck), Cerconota, 21-900
maroniella (Busck), Eomichla, 21-1569
martini Amsel, Antaeotricha, 21-611
mastictor Bradley, Cnissostages, 17-3
mastodes Meyrick, Baeonoma, 21-836
mathewi Zeller, Heliostibes, 21-1579
maturescens (Meyrick), Timocratica, 21-1494
maxima (Meyrick), Atoposea, 32-7
meadi H. Edwards, Thyridopteryx, 16-59
meadii Kirby, Thyridopteryx, 16-59
meator Hodges, Batrachedra, 24-26
meconitis (Meyrick), Onebala, 29-627
mediana Walker, Cotaena, 34-1
mediatrix Zeller, Sophronia, 29-248
mediella Busck, Ethmia, 21-312
mediocris Walsingham, Pigritia, 23-111
mediofuscella (Clemens), Chionodes, 29-77
medioliniella (Kearfott), Acrolophus, 15-92
medulella Busck, Telphusa, 29-288
medullata Meyrick, Scythris, 28-8
megaleuca (Meyrick), Timocratica, 21-1495
megaspilella (Walker), Stenoma, 21-1249
meibomiella Forbes, Anacampsis, 29-323
melaleuca Clarke, Anchimompha, 25-1
melanactis Meyrick, Acrocercops, 20-88
melanamba Meyrick, Battaristis, 29-351
melanarma Meyrick, Antaeotricha, 21-665
melanarthra (Lower), Sitotroga, 29-513
melanesia Meyrick, Stenoma, 21-1318
melanixa Meyrick, Stenoma, 21-1319
melanobathra Meyrick, Untomia, 29-498
melanocampa (Meyrick), Scrobipalpula, 29-229

melanocosma Meyrick, Acrocercops, 20-89
melanocosta Becker, Timocratica, 21-1496
melanocrypta Meyrick, Stenoma, 21-1195
melanolepis (Clarke), Scrobipalpula, 29-243
melanoleuca Walsingham, Telphusa, 29-289
melanoma Clarke, Aniuta, 21-1524
melanometra (Meyrick), Chlamydastis, 21-951
melanonca (Meyrick), Chlamydastis, 21-952
melanophaea (Forbes), Compsolechia, 29-414
melanopthalma (Meyrick), Psilocorsis, 21-277
melanopis Meyrick, Antaeotricha, 21-666
melanoplintha (Meyrick), Symmetrischema, 29-266
melanostictella (Zeller), Recurvaria, 29-208
melanostriga Becker, Timocratica, 21-1497
melanota (Walsingham), Cymotricha, 29-542
melantherella Busck, Acrocercops, 20-90
melema (Walsingham), Cerconota, 21-877
meliacella Becker, Phyllocnistis, 20-142
meliacta (Meyrick), Inga, 21-1632
melichrosta (Meyrick), Atticonviva, 15-213
meligrapta Meyrick, Stenoma, 21-1320
melinopa (Meyrick), Antaeotricha, 21-667
melissia (Walsingham), Cymotricha, 29-543
melitaula Meyrick, Argyresthia, 37-5
melithrepta Meyrick, Glyphidocera, 23-21
melitoptila (Meyrick), Arogalea, 29-56
melixesta Meyrick, Stenoma, 21-1321
melobaphes Walsingham, Gonionota, 21-163
memnonia (Meyrick), Cymotricha, 29-544
mendax (Zeller), Antaeotricha, 21-668
mendoron (Busck), Chlamydastis, 21-953
mendozella Kieffer & Jörgensen, Tecia, 29-822
menestella (Walsingham), Antaeotricha, 21-574
menidias Meyrick, Xylesthia, 15-444
meniscogramma Clarke, Philomusaea, 21-1697
mentigera Meyrick, Antaeotricha, 21-664
menura Clarke, Gonionota, 21-164
mercata (Meyrick), Inga, 21-1633
merida (Strand), Urodus, 36-63
meridiana (Meyrick), Antaeotricha, 21-607
meridionalis Becker, Timocratica, 21-1498
meridionalis Walsingham, Amydria, 15-206
meridogramma Meyrick, Stenoma, 21-1322
merismatella (Zeller), Recurvaria, 29-209
merocoma Meyrick, Acrolophus, 15-105
mesodelta Meyrick, Compsolechia, 29-415
mesogramma Meyrick, Antispila, 14-3
mesonyctia Meyrick, Zaratha, 26-41
mesosaris (Meyrick), Antaeotricha, 21-669
mesosceptra (Meyrick), Exaeretia, 21-5
mesostrota Meyrick, Antaeotricha, 21-670
metachlora Meyrick, Imma, 41-30
metacymba (Meyrick), Chlamydastis, 21-954
metacystis (Meyrick), Chlamydastis, 21-955
metadupa (Walsingham), Compsolechia, 29-416
metallicella Busck, Heliodines, 40-21
metallifera (Walsingham), Mompha, 25-15
metamochla (Meyrick), Chlamydastis, 21-956
metanastes Walsingham, Opogona, 15-316
metellus (Druce), Druceiella, 5-32
methystica Meyrick, Stenoma, 21-1323
metochra Meyrick, Commatica, 29-739
metonella Pierce & Metcalfe, Tinea, 15-416
metricus (Pfitzner, 1914), Druceiella, 5-33
metricus (Pfitzner, 1938), Druceiella, 5-33

metrodoxa Meyrick, Opsodoca, 15-326
metroleuca Meyrick, Stenoma, 21-1324
mexica Walsingham, Drastea, 15-250
mexicana Bastida, Tegeticula, 13-2
mexicana Walsingham, Dichomeris, 29-589
mexicanellus Beutenmüller, Acrolophus, 15-64
mexicanensis (Viette), Aepytus, 5-66
mexicanus Gaede, Oiketicus, 16-55
meyeriana (Stoll), Stenoma, 21-1325
michaeli (Pfitzner), Aplatissa, 5-57
michaelis (Pfitzner), Aplatissa, 5-57
micranthis Meyrick, Synactias, 29-817
microglyptis Meyrick, Cryptolechia, 21-58
microlepta Meyrick, Opostega, 8-7
micromacha Meyrick, Acrolophus, 15-106
microphis Meyrick, Acrocercops, 20-91
microphthalma Meyrick, Petalothyrsa, 21-1083
microptera (Schaus), Animula, 16-43
micropteroides Kristensen & Nielsen, Hypomartyria, 1-1
microsacta Meyrick, Acartophila, 21-1516
microsticta Walsingham, Atteva, 36-6
microtypa (Meyrick), Antaeotricha, 21-671
micrura Walsingham, Parornix, 20-40
migueli Nielsen & Robinson, Calada, 5-23
milichodes Meyrick, Stenoma, 21-1326
milictis Meyrick, Antaeotricha, 21-672
militaris (Meyrick), Gonionota, 21-165
miltopa (Meyrick), Hypercallia, 21-220
miltopeza Clarke, Doshia, 21-84
miltophragma (Meyrick), Cymotricha, 29-545
mimasalis (Walker), Acrolophus, 15-107
mimetis Meyrick, Cosmopterix, 27-9
mimobathra (Meyrick), Inga, 21-1634
mimulina (Butler), Gonionota, 21-166
minerva (Meyrick), Psilocorsis, 21-278
miniata (Dognin), Hypercallia, 21-221
minima (Walsingham), Acrolophus, 15-108
minimella Busck, Scardia, 15-355
minimella Forbes, Oenoe, 15-309
minna (Busck), Cerconota, 21-878
minnetta (Butler), Borkhausenia, 21-1537
minor Busck, Stenoma, 21-1327
minuscula (Walsingham), Erechthias, 15-259
minuta Busck, Decantha, 21-1546
minuta Clarke, Porphyrosela, 20-136
minuta Sattler, Deltophora, 29-89
minutanus Brèthes, Cecidoses, 11-2
minutella (Fabricius), Tinea, 15-396
mirabilinella Comstock, Lithariapteryx, 40-34
miranda Clarke, Pseudarla, 29-196
mirella (Möschler), Urodus, 36-64
misema Walsingham, Acrolophus, 15-109
miseta (Walsingham), Cerconota, 21-879
mistipalpis (Walsingham), Trichotaphe, 29-666
mitis (Meyrick), Gonionota, 21-167
mitratella (Busck), Antaeotricha, 21-673
mitrodeta (Meyrick), Moca, 41-5
mixadelpha (Meyrick), Inga, 21-1635
mixophanes Meyrick, Lithopsaestis, 15-296
mixotypa (Meyrick), Acrolepiopsis, 39-11
mnesicentra Meyrick, Xyrosaris, 36-85
mnesicosma Meyrick, Ethmia, 21-366
mniocosma Meyrick, Compsolechia, 29-417
mniodora Meyrick, Stenoma, 21-1328
mniograpta (Meyrick), Moca, 41-6

mochlopa (Meyrick), Chlamydastis, 21-957
mochlopis Meyrick, Cymotricha, 29-546
modesta Druce, Urodus, 36-65
modestus Busck, Acrolophus, 15-110
modulata (Meyrick), Antaeotricha, 21-674
molifica (Meyrick), Inga, 21-1636
molinella (Stoll), Chlamydastis, 21-958
mollipedella Clemens, Plutella, 35-18
molybdaspis Meyrick, Psittacastis, 21-289
molybdina (Walsingham), Compsolechia, 29-418
molybditis (Zeller), Stigmella, 7-11
molybdopa (Meyrick), Inga, 21-1637
momus (Druce), Druceiella, 5-33
monastra (Meyrick), Chlamydastis, 21-959
monoargenteus Viette, Aepytus, 5-93
monochromella (Walker), Compsolechia, 29-419
monoclona Busck, Antaeotricha, 21-675
monocolona Meyrick, Antaeotricha, 21-675
monoctenis Meyrick, Acrolophus, 15-111
monopa Duckworth, Lethata, 21-1053
monosaris (Meyrick), Antaeotricha, 21-676
monosperma Meyrick, Opostega, 8-8
monotona (Amsel), Durrantia, 21-407
monotonia (Strand), Timocratica, 21-1499
montezuma Meyrick, Tinea, 15-397
monticola (Maassen), Pfitzneriella, 5-129
montium (Walsingham), Scardia, 15-356
monura Herrich-Schäffer, Urodus, 36-66
moorei Busck, Arauzona, 21-1722
morata Meyrick, Machimia, 21-242
morbida (Zeller), Chlamydastis, 21-960
morbidula Meyrick, Acrolophus, 15-112
morrisoni (Walsingham), Acrolophus, 15-138
mortonjonesi Vázquez, Oiketicus, 16-52
motasi Povolný, Gnorimoschema, 29-163
motasi Povolný, Scrobipalpula, 29-244
mucida Duckworth, Lethata, 21-1054
mulciber (Meyrick), Alucita, 31-10
mulleri Busck, Ethmia, 21-367
mülleri Busck, Ethmia, 21-367
multidentatus Vázquez, Oiketicus, 16-55
multimaculella (Chambers), Setomorpha, 15-359
multinotata (Meyrick), Anacampsis, 29-324
multipunctella (Chambers), Ethmia, 21-391
mummea Pfitzner, "Dalaca", 5-127
mummia Schaus, "Dalaca", 5-127
mundana (Meyrick), Haplochela, 29-503
mundella (Walker), Antaeotricha, 21-677
mundula Meyrick, Stenoma, 21-1329
munona Schaus, Aepytus, 5-76
murariella Staudinger, Tinea, 15-398
murenula Meyrick, Choropleca, 15-226
muricolor Walsingham, Amydria, 15-207
murina Hodges, Teladoma, 27-43
murinella (Walker), Antaeotricha, 21-678
muscula Zeller, Stenoma, 21-1330
musota Meyrick, Mompha, 25-16
mustella (Walsingham), Antaeotricha, 21-679
mutabilis Meyrick, Dissoptila, 29-758
muysca (Pfitzner), Aepytus, 5-68
mydopis (Meyrick), Inga, 21-1638
mylicopa Meyrick, Zymologa, 15-454
myopina (Zeller), Lethata, 21-1055
myrochroa (Meyrick), Lethata, 21-1056
myrodora (Meyrick), Cerconota, 21-880

myrrhinopa Meyrick, Stenoma, 21-1331
mysteriodes Meyrick, Thalassonympha, 35-19
mysticopis (Meyrick), Chlamydastis, 21-961
myura Meyrick, Lyonetia, 19-5
nana Dietz, Holcocera, 23-93
nannophyes Pfitzner, "Dalaca", 5-116
nasuta Zeller, Alucita, 31-11
nasutitermina Silvestri, Ectinocampa, 15-254
navicularis (Meyrick), Antaeotricha, 21-680
navigatrix (Meyrick), Peleopoda, 21-415
naxia Meyrick, Aristotelia, 29-28
neanica Walsingham, Stenoma, 21-1310
neastra (Meyrick), Orphnolechia, 21-1080
neblina Fleming, Alinguata, 31-1
nebras (Meyrick), Neomachlotica, 34-19
nebrita Walsingham, Stenoma, 21-1332
nebula Hodges, Teladoma, 27-44
nebulombra Powell, Ethmia, 21-392
negotiosa Meyrick, Stenoma, 21-1333
negreai Căpuse & Georgesco, Niditinea, 15-307
nemeseta Meyrick, Spanioptila, 20-113
neochorda Meyrick, Glyphipterix, 34-50
neocompsa Meyrick, Syntetrernis, 26-38
neocrossa (Meyrick), Antaeotricha, 21-681
neoleuca Meyrick, Monachozela, 14-13
neopercna Meyrick, Stenoma, 21-1334
neoptila Meyrick, Stenoma, 21-1335
neospila (Meyrick), Inga, 21-1639
neozona (Meyrick), Blastobasis, 23-77
nephalia Walsingham, Holcocera, 23-94
nephallactis (Meyrick), Moca, 41-7
nephelaegis Meyrick, Plutella, 35-15
nephelocyma (Meyrick), Antaeotricha, 21-682
nepheloleuca Meyrick, Stenoma, 21-1430
nephelophracta Meyrick, Gelechia, 29-135
nephelozyga Meyrick, Tinaegeria, 21-1729
neptes Walsingham, Valentinia, 23-107
neptica Walsingham, Gelechia, 29-136
nerterodes Meyrick, Commatica, 29-740
nerteropa Meyrick, Antaeotricha, 21-683
nesitis (Walsingham), Parectopa, 20-28
nessica (Walsingham), Trichotaphe, 29-667
nestes (Busck), Chlamydastis, 21-962
neurocentra Meyrick, Stenoma, 21-1336
neurographa Meyrick, Antaeotricha, 21-684
neurophora Meyrick, Compsolechia, 29-420
neuroscia Meyrick, Machimia, 21-243
neurotona (Meyrick), Stenoma, 21-1337
nictitans (Zeller), Antaeotricha, 21-685
niepelti Pfitzner, "Dalaca", 5-121
nigratomella (Clemens), Battaristis, 29-352
nigricans (Busck), Stenoma, 21-1338
nigriceps Zeller, Tinea, 15-399
nigricornis Walker, Dalaca, 5-5
nigrifoldella Gregson, Tinea, 15-403
nigripalpis Walsingham, Tinea, 15-400
nigripectus Walsingham, Gelechia, 29-137
nigriplaga Dognin, Coptotelia, 21-42
nigritaenia Powell, Ethmia, 21-368
nigritella Busck, Dorata, 15-248
nigroatomella (Dietz), Nemapogon, 15-305
nigrovenosalis (Viette), Aepytus, 5-90
nigrovittata (Walsingham), Xystrologa, 15-451
nimbata Meyrick, Antaeotricha, 21-686
nimbosa (Zeller), Cerconota, 21-881

niobella (R. Felder & Rogenhofer), Compsolechia, 29–421
niphacma Meyrick, Stenoma, 21-1339
nipharcha (Meyrick), Moca, 41-8
niphatma Meyrick, Urodus, 36-67
niphocentra Meyrick, Compsolechia, 29-422
niphochlaena (Meyrick), Stenoma, 21-1340
niphocycla Meyrick, Hypercallia, 21-222
niphosperma (Meyrick), Acrolepiopsis, 39-12
niphostoma (Meyrick), Moca, 41-9
nitella Voute, Setomorpha, 15-359
nitens (Butler), Cerconota, 21-882
nitescens Meyrick, Antaeotricha, 21-687
nitidorella (Walker), Antaeotricha, 21-688
nitrota Meyrick, Antaeotricha, 21-689
nivea Becker, Timocratica, 21-1500
niveipunctata Walsingham, Acrolophus, 15-113
niveisignella (Zeller), Parastega, 29-792
niveiventris R. Felder & Rogenhofer, Zaratha, 26-42
niveosella (Busck), Ethmia, 21-369
niviliturella (Walker), Antaeotricha, 21-574
nivosa (R. Felder & Rogenhofer), Tiquadra, 15-431
nivosella (Walker), Ethmia, 21-369
noctivaga (Walsingham), Acrolophus, 15-114
noctuella (Madinier), Perileucoptera, 19-8
noctuides Pfitzner, Dalaca, 5-3
noctuina (Walsingham), Acrolophus, 15-115
nocturna (Meyrick), Anacampsis, 29-325
nolickeniella (Zeller), Acrocercops, 20-92
nona Hodges, Friseria, 29-110
nonagriella (Walker), Anadasmus, 21-434
normalis Meyrick, Auximobasis, 23-55
notandella (Busck), Eomichla, 21-1570
notatella (Walker), Ethmia, 21-370
notaula Meyrick, Ilarches, 29-775
notella (Busck), Machimia, 21-244
nothostigma Meyrick, Recurvaria, 29-210
notifera (Meyrick), Gonioterma, 21-1027
notodontella Zeller), Gluphidocera, 23-22
notogramma (Meyrick), Antaeotricha, 21-690
notolopha Meyrick, Glyphidocera, 23-22
notomurinella Powell, Ethmia, 21-371
notopyrsa Meyrick, Copocentra, 40-5
notorrhoa Meyrick, Scythris, 28-9
notosaris (Meyrick), Antaeotricha, 21-691
notosemia (Zeller), Antaeotricha, 21-692
notospila (Meyrick), Taygete, 29-277
noverca (Meyrick), Cerconota, 21-883
n-signatus Weymer, Phassus, 5-38
nubifer Walsingham, Acrolophus, 15-116
nubifera (Meyrick), Alucita, 31-12
nubilella Amsel, Tiquadra, 15-432
nubilella Möschler, Diastoma, 21-992
nubilipennella (Clemens), Niditinea, 15-306
nucifer Walsingham, Gelechia, 29-138
nucifera Meyrick, Gelechia, 29-138
nuciferae Hodges, Batrachedra, 24-27
nucifraga Meyrick, Tiquadra, 15-433
nuclearis Meyrick, Antaeotricha, 21-693
nugella R. Felder & Rogenhofer, Glyphipterix, 34-51
numeratrix Meyrick, Atteva, 36-7
numidia (Druce), Acrolophus, 15-117
nummulata Meyrick, Eomichla, 21-1571
nuntia (Meyrick), Antaeotricha, 21-460
nuptella (R. Felder & Rogenhofer), Compsolechia, 29-423

nycteropa Meyrick, Stenoma, 21-1341
nyctiphanes Meyrick, Cosmopterix, 27-19
nymphas (Meyrick), Cerconota, 21-884
nymphotima Meyrick, Stenoma, 21-1413
obelodes Meyrick, Stenoma, 21-1125
oberthuri (Viette), Parapielus, 5-26
obligata Busck, Telphusa, 29-290
obligatella Möschler, Euarne, 36-28
obliquella Dietz, Amydria, 15-208
obliquestrigata Strand, "Dalaca", 5-122
obliquistriga Dognin, Hypercallia, 21-223
oblita (Butler), Stenoma, 21-1342
oblitella (R. Felder & Rogenhofer), Taruda, 21-303
obmutescens (Meyrick), Anadasmus, 21-435
obnubila Busck, Catarata, 21-840
obnupta (Meyrick), Chlamydastis, 21-963
obolarcha Meyrick, Lamprolophus, 40-30
obovata Meyrick, Stenoma, 21-1343
obscura Duckworth, Lethata, 21-1057
obscurella (Beutenmüller), Ethmia, 21-312
obsessa (Walsingham), Mompha, 25-17
obsordescens (Meyrick), Cerconota, 21-885
obstrciat Meyrick, Auximobasis, 23-56
obstructa Meyrick, Prochola, 26-27
obtusa (Meyrick), Antaeotricha, 21-694
obversa Meyrick, Acrocercops, 20-93
obydella (R. Felder & Rogenhofer), Taruda, 21-299
occultum (Walsingham), Acrolophus, 15-118
oceanica (Tindale), Dalaca, 5-3
oceanitis (Meyrick), Cerconota, 21-886
ocellata (Busck), Stenoma, 21-1346
ocellea Forbes, Mothonica, 21-1074
ocellifer (Walsingham), Antaeotricha, 21-695
ocelligera (Butler), Compsolechia, 29-424
ochleria Walsingham, Cosmopterix, 27-20
ochlodes Walsingham, Stenoma, 21-1344
ochoterenai Vázquez, Oiketicus, 16-57
ochracea (Möschler), Acrolophus, 15-119
ochracea Walker, Tinaegeria, 21-1730
ochracea (Zeller), Utilia, 21-1719
ochracma Meyrick, Cryphiotechna, 15-243
ochreostrigella (Chambers), Scrobipalpula, 29-240
ochrescens (Meyrick), Sitotroga, 29-513
ochricollis Zeller, Stenoma, 21-1345
ochricostata Zeller, Falculina, 21-1000
ochridorsis Zeller, Argyresthia, 37-6
ochrifoliata Walsingham, Telphusa, 29-291
ochrobasis (Duckworth), Rectiostoma, 21-1109
ochrobathra Meyrick, Blastobasis, 23-78
ochrocrossa Meyrick, Dinotropa, 21-1549
ochrodyta Meyrick, Ceromitia, 12-8
ochrogypsa Meyrick, Auxotricha, 21-17
ochrolepra Powell, Pyramidobela, 21-1704
ochroleuca Clarke, Aniuta, 21-1525
ochromicta Meyrick, Prochola, 26-28
ochropa Walsingham, Stenoma, 21-1346
ochrophanes (Meyrick), Dichomeris, 29-565
ochropis Meyrick, Parastega, 29-790
ochropyga (Walsingham), Cymotricha, 29-552
ochrosaris (Meyrick), Gonioterma, 21-1030
ochrosemia (Zeller), Mompha, 25-18
ochrospora Meyrick, Atelosticha, 21-1528
ochrothicata Busck, Stenoma, 21-1347
ochrothicta Meyrick, Stenoma, 21-1347
ochrotoma (Meyrick), Carticeja, 29-59

octacentra Meyrick, Stenoma, 21-1169
oculosa Duckworth, Lethata, 21-1058
oecophila (Staudinger), Oecia, 23-41
oenoes Meyrick, Phelotropa, 21-1086
oenotheraeella (Chambers), Mompha, 25-8
ogmodes Meyrick, Clinograptis, 15-235
ogmolopha (Meyrick), Antaeotricha, 21-696
ogmosaris (Meyrick), Antaeotricha, 21-697
oleagina Zeller, Cnissostages, 17-4
olga Meyrick, Zelosyne, 29-828
oligarcha (Meyrick), Gonionota, 21-169
oligastra Meyrick, Glyphipterix, 34-52
olivescens (Pfitzner), Pfitzneriana, 5-59
olivescens (Pfitzner), Pfitzneriana, 5-59a
olyranta Meyrick, Ussara, 34-30
olyritis (Meyrick), Stigmella, 7-12
omagua (Pfitzner), Aepytus, 5-104
omega Powell, Ethmia, 21-372
ommatopa (Meyrick), Chlamydastis, 21-964
omphacopa (Meyrick), Cerconota, 21-855
omphalopa (Meyrick), Evippe, 29-99
oncotera (Walsingham), Eunebristis, 29-606
onychias Meyrick, Lindera, 15-292
operculella (Zeller), Phthorimaea, 29-185
operosella Zeller, Setomorpha, 15-359
ophiaula Meyrick, Gelechia, 29-139
ophiomorpha Meyrick, Gelechia, 29-140
ophiopa (Meyrick), Chlamydastis, 21-965
ophitis Walsingham, Simoneura, 29-808
ophrysta Meyrick, Antaeotricha, 21-698
oppidana Meyrick, Prochola, 26-29
opsonoma (Meyrick), Trichotaphe, 29-668
opticosema Meyrick, Urodus, 36-68
optima Duckworth, Lethata, 21-1059
opulenta Walsingham, Pseudastasia, 40-35
orasialis Walsingham, Acrolophus, 15-120
orasiusalis (Walker), Acrolophus, 15-120
ordinata Walsingham, Comodica, 15-238
oreas (Schaus), Aepytus, 5-85
orgadopa (Meyrick), Antaeotricha, 21-699
orgilopis Meyrick, Telphusa, 29-292
oriarcha (Meyrick), Caloptilia, 20-10
oribatis Meyrick, Aristotelia, 29-29
originalis Meyrick, Protonyctia, 38-1
orion (Busck), Chlamydastis, 21-966
oriphanta (Meyrick), Gonionota, 21-170
orizabae (Dyar), Acrolophus, 15-121
orizavae Schaus, Oiketicus, 16-57
ornata Busck, Ethmia, 21-380b
ornata (Walsingham), Acrolophus, 15-122
ornata Walsingham, Eucosmophora, 20-20
ornatipalpella (Walsingham), Recurvaria, 29-211
orneopis Meyrick, Stenoma, 21-1348
orphnaea (Meyrick), Afdera, 21-10
orphnopa Meyrick, Prochola, 26-30
orphnopis Meyrick, Doliotechna, 21-1557
orthobasis Meyrick, Prochola, 26-31
orthocampta Meyrick, Battaristis, 29-353
orthocapna Meyrick, Stenoma, 21-1349
orthochaeta (Meyrick), Hypercallia, 21-224
orthoctenis Meyrick, Glyphidocera, 23-23
orthodelta Meyrick, Antispila, 14-4
orthodeta Meyrick, Glyphipterix, 34-53
orthodoxa (Meyrick), Inga, 21-1640
orthographa Meyrick, Stenoma, 21-1350

ortholampra Meyrick, Stenoma, 21-1351
orthopa Meyrick, Stenoma, 21-1467
orthophaea Meyrick, Antaeotricha, 21-700
orthophracta (Meyrick), Compsolechia, 29-425
orthophragma (Meyrick), Inga, 21-1641
orthophrontis Meyrick, Holcocera, 23-95
orthotenes Meyrick, Glyphidocera, 23-24
orthotona Meyrick, Antaeotricha, 21-701
orthozona Meyrick, Baeonoma, 21-837
orthridia (Meyrick), Cerconota, 21-878
orthriopa Meyrick, Antaeotricha, 21-702
orthroptila Meyrick, Stenoma, 21-1396
ostariella (Walsingham), Recurvaria, 29-212
osteacma Meyrick, Orthenches, 35-8
ostensella (Walker), Cymotricha, 29-547
ostiaria Meyrick, Tinea, 15-401
ostodes (Walsingham), Antaeotricha, 21-703
oterosella Busck, Ethmia, 21-373
otiosa Walsingham, Eritarbes, 27-73
ovata (Zeller), Urodus, 36-69
ovatella (Walker), Stenoma, 21-1352
oviformis Köhler, Oiketicus, 16-48
ovulifera (Meyrick), Antaeotricha, 21-704
oxinopa Meyrick, Eupragia, 21-91
oxybela Clarke, Gonionota, 21-171
oxybela Meyrick, Machimia, 21-245
oxycentra Meyrick, Antaeotricha, 21-705
oxydecta (Meyrick), Antaeotricha, 21-706
oxyglypta (Meyrick), Glyphipterix, 34-54
oxygrapta (Meyrick), Phyllonorycter, 20-129
oxyloba (Meyrick), Pachygeneia, 29-789
oxymora Meyrick, Tinea, 15-402
oxyplaca (Meyrick), Chlamydastis, 21-967
oxyschista (Meyrick), Antaeotricha, 21-707
oxyscia Meyrick, Stenoma, 21-1353
oxytypa Meyrick, Sclerograptis, 29-807
pacalis Meyrick, Blastobasis, 23-79
pacatum Walsingham, Gonioterma, 21-1028
pachybathra (Meyrick), Inga, 21-1642
pachynta Meyrick, Acrolophus, 15-123
pactota Meyrick, Antaeotricha, 21-708
paedisca Walsingham, Hybroma, 15-276
paenitens Meyrick, Pachysaris, 29-637
pagana (Meyrick), Inga, 21-1643
pagella Hodges, Lita, 29-173
pagidotis (Meyrick), Inga, 21-1644
pala Powell, Ethmia, 21-374
palaestrica (Butler), Trichotaphe, 15-441
palaestrias Meyrick, Antaeotricha, 21-709
palearis (Meyrick), Sitotroga, 29-513
palirrhoa Meyrick, Commatica, 29-741
pallens (Blanchard), Dalaca, 5-3
pallescentella Stainton, Tinea, 15-403
palliata (Walsingham), Cerconota, 21-887
pallicosta (R. Felder & Rogenhofer), Antaeotricha, 21-710
pallida Walsingham, Pammeces, 26-14
pallidicostella (Walsingham), Urodus, 36-70
pallidorsella Zeller, Tinea, 15-404
pallidovenata Grossbeck, Thyridopteryx, 16-60
pallidus Möschler, Acrolophus, 15-124
pallulella (Busck), Antaeotricha, 21-813
palmar Viette, Phialuse, 5-113
palpalis (Zeller), Timocratica, 21-1501
palpella Forbes, Infurcitinea, 15-280

palpella (Walsingham), Cronicombra, 34-10
palpiannulella (Chambers), Monochroa, 29-3
palpigera (Walsingham), Brachyacma, 29-524
panamae Busck, Acrolophus, 15-125
pancrypta (Meyrick), Ascalenia, 27-67
pandora Meyrick, Plocamosaris, 29-641
pannephela Meyrick, Acrolophus, 15-126
panolbia (Walsingham), Alucita, 31-13
panscia Meyrick, Choropleca, 15-227
pantalaena (Walsingham), Aristotelia, 29-30
pantogenes Meyrick, Stenoma, 21-1354
paphia Meyrick, Aristotelia, 29-31
paphlactis (Meyrick), Friseria, 29-111
papiella Powell, Ethmia, 21-375
paracapna Meyrick, Stenoma, 21-1355
paracrypta Meyrick, Antaeotricha, 21-711
paracta (Meyrick), Antaeotricha, 21-712
paradisea Walsingham, Glyphipterix, 34-55
paradoxica (Chambers), Prodoxus, 13-4
paradromis (Meyrick), Chlamydastis, 21-968
paragrapta (Meyrick), Cuphodes, 20-110
paralagneia Clarke, Doina, 21-78
paramochla Meyrick, Stenoma, 21-1314
paraplecta Meyrick, Stenoma, 21-1356
paraplutella (Busck), Aroga, 29-52
parastis Gyen, Antaeotricha, 21-713
paratma (Meyrick), Moca, 41-10
paravexillata Clarke, Gonionota, 21-172
parazela Meyrick, Battaristis, 29-354
parcens Meyrick, Basanasca, 15-217
pardalodes Meyrick, Stenoma, 21-1357
pardella Busck, Moriloma, 27-37
parephoria Clarke, Aristotelia, 29-32
parishi Gaedike, Parochromolopis, 33-1
paritor Hodges, Batrachedra, 24-28
parmata Meyrick, Compsolechia, 29-426
parmifera Meyrick, Comotechna, 21-22
parmulata Meyrick, Commatica, 29-742
paromias (Meyrick), Opostega, 8-9
paropta Meyrick, Stenoma, 21-1358
paropus (Druce), Aepytus, 5-70
particeps Meyrick, Acrolophus, 15-127
particularis (Zeller), Antaeotricha, 21-714
parvella (Fabricius), Taygete, 29-278
parvifusca Becker, Timocratica, 21-1502
parviguttata (Bryk), Dalaca, 5-3
parvileuca Becker, Timocratica, 21-1503
parvipulvella (Chambers), Brachmia, 29-520
parvus (Walsingham), Acrolophus, 15-128
passiflorae Clarke, Odonna, 21-1689
pastranai (Bourquin), Caloptilia, 20-11
pastulella (Fabricius), Atteva, 36-8
patagonica Povolný, Scrobipalpula, 29-245
patellifera (Meyrick), Zetesima, 21-1513
patens Meyrick, Stenoma, 21-1359
paterata Meyrick, Aristotelia, 29-33
pateropa Meyrick, Cryptolechia, 21-59
patria (Meyrick), Alucita, 31-14
patriciae Nielsen & Robinson, Dalaca, 5-6
patula Meyrick, Stenoma, 21-1342
paucella (Walker), Ethmia, 21-376
pauculella (Walker), Amydria, 15-209
pauper Walsingham, Acrolophus, 15-129
pauperatella (Walker), Gonioterma, 21-1022
paurocentra (Meyrick), Anadasmus, 21-436

pauroconis (Meyrick), Antaeotricha, 21-461
peccans (Butler), Stenoma, 21-1360
pecten Clarke, Coptotelia, 21-43
pectinella (Forbes), Cymotricha, 29-548
peculella (Busck), Compsolechia, 29-427
pedipogon Strand, Phassus, 5-36
pegaea Meyrick, Hybroma, 15-277
pelasta (Meyrick), Ithome, 27-81
pelinitis (Meyrick), Anadasmus, 21-437
pelinoma Meyrick, Opogona, 15-317
pelinopis Meyrick, Coleophora, 24-9
pellocoma (Meyrick), Antaeotricha, 21-715
pelodes (Walsingham), Anadasmus, 21-438
pelomacta (Meyrick), Moca, 41-11
peloptila (Meyrick), Anacampsis, 29-326
penai Davis & Nielsen, Apoplania, 3-2
penetrans Meyrick, Recurvaria, 29-213
penetratrix Meyrick, Telphusa, 29-293
penicillata (Walsingham), Aristotelia, 29-34
pentachorda Meyrick, Cosmopterix, 27-21
pentadora (Meyrick), Chionodes, 29-78
pentagramma Meyrick, Calliprora, 29-64
pentalitha Meyrick, Antispila, 14-5
pentapyrga (Meyrick), Walshia, 27-109
pentastra Meyrick, Compsolechia, 29-428
pentathlopa Meyrick, Cryptolechia, 21-60
penthica Walsingham, Ethmia, 21-377
penumbra Walsingham, Acrolophus, 15-130
peperita (Walsingham), Machimia, 21-246
peplophanes Meyrick, Calliathla, 35-1
peragrapta Meyrick, Crasimorpha, 29-501
perceptella (Busck), Teuchophanes, 29-648
percnacma Meyrick, Vazugada, 29-685
percnocarpa (Meyrick), Antaeotricha, 27-716
percnocoma Meyrick, Cryptolechia, 21-61
percnogona Meyrick, Antaeotricha, 21-717
percnoleuca Meyrick, Glyphidocera, 23-25
percnopholis Walsingham, Dichomeris, 29-590
percnorma (Meyrick), Inga, 21-1645
percnoscia Meyrick, Holcocera, 23-96
percnospila (Meyrick), Compsolechia, 29-429
percnotoxa Meyrick, Atelosticha, 21-1529
percussella Zeller, Argyresthia, 37-7
perdigna Walsingham, Opostega, 8-10
perducta (Meyrick), Chlamydastis, 21-969
perfidiosa (Meyrick), Gnorimoschema, 29-164
perforata Nielsen & Robinson, Callipielus, 5-14
perfossa Meyrick, Aristotelia, 29-35
perfracta Meyrick, Glyphipterix, 34-56
perfusa Meyrick, Antaeotricha, 21-579
perianthes Meyrick, Machimia, 21-247
periapta Walsingham, Mothonica, 21-1075
periaula Meyrick, Stenoma, 21-1361
perichalca Meyrick, Heliodines, 40-22
pericyclota (Meyrick), Inga, 21-1646
peridesma Meyrick, Stenoma, 21-1362
perinaeta (Walsingham), Battaristis, 29-355
perioditis (Meyrick), Inga, 21-1647
periphereia Clarke, Gonionota, 21-173
periphrictis (Meyrick), Antaeotricha, 21-718
perirrhoa Meyrick, Stenoma, 21-1363
periscelta (Meyrick), Gonioterma, 21-1029
perischias Meyrick, Urodus, 36-71
perisepta Meyrick, Tinea, 15-405
perissarcha Meyrick, Acrolophus, 15-131

perissosema (Meyrick), Chionodes, 29-79
peritheta Walsingham, Penica, 20-38
peritura Meyrick, Calliprora, 29-65
perjecta Meyrick, Stenoma, 21-1364
perjura (Meyrick), Anadasmus, 21-421
perlatella (Walker), Compsolechia, 29-430
permixtella Walsingham, Dialectica, 20-42
permota Meyrick, Mompha, 25-19
permundella (Walker), Cymotricha, 29-549
pernigrella (Forbes), Ithome, 27-82
pernota (Walsingham), Mompha, 25-20
perobscura Walsingham, Glyphidocera, 23-26
perobtusa (Meyrick), Ifeda, 21-1582
peronacma Meyrick, Otochares, 15-328
peronia Busck, Stenoma, 21-1365
perophora Meyrick, Stenoma, 21-1291
perpetua Meyrick, Acrolophus, 15-132
perplexa Clarke, Aristotelia, 29-36
perpulchra Walsingham, Ethmia, 21-378
perquisita Meyrick, Anacampsis, 29-327
perrensella (Walsingham), Acrolophus, 15-133
perrensi (Druce), Acrolophus, 15-134
perseae (Busck), Caloptilia, 20-12
perseaphaga Clarke, Coptotelia, 21-44
persimilella Walsingham, Auximobasis, 23-57
persistis (Meyrick), Gonionota, 21-174
persita Meyrick, Stenoma, 21-1366
perspicilla (Stoll), Arrhenophanes, 17-1
perspicua (Walsingham), Telphusa, 29-294
pertinax Meyrick, Stenoma, 21-1367
pertinens Meyrick, Gelechia, 29-141
perturbata Meyrick, Acrocercops, 20-94
peruviella Busck, Filinota, 21-96
pervallata Meyrick, Prochola, 26-32
petasodes (Meyrick), Inga, 21-1648
petraea Walsingham, Gelechia, 29-142
petrina Walsingham, Stenoma, 21-1203
petrographa Meyrick, Anacampsis, 29-328
petromorpha Meyrick, Compsolechia, 29-431
petropolisiensis Viette, Aepytus, 5-77
petulans Meyrick, Hapalosaris, 29-167
pexa Meyrick, Opostega, 8-11
phaeobathra (Meyrick), Cronicombra, 34-11
phaeoceros Meyrick, Ceromitia, 12-9
phaeocrossa (Meyrick), Inga, 21-1649
phaeomalla Meyrick, Acrolophus, 15-135
phaeomystis Meyrick, Stenoma, 21-1368
phaeonephela Meyrick, Tinea, 15-406
phaeoneura (Meyrick), Antaeotricha, 21-719
phaeopa Meyrick, Stomopteryx, 29-258
phaeophanes (Meyrick), Cerconota, 21-888
phaeoplintha (Meyrick), Antaeotricha, 21-720
phaeoptera Forbes, Stilbosis, 27-103
phaeosaris Meyrick, Antaeotricha, 21-721
phaeospila Meyrick, Syncraternis, 15-361
phaeostrota Meyrick, Rhynchotona, 29-805
phaeotoxa Meyrick, Compsolechia, 29-432
phalacra (Walsingham), Stegasta, 29-254
phalacropa Meyrick, Stenoma, 21-1369
phalaropis Meyrick, Mompha, 25-21
phalerus Druce, Phassus, 5-39
phaneropis (Meyrick), Marmara, 20-119
phanerozona Meyrick, Empedaula, 29-762
phanocrossa Meyrick, Commatica, 29-743
phantasmella Walsingham, Leucophasma, 15-290

phaobregna Clarke, Doina, 21-79
pharus (Druce), Phassus, 5-44
phaselodes (Meyrick), Antaeotricha, 21-722
phaseolodes (Busck), Antaeotricha, 21-722
phaula (Walsingham), Antaeotricha, 21-723
phepsalitis Meyrick, Compsolechia, 29-433
phiaropis (Meyrick), Caloptilia, 20-13
philantha Meyrick, Triclonella, 27-56
philiponi Viette, Aepytus, 5-103
phillita Clarke, Profilinota, 21-271
philochrysa Meyrick, Macarocosma, 21-1686
philomela (Meyrick), Timocratica, 21-1504
phlebotes (Walsingham), Hypercallia, 21-225
phlogophora Walsingham, Pammeces, 26-15
phlyctaenopa Meyrick, Rhopalosetia, 30-10
phococara Clarke, Dita, 21-1551
phocodes Meyrick, Gonionota, 21-175
phoebe (Busck), Antaeotricha, 21-748
phoenicura Meyrick, Ethmia, 21-379
phoenissa (Butler), Pachyphoenix, 21-1691
phoenogramma Meyrick, Vazugada, 29-686
phollicodes (Meyrick), Antaeotricha, 21-724
phormophora (Meyrick), Erysiptila, 21-308
phortax Meyrick, Gonioterma, 21-1030
phosphorodes Meyrick, Gonada, 21-109
phosphoropa (Meyrick), Beltheca, 29-706
phryactis Meyrick, Antaeotricha, 21-725
phthiochroma Clarke, Gonionota, 21-176
phthorosema (Meyrick), Moca, 41-12
phycitana Walsingham, Carposina, 32-2
phylacis Walsingham, Ethmia, 21-380
phylacis Walsingham, Ethmia, 21-380a
phylacops Powell, Ethmia, 21-381
phyllocosma Meyrick, Stenoma, 21-1279
phylloxantha Meyrick, Stenoma, 21-1370
physotricha (Meyrick), Cerconota, 21-860
phytomiella Busck, Anacampsis, 29-329
phytoptera (Busck), Chlamydastis, 21-970
pialea (Meyrick), Gonionota, 21-177
picolella Busck, Beltheca, 29-707
picrantis Meyrick, Stenoma, 21-1371
picrogramma Meyrick, Gelechia, 29-143
picta (Zeller), Stenoma, 21-1372
picticollis (Walsingham), Iphimachaera, 29-776
picticornis Walsingham, Coleophora, 24-10
picticornis (Walsingham), Compsolechia, 29-434
pictoria Meyrick, Triclonella, 27-57
pictrix Meyrick, Psittacastis, 21-290
pictus Walsingham, Phyllonorycter, 20-130
picula Walsingham, Recurvaria, 29-214
pignerata (Meyrick), Ithome, 27-83
piligera Meyrick, Acrocercops, 20-95
pinae Amsel, Holcocera, 23-97
pinnifera Meyrick, Acrolophus, 15-136
piperata (Walsingham), Dichomeris, 29-591
piperatella Busck, Plumana, 16-7
piperatella (Walsingham), Mompha, 25-22
piperella Powell, Ethmia, 21-382
pircuniae (Zeller), Tiquadra, 15-434
pisoniae Busck, Nealyda, 29-8
pisoniella Busck, Scelorthus, 40-37
pistopis Meyrick, Zelleria, 36-90
pithecolobiella Busck, Neurostrota, 20-22
pittionii (Viette), Aepytus, 5-88
pizote (Schaus), Cryptothelea, 16-24

placoterma Meyrick, Commatica, 29-744
plaesiosema (Turner), Symmetrischema, 29-266
plagifera Meyrick, Tischeria, 9-6
plagosa (Zeller), Stenoma, 21-1373
planicoma (Meyrick), Antaeotricha, 21-726
planistes Forbes, Aphanosara, 27-5
platensis Berg, Oiketicus, 16-58
platiastis Meyrick, Compsolechia, 29-435
platyaula Meyrick, Crambodoxa, 29-752
platycolpa (Meyrick), Gonioterma, 21-1033
platydesma Meyrick, Antaeotricha, 21-727
platydoxa Meyrick, Gelechia, 29-144
platyochra Meyrick, Glyphipterix, 34-57
platyphylla Meyrick, Stenoma, 21-1374
platysaris Meyrick, Tinea, 15-407
platysoma (Walsingham), Taygete, 29-279
platyspora (Meyrick), Chlamydastis, 21-971
platyterma Meyrick, Stenoma, 21-1375
platyxantha (Meyrick), Triclonella, 27-58
platyxipha Meyrick, Calliprora, 29-66
plaumanni Powell, Ethmia, 21-383
playa Powell, Ethmia, 21-384
plebicola (Meyrick), Anadasmus, 21-439
plectanopa Meyrick, Colonanthes, 29-726
plectanota (Meyrick), Inga, 21-1650
plejadella (R. Felder & Rogenhofer), Compsolechia, 29-451
plenella (Busck), Cotaena, 34-2
pleonastes (Meyrick), Gonioterma, 21-1031
pleroma Walsingham, Gelechia, 29-145
plerotis Meyrick, Antaeotricha, 21-728
plesistia (Meyrick), Antaeotricha, 21-729
pleurosaris Meyrick, Recurvaria, 29-215
pleurotella Walsingham, Aerotypia, 29-692
pleurotricha (Meyrick), Anadasmus, 21-435
plexigrmma Meyrick, Dichomeris, 29-592
pleximorpha (Meyrick), Anadasmus, 21-443
plicata Walsingham, Ptilostonychia, 29-804
plocogastra Meyrick, Scythris, 28-10
plocogramma (Meyrick), Chlamydastis, 21-972
plumata (Meyrick), Evippe, 29-100
plumbata (Meyrick), Ascalenia, 27-68
plumbeella (Beutenmüller), Ethmia, 21-391
plumbeolata (Walsingham), Compsolechia, 29-436
plumella (Walsingham), Protodarcia, 15-346
plumosa (Busck), Antaeotricha, 21-730
pluriargenteus (Viette), Aepytus, 5-95
plurima Walsingham, Stenoma, 21-1376
plusia (Herrich-Schäffer), Phassus, 5-47
plutella (Chambers), Evippe, 29-98
pneumatica (Meyrick), Caloptilia, 20-14
poecilia Clarke, Gonionota, 21-178
poecilosoma Walsingham, Zelosyne, 29-829
poecilta Walsingham, Choropleca, 15-228
poeyi Lucas, Oiketicus, 16-57
poeyi Walsingham, Acrolophus, 15-137
pogonites Walsingham, Amydria, 15-210
polingi (Barnes & Benjamin), Astala, 16-36
poliombra Meyrick, Anacampsis, 29-330
poliopa (Meyrick), Chlamydastis, 21-973
poliopis (Meyrick), Acrolepiopsis, 39-13
politica Meyrick, Arotromima, 29-705
pollenifera Davis, Parategeticula, 13-3
poltrona (Schaus), Aepytus, 5-97
polychroma Meyrick, Erithyma, 21-90

polydora Meyrick, Phanerozela, 14-14
polyglypta (Meyrick), Antaeotricha, 21-731
polygoni (Zeller), Mompha, 25-23
pomaceella (Walker), Anacampsis, 29-331
pometella (Harris), Dichomeris, 29-586
pompeiana Meyrick, Timocratica, 21-1505
ponderata (Meyrick), Chlamydastis, 21-936
pontifex Meyrick, Opostega, 8-12
pontifica Forbes, Acrocercops, 20-96
pontifica Meyrick, Tiquadra, 15-435
popeanellus (Clemens), Acrolophus, 15-138
popperi (Pfitzner), Parapielus, 5-25
porinodes Meyrick, Stenoma, 21-1421
porphyraspis (Meyrick), Charistica, 29-720
porphyrastis Meyrick, Stenoma, 21-1377
porphyrina Meyrick, Urodus, 36-72
porphyrogramma Meyrick, Trichotaphe, 29-669
porphyropis Meyrick, Copocentra, 40-6
porphyrospila (Meyrick), Cronicombra, 34-12
porpotis (Meyrick), Inga, 21-1651
porrectella (Linnaeus), Plutella, 35-16
portentosa Busck, Zetesima, 21-1514
postpallescens (Walsingham), Stegasta, 29-255
postscripta Meyrick, Antispila, 14-6
postvariabilis Nielsen & Robinson, Dalaca, 5-9
potosi Busck, Metopleura, 29-783
practica Meyrick, Acrolophus, 15-139
praecauta Meyrick, Stenoma, 21-1378
praeceps (Meyrick), Anadasmus, 21-422
praecincta Meyrick, Antispila, 14-7
praecisa Meyrick, Antaeotricha, 21-732
praeclivis (Meyrick), Gonionota, 21-179
praediata Meyrick, Ascalenia, 27-69
praefinita (Meyrick), Telphusa, 29-282
praefulgens Meyrick, Lamprozela, 14-12
praenivea (Meyrick), Compsolechia, 29-437
praenubila (Meyrick), Chlamydastis, 21-974
praerupta Meyrick, Antaeotricha, 21-733
praesignis (Meyrick), Semophylax, 29-512
praestabilis Meyrick, Tinea, 15-408
praestans (Walsingham), Opogona, 15-318
praesul Meyrick, Borkhausenia, 21-1538
praetexta Clarke, Darlia, 29-753
praetextata (Meyrick), Urodus, 36-73
praetusalis Walsingham, Acrolophus, 15-140
praeumbrata (Meyrick), Niditinea, 15-307
praevecta Meyrick, Cryptolechia, 21-62
prasina Meyrick, Anacampsis, 29-332
prasinantha Meyrick, Porpodryas, 29-510
prasinaula (Meyrick), Acrolepiopsis, 39-14
prasinoptila Meyrick, Alsodryas, 29-697
prasinospora Meyrick, Imma, 41-31
prasoleuca (Meyrick), Cerconota, 21-870
prasophanes Meyrick, Prochola, 26-33
pratifera (Meyrick), Antaeotricha, 21-734
prattiella Busck, Ethmia, 21-385
prensans Meyrick, Dichomeris, 29-593
prensoria Meyrick, Tinea, 15-409
prepodes Walsingham, Acrolophus, 15-141
pretiosus (Herrich-Schäffer), Phassus, 5-47
pretusalis (Walker), Acrolophus, 15-140
primigenia Meyrick, Anacampsis, 29-333
princeps (Busck), Lita, 29-174
priscella Busck, Tocasta, 24-13
prismatica Walsingham, Lipomerinx, 15-295

prismatopa Meyrick, Battaristis, 29-356
pristinella (Walker), Acrolophus, 15-142
proagorella (Zeller), Holcocera, 23-98
probata Walsingham, Dysoptus, 15-252
problema Walsingham, Pammeces, 26-16
probolopis Meyrick, Aristotelia, 29-37
procritica Meyrick, Stenoma, 21-1379
procyphodes (Meyrick), Cymotricha, 29-550
producta (Walsingham), Tricherodes, 21-306
projecta (Meyrick), Gonioterma, 21-1032
prolectans (Meyrick), Gonionota, 21-180
prolixa Meyrick, Auximobasis, 23-58
promeces Walsingham, Ectaga, 21-88
prometopias (Gyen), Monopis, 15-300
prominula Meyrick, Coptotelia, 21-42
promotella Zeller, Stenoma, 21-1380
propriella (Walker), Psittacastis, 21-291
propriella Zeller, Psilocorsis, 21-279
proseni (Pastrana), Alucita, 31-15
prosopus (Druce), Pfitzneriana, 5-62
prosora (Walsingham), Antaeotricha, 21-735
prostylias Meyrick, Cranaodes, 15-240
protocrossa Meyrick, Imma, 41-32
protomochla Meyrick, Opostega, 8-13
protonistis Meyrick, Episyrta, 15-257
protosaris Meyrick, Antaeotricha, 21-736
protozona Meyrick, Gelechia, 29-146
proximella Busck, Ethmia, 21-386
prozona Meyrick, Dissoptila, 29-759
prudentula (Meyrick), Chlamydastis, 21-975
pruniramiella Clemens, Xylesthia, 15-445
prytanes (Schaus), Aepytus, 5-100
psalmographa Meyrick, Stenoma, 21-1381
psammolitha Meyrick, Glyphidocera, 23-27
psammophila Meyrick, Acrolophus, 15-143
pselaphotis Meyrick, Parectopa, 20-29
psephophragma Meyrick, Athrinacia, 21-14
pseudacma (Meyrick), Cerconota, 21-891
pseuderiocrania Kristensen & Nielsen, Heterobathmia,
    2-2
pseudochyta Meyrick, Antaeotricha, 21-737
pseudodimiata (Paclt), Dalaca, 5-3
pseudohirsutus Hasbrouck, Acrolophus, 15-144
pseudonoma Meyrick, Acrolophus, 15-145
pseudophile (Dyar), Lactura, 36-17
pseudospretella (Stainton), Hofmannophila, 21-1580
pseudostoma Meyrick, Glyphipterix, 34-58
psidiella Busck, Chilocampyla, 20-41
psidii (Sepp), Lethata, 21-1060
psilodoxa Meyrick, Sorotacta, 29-811
psilomorpha Meyrick, Stenoma, 21-1382
psittacanthus Heppner, Parochromolopis, 33-2
psittacopa (Meyrick), Hypercallia, 21-226
psoloessa Meyrick, Acrolophus, 15-146
psoricopterella (Walsingham), Acompsia, 29-516
pterodactylella Walker, Zaratha, 26-42
pterygota Meyrick, Commatica, 29-745
ptilallactis (Meyrick), Antaeotricha, 21-729
ptilocompa (Meyrick), Cymotricha, 29-551
ptilocrates Meyrick, Antaeotricha, 21-569
ptilopa (Meyrick), Chlamydastis, 21-976
ptilosema Meyrick, Cerconota, 21-889
ptilostoma Meyrick, Glyphidocera, 23-28
ptisanopa Meyrick, Syncerastis, 36-31
ptochogramma Meyrick, Compsolechia, 29-438

ptychobathra Meyrick, Stenoma, 21-1383
ptychocentra Meyrick, Stenoma, 21-1384
ptychocryptis Meyrick, Glyphidocera, 23-29
ptychophthalma Meyrick, Stenoma, 21-1385
ptychoptila Meyrick, Atopotorna, 21-1531
ptycta (Walsingham), Antaeotricha, 21-738
pucaraensis Pastrana, Ceromitia, 12-10
pudibundella.- (Chambers), Aristotelia, 29-42
pudibundella (Zeller), Aristotelia, 29-38
pugionaria Meyrick, Holcocera, 23-99
pugnax Walsingham, Durrantia, 21-411
pulchricornis Walsingham, Coleophora, 24-11
pulicella Walsingham, Aristotelia, 29-39
pulla (Esper), Epichnopterix, 16-11
pulverea Walsingham, Tischeria, 9-7
pulverella (Walsingham), Parectopa, 20-30
pulvinata Meyrick, Urodus, 36-74
pumicea Walsingham, Acrolophus, 15-147
pumila Walsingham, Opostega, 8-15
pumiliella (Walsingham), Oenoe, 15-310
pumilis (Busck), Antaeotricha, 21-739
punctata (Druce), Acrolophus, 15-148
punctatella (Walker), Dichomeris, 29-594
punctella (Cramer), Atteva, 36-8
punctessa Powell, Ethmia, 21-387
punctiferella (Busck), Obithome, 27-85
punctiferella Walker, Alucita, 31-16
punctipennella Clemens, Anorthosia, 29-519
pungens Meyrick, Compsolechia, 29-439
punicea (Meyrick), Cerconota, 21-843
purpurascens Forbes, Perimede, 27-89
purpurascens Walsingham, Psilocorsis, 21-280
purulenta Zeller, Antaeotricha, 21-740
pusilla Dietz, Holcocera, 23-100
pusilla (Zeller), Acrolophus, 15-149
pustulatella (Walker), Stenoma, 21-1386
pustulella (Fabricius), Atteva, 36-8
putella Busck, Recurvaria, 29-216
pygmaea Busck, Urangela, 27-66
pygmaea (Walsingham), Acrolophus, 15-150
pylonias Meyrick, Acrocercops, 20-97
pyramellus (Barnes & McDunnough), Acrolophus, 15-151
pyramidea Walsingham, Stenoma, 21-1387
pyrenodes (Meyrick), Lethata, 21-1061
pyretodes Meyrick, Semiomeris, 29-645
pyrgota (Meyrick), Antaeotricha, 21-741
pyrobathra (Meyrick), Antaeotricha, 21-742
pyrocalyx Meyrick, Machimia, 21-248
pyrocausta (Meyrick), Gonionota, 21-181
pyrodercia Walsingham, Aristotelia, 29-40
pyrographa Walsingham, Opogona, 15-319
pyrograpta Meyrick, Machimia, 21-249
pyromantis Meyrick, Tinaegeria, 21-1731
pyronota Meyrick, Gonada, 21-110
pyropis (Meyrick), Plocamosaris, 29-642
pyrothyris (Meyrick), Inga, 21-1652
pyrozela Meyrick, Cosmopterix, 27-22
pyrrhias Meyrick, Stenoma, 21-1388
pyrrhonota Meyrick, Stenoma, 21-1389
pyrrhophthalma Meyrick, Phytomimia, 21-269
pyrrhopis Meyrick, Prophoraula, 29-644
pyrrhotrota (Meyrick), Gonionota, 21-182
pyrrhoxantha (Meyrick), Inga, 21-1653
pyrsophanes Meyrick, Psittacastis, 21-292
pythonaea Meyrick, Antaeotricha, 21-743

quadratella (Walker), Anadasmus, 21-440
quadricornis Nielsen & Robinson, Dalaca, 5-4
quadridentata Davis, Lumacra, 16-30
quadrifascia (Walker), Compsolechia, 29-440
quadrimaculata Walsingham, Holophycis, 29-483
quadrimaculella (Chambers), Brymblia, 21-1541
quadristrigella (Zeller), Parectopa, 20-31
quadrivittana (Walker), Imma, 41-33
quatiens Meyrick, Antaeotricha, 21-664
quercipominella (Chambers), Dichomeris, 29-586
quiescens (Meyrick), Antaeotricha, 21-744
quinquedentata (Walsingham), Telphusa, 29-295
quinqueguttata Walsingham, Heliodines, 40-23
quinquepunctata (Forbes), Ithome, 27-78
quinquepunctella Walsingham, Anacampsis, 29-334
quinquepunctellus (Chambers), Prodoxus, 13-4
quinquicristatella (Chambers), Stilbosis, 27-105
radicalis (Zeller), Antaeotricha, 21-745
radicata Meyrick, Aristotelia, 29-35
radicicola Meyrick, Antaeotricha, 21-746
ramigera Meyrick, Acrocercops, 20-98
raptans (Meyrick), Palinorsa, 21-262
raricilia Meyrick, Antaeotricha, 21-808
rastricornis Meyrick, Acrolophus, 15-152
rebeli (Köhler), Curtorama, 16-33
rebellis Meyrick, Leistogenes, 29-778
recens Hodges, Lita, 29-175
receptella (Walker), Stenoma, 21-1390
reciprocella (Walker), Antaeotricha, 21-747
recondita Meyrick, Stenoma, 21-1391
recta Meyrick, Compsolechia, 29-441
recticostella Walsingham, Glyphidocera, 23-30
rectificata Meyrick, Stenoma, 21-1181
rectistrigella (Barnes & Busck), Lita, 29-176
rectivittella Zeller, Plutella, 35-17
recurrens (Meyrick), Cerconota, 21-890
recurvella (Walker), Cerconota, 21-891
redintegrata Meyrick, Stenoma, 21-1392
reductella (Walker), Antaeotricha, 21-748
redundans Walsingham, Phytomimia, 21-267
reduplicata (Walsingham), Lepyrotica, 15-287
reedella (Chambers), Dichomeris, 29-586
reedi (Ureta), Parapielus, 5-28
reflexa (Fabricius), Acrolophus, 15-153
refracta (Meyrick), Compsolechia, 29-442
refractella Zeller, Glyphipterix, 34-59
refractrix Meyrick, Antaeotricha, 21-749
refuga (Meyrick), Inga, 21-1654
refulgens Meyrick, Parectopa, 20-32
regesta Meyrick, Stenoma, 21-1393
regia (Meyrick), Satrapodoxa, 29-806
regiella (Busck), Eomichla, 21-1572
regifica Meyrick, Filinota, 21-101
relata Meyrick, Stenoma, 21-1394
relegata (Meyrick), Exaeretia, 21-6
relicta (Meyrick), Inga, 21-1665
religata Meyrick, Compsolechia, 29-443
remivola Meyrick, Trierostola, 15-442
remorsa (Meyrick), Antaeotricha, 21-780
remota (Pfitzner), Pfitzneriella, 5-131
remotella (Staudinger), Cryptolechia, 21-63
renascens (Walsingham), Trichotaphe, 29-670
rendalli Walsingham, Dialectica, 20-43
renitens Meyrick, Myrmecozela, 15-302
renovata (Meyrick), Amiantastis, 16-3

renselariana (Stoll), Antaeotricha, 21-750
repandella (Walker), Compsolechia, 29-444
reparabilis, Walsingham, Glyphidocera, 23-31
repentina (Walsingham), Friseria, 29-112
repertor Hodges, Batrachedra, 24-29
repletana (Walker), Glyphipterix, 34-60
reprehensa Meyrick, Antaeotricha, 21-751
residuella (Zeller), Stenoma, 21-1395
resiliens Meyrick, Antaeotricha, 21-752
resona Meyrick, Exoncotis, 15-262
resurgens Walsingham, Durrantia, 21-412
reticulata Davis, Paucivena, 16-12
retracta Meyrick, Trichotaphe, 29-671
revecta Meyrick, Prochola, 26-34
revelata Meyrick, Ascalenia, 27-70
reversella (Walker), Tiquadra, 15-436
rhabdodes Walsingham, Anacampsis, 29-335
rhabdophora Forbes, Triclonella, 27-59
rhacina Walsingham, Gonionota, 21-183
rhedaria (Meyrick), Chionodes, 29-77
rhicnota Walsingham, Recurvaria, 29-217
rhipidaula (Meyrick), Antaeotricha, 21-753
rhizogramma Meyrick, Ageliarches, 29-694
rhizophorae Walsingham, Coptodisca, 14-9
rhodanopa Meyrick, Sericostola, 34-31
rhodanthes Meyrick, Loxotoma, 21-1067
rhodochra Clarke, Nedenia, 21-257
rhodoclea Meyrick, Nicanthes, 26-9
rhodoclista (Meyrick), Inga, 21-1655
rhodocolpa Meyrick, Stenoma, 21-1396
rhodocosma (Meyrick), Empedaula, 29-763
rhodocrossa (Meyrick), Erythriastis, 29-766
rhodogramma Meyrick, Calliprora, 29-67
rhodograpta Meyrick, Filinota, 21-102
rhodopetala (Meyrick), Charistica, 29-721
rhodosarca (Walsingham), Hypercallia, 21-227
rhodosarea Meyrick, Hypercallia, 21-227
rhodosema (Meyrick), Despina, 21-1548
rhomaeopa (Meyrick), Chlamydastis, 21-977
rhombica Meyrick, Compsolechia, 29-445
rhombifera Meyrick, Choropleca, 15-229
rhombophora Meyrick, Bucculatrix, 19-15
rhombophorella (Zeller), Recurvaria, 29-218
rhothiodes Meyrick, Stenoma, 21-1397
rhynchacma Meyrick, Opogona, 15-320
rhynchograpta Meyrick, Acrocercops, 20-99
rhypara Walsingham, Glyphidocera, 23-32
rhypodes Walsingham, Gelechia, 29-147
rhythmodes Meyrick, Battaristis, 29-357
ribbei Zeller, Antaeotricha, 21-754
ribeella (Zeller), Onebala, 29-628
ridicula Meyrick, Acrolophus, 15-154
ridingsella Clemens, Adela, 12-17
riggii (Orfila), Alucita, 31-17
rileyi (Viette), Aepytus, 5-91
rileyi (Walsingham), Pyroderces, 27-39
ripula Walsingham, Telphusa, 29-296
rita (Busck), Stenoma, 21-1415
rivulella (Möschler), Stegasta, 29-251
rixator Hodges, Batrachedra, 24-30
robertus Busck, Acrolophus, 15-51
robiginosa (Meyrick), Cerconota, 21-892
robustella (Walker), Stegasta, 29-251
rogifera Meyrick, Machimia, 21-250
rosa (Busck), Stenoma, 21-1398

rosacea (Butler), Cerconota, 21-893
rosacea (Forbes), Gonionota, 21-184
rosapicella Busck, Rhindoma, 21-296
roscida (Meyrick), Moca, 41-13
rosea (Meyrick), Inga, 21-1656
roseicorpus (Dognin), Philtronoma, 21-265
roseomarginella (Busck), Inga, 21-1657
roseosuffusella (Clemens), Aristotelia, 29-41
rostella (R. Felder & Rogenhofer), Commatica, 29-736
rostriformis (Meyrick), Antaeotricha, 21-755
rostrigera Meyrick, Elasiprora, 29-761
rosulentus Weymer, Phassus, 5-45
rotans Meyrick, Phyllocnistis, 20-143
rotigera Meyrick, Parectopa, 20-33
rotundata Walsingham, Untomia, 29-499
ruba Duckworth, Lethata, 21-1062
rubella (Blanchard), Heliodines, [20-144] 40-23.1
rubens Meyrick, Gonada, 21-111
rubensella (Chambers), Aristotelia, 29-42
rubentula (Meyrick), Erythriastis, 29-767
rubidella (Clemens), Aristotelia, 29-42
rubiginosella (Walker), Cymotricha, 29-552
rubristricta (Walsingham), Exaeretia, 21-7
ruderella (Chambers), Dichomeris, 29-586
ruderella (R. Felder & Rogenhofer), Tiquadra, 15-421
ruderella Zeller, Setomorpha, 15-359
rufispinis (Meyrick), Chlamydastis, 21-966
rufitecta Meyrick, Molopostola, 29-785
rugosella (Busck), Moca, 41-14
rugosus (Köhler), Curtorama, 16-33
rupestris (Walsingham), Acrolophus, 15-155
rupicella Zeller, Setomorpha, 15-359
ruricola (Meyrick), Inga, 21-1658
rurigena Meyrick, Pachysaris, 29-638
rusposoria Povolný, Keiferia, 29-171
rustica Clarke, Retha, 21-1706
rusticella (Walker), Onebala, 29-629
rusticus (Walsingham), Dichomeris, 29-565
rutella Zeller, Setomorpha, 15-359
rutila (Bartlett-Calvert), Lactura, 36-18
sabinella Forbes, Phaeoses, 15-334
saccharella (Busck), Dicranoctetes, 22-1
sacculata Meyrick, Opostega, 8-15
sachari Busck, Acrolophus, 15-156
sacra (Meyrick), Timocratica, 21-1510a
sacricola (Meyrick), Cerycangela, 29-713
sagaritis Meyrick, Acrolophus, 15-157
sagax Busck, Stenoma, 21-1399
sagifera Meyrick, Glyphidocera, 23-33
sagittella Busck, Dorata, 15-249
saguanmachica (Pfitzner), Aepytus, 5-110
salasi Robinson, Callipielus, 5-13
salebrosa Meyrick, Compsolechia, 29-446
salicis Walsingham, Leucoptera, 19-7
salinae Walsingham, Glyphidocera, 23-34
salome Busck, Stenoma, 21-1400
saltatoria Meyrick, Copocentra, 40-7
saltatrix Walsingham, Opostega, 8-16
salubris Meyrick, Stenoma, 21-1401
salutaris (Butler), Cerconota, 21-900
salva (Meyrick), Chionodes, 29-80
salvini (Druce), Acrolophus, 15-158
sana Meyrick, Antaeotricha, 21-756
sanctaecrucis Walsingham, Dialectica, 20-44
sanctipaulensis (Strand), Urodus, 36-75

sanctivicentii Walsingham), Cosmopterix, 27-23
sandaracota (Meyrick), Charistica, 29-722
sandra Powell, Ethmia, 21-388
sanguinea Butler, Pachyphoenix, 21-1692
sanidopa Meyrick, Cathelotis, 30-1
saphirinella (Chambers), Gnorimoschema, 29-165
sapphiritis Meyrick, Strobisia, 29-490
sarcinata Meyrick, Antaeotricha, 21-757
sarcitella (Linnaeus), Endrosis, 21-1563
sarcitrella (Linnaeus), Endrosis, 21-1563
sarcochlora (Meyrick), Friseria, 29-107
sarcodes Walsingham, Aristotelia, 29-43
sardania Meyrick, Antaeotricha, 21-758
sarissa Clarke, Alynda, 21-1522
sarista Meyrick, Acrolophus, 15-159
sartor Walsingham, Recurvaria, 29-219
satelles Meyrick, Stenoma, 21-1402
satrapis (Meyrick), Gonionota, 21-185
satura (Meyrick), Inga, 21-1659
saturata Meyrick, Dichomeris, 29-595
saturnina Meyrick, Aristotelia, 29-44
satyrisca Meyrick, Acrolophus, 15-160
satyropa (Meyrick), Lethata, 21-1063
saulopis Meyrick, Gonionota, 21-186
saxea Meyrick, Recurvaria, 29-220
scalata (Meyrick), Anacampsis, 29-336
scandalitis Meyrick, Homosetia, 15-269
scapularis (Meyrick), Antaeotricha, 21-759
scardamyctis Meyrick, Lepyrotica, 15-288
scardina (Zeller), Acrolophus, 15-138
scariphista (Meyrick), Doina, 21-80
sceptrifera (Meyrick), Gonioterma, 21-1033
schajovskoii Pastrana, Ceromitia, 12-11
schausi (Viette), Aepytus, 5-89
schausia Busck, Lactura, 36-19
schenoxantha (Schaus), Lactura, 36-20
schistodes Meyrick, Acrolophus, 15-161
scholias Meyrick, Compsolechia, 29-423
schulzella (Fabricius), Heliodines, 40-24
schwarziella Busck, Euclemensia, 27-2
sciactis Meyrick, Steremniodes, 29-813
sciaphilina (Zeller), Cerconota, 21-894
sciastes Walsingham, Dichomeris, 29-596
scintillula (Walsingham), Parelectroides, 29-796
sciocnesta Meyrick, Stenoma, 21-1330
sciocrates (Meyrick), Inga, 21-1660
sciodeta Meyrick, Cryptolechia, 21-64
sciogama Meyrick, Stenoma, 21-1403
sciographa Meyrick, Ceromitia, 12-12
sciomima Meyrick, Compsolechia, 29-447
sciophanes Walsingham, Imma, 41-34
sciophanta Meyrick, Phyllocnistis, 20-145
sciophthalma (Meyrick), Lethata, 21-1064
sciospila (Meyrick), Antaeotricha, 21-760
sciospora (Meyrick), Amiantastis, 16-2
sciotoxa (Meyrick), Inga, 21-1661
scitella (Walker), Compsolechia, 29-448
scitiorella (Walker), Stenoma, 21-1404
sciurella (Walsingham), Thiotricha, 29-304
scoliandra (Meyrick), Zetesima, 21-1515
scolopacina (Walsingham), Cerconota, 21-895
scopodes Meyrick, Acrolophus, 15-162
scopulata (Meyrick), Compsolechia, 29-449
scoriodes (Meyrick), Stenoma, 21-1405
scortea (Meyrick), Anadasmus, 21-438

scotera Walsingham, Acrolophus, 15-163
scoteropis Meyrick, Stegasta, 29-256
scotina (Walsingham), Acrolophus, 15-164
scotocleptes Meyrick, Tinea, 15-410
scotodes Walsingham, Gelechia, 29-148
scrupulata (Meyrick), Acrolophus, 15-165
scrutatricella Zeller, Tinea, 15-411
scutella (Zeller), Compsolechia, 29-450
scutellata (Meyrick), Chlamydastis, 21-978
scutula Powell, Ethmia, 21-389
scutulata Meyrick, Comotechna, 21-23
scythrochalca Meyrick, Urodus, 36-76
scythropa Walsingham, Ethmia, 21-390
scythropiella (Walsingham), Lepyrotica, 15-285
secretella (Walker), Compsolechia, 29-451
sectella (Walker), Compsolechia, 29-366
sectifera Meyrick, Mompha, 25-24
secundata Meyrick, Stenoma, 21-1406
secundella (Walker), Compsolechia, 29-430
sedata (Butler), Phthorimaea, 29-185
seducta (Meyrick), Cerconota, 21-896
seductella (Walker), Compsolechia, 29-452
segmentata (Meyrick), Antaeotricha, 21-761
seitzi (Gaede), Animula, 16-45
sejunctella (Walker), Machimia, 21-251
selectella (Walker), Acrolepiopsis, 39-20
selectella (Walker), Parelectroides, 29-797
selene Clarke, Gonionota, 21-187
sellifera Meyrick, Antaeotricha, 21-762
sematopa Meyrick, Stenoma, 21-1407
semiacma Meyrick, Ethirostoma, 29-769
semialbata Meyrick, Prochola, 26-35
semiberbis Meyrick, Comotechna, 21-24
semibrunnea Dognin, Cryptolechia, 21-65
semicinerea Zeller, Antaeotricha, 21-763
semiclausa (Meyrick), Caloptilia, 20-15
semicretata Meyrick, Orthenches, 35-9
semicuprata Meyrick, Trichotaphe, 29-672
semiglobata Meyrick, Tiquadra, 15-437
semilugens (Zeller), Ethmia, 21-391
seminigera Meyrick, Acrolophus, 15-166
seminigrescens Meyrick, Loxotoma, 21-1068
seminolella (Beutenmüller), Praeacedes, 15-341
semiombra Dyar, Ethmia, 21-392
semiombra Dyar, Ethmia, 21-392a
semiopaca (Grote, 1881), Ethmia, 21-391
semiotarsa Meyrick, Struthoscelis, 21-1710
semiovata Meyrick, Antaeotricha, 21-764
semisignella (Walker), Antaeotricha, 21-765
semisiquella (Busck), Antaeotricha, 21-765
semitenebrella Dyar, Ethmia, 21-393
semnodoxa Meyrick, Harpella, 21-1578
semocrossa Meyrick, Peleopoda, 21-416
semotella (Walker), Inga, 21-1662
senariella (Zeller), Recurvaria, 29-221
senecta Walsingham, Arogalea, 29-57
sensilis Meyrick, Borkhausenia, 21-1539
separabilis Walsingham, Diastaltica, 29-754
separatella (Walker), Inga, 21-1663
sepias Meyrick, Loxotrochis, 41-17
seppiana (Stoll), Gonioterma, 21-1033
septemstrigella (Chambers), Augolychna, 15-214
septemstrigella Zeller, Glyphipterix, 34-61
sepulcralis (Meyrick), Acrolophus, 15-167
sequella Busck, Triclonella, 27-60

sequens Meyrick, Tinea, 15-412
sequestra Meyrick, Stenoma, 21-1408
serangodes Meyrick, Antaeotricha, 21-766
seraphica (Meyrick), Acrolepiopsis, 39-15
serarcha Meyrick, Antaeotricha, 21-767
serena (Meyrick), Cymotricha, 29-553
sericata (Butler), Stenoma, 21-1409
sericella Forbes, Taeniodictys, 15-366
serratus Hasbrouck, Acrolophus, 15-168
serrigera Meyrick, Acrocercops, 20-100
serta (Schaus), Aepytus, 5-64
serta Walsingham, Opogona, 15-321
sertifera (Meyrick), Alucita, 31-18
sertigera (Meyrick), Onebala, 29-360
servilis (Walsingham), Onebala, 29-633
servula Meyrick, Commatica, 29-746
sesamodes Meyrick, Compsolechia, 29-453
sesquitertia (Zeller), Stenoma, 21-1410
setiacma Meyrick, Acrolophus, 15-169
sevactella (Walker), Ilingiotis, 29-612
severella (Walker), Battaristis, 29-349
sexangula Meyrick, Phyllocnistis, 20-146
sexmaculata (Dognin), Stenoma, 21-1411
sexpunctella Walsingham, Heliodines, 40-25
siderea (Walsingham), Atteva, 36-9
siderophaea (Walsingham), Compsolechia, 29-454
sigmoidella Dietz, Setomorpha, 15-359
signatus Busck, Acrolophus, 15-170
signifera Meyrick), Inga, 21-1664
significa (Meyrick), Exaeretia, 21-8
signigera Meyrick, Philaustera, 35-10
silvibasis (Duckworth), Rectiostoma, 21-1110
silvicolor Meyrick, Phytomimia, 21-267
similatella Busck, Ethmia, 21-394
similatella (Zeller), Caloptilia, 20-16
similis (Busck), Antaeotricha, 21-768
similis Walsingham, Cosmopterix, 27-24
similis (Zukowsky), Pfitzneriella, 5-130
simplex Busck, Stenoma, 21-1412
simplex (Viette), Aepytus, 5-83
simplex (Walsingham), Opogona, 15-322
simulella (Dietz), Lindera), 15-294
sinaloanus Vázquez, Oiketicus, 16-57
siraphora (Meyrick), Cerconota, 21-897
sladeni (Hampson), Aepytus, 5-99
smaragdopis Meyrick, Telphusa, 29-297
smaragdulella (Walker), Compsolechia, 29-430
smileuta Meyrick, Antaeotricha, 21-769
smithi Druce, Phassus, 5-50
smodicopa (Meyrick), Chlamydastis, 21-979
socia Meyrick, Panclintis, 26-17
socialis (Beutelspacher), Amydria, 15-202
sodalis (Walsingham), Inga, 21-1665
solanella (Boisduval), Phthorimaea, 29-185
solani (E. M. Hering), Phyllonorycter, 20-131
solanivora (Povolný), Ptycerata, 29-198
solella (Walker), Promenesta, 21-1096
solenobiella Walsingham, Tinea, 15-413
solida Walsingham, Eupragia, 21-92
solidella (Walker), Compsolechia, 29-455
sollers Meyrick, Prochola, 26-36
sommerella (Zeller), Stenoma, 21-1413
somnulentella (Zeller), Bedellia, 19-4
songoensis (Pfitzner), Druceiella, 5-30
sonorensis Walsingham, Gelechia, 29-149

sordidata (Zeller), Urodus, 36-77
soronella Busck, Arogalea, 29-58
sororia (Zeller), Anadasmus, 21-441
sortifera Meyrick, Antaeotricha, 21-770
sortis Meyrick, Acrocercops, 20-101
sparganota Meyrick, Antaeotricha, 21-771
sparsiciliella (Clemens), Inga, 21-1666
spathista Meyrick, Acrolophus, 15-171
spatula (Busck), Lactura, 36-21
speciosella (Walker), Compsolechia, 29-458
spectata Meyrick, Philonome, 19-3
specter Schaus, Oiketicus, 16-54
spectrophthalma (Meyrick), Chlamydastis, 21-980
speculans Meyrick, Glyphipterix, 34-62
speculatrix (Meyrick), Inga, 21-1667
sperata (Busck), Antaeotricha, 21-454
spermatias Meyrick, Syrmologa, 15-364
spermatopis Meyrick, Zemiocrita, 21-307
spermidias Meyrick, Stenoma, 21-1317
spermolitha (Meyrick), Antaeotricha, 21-772
sphecophila (Meyrick), Taygete, 29-280
sphenisca Powell, Ethmia, 21-395
sphenodelta Meyrick, Battaristis, 29-358
sphenogramma Clarke, Gonionota, 21-188
sphenophora (Walsingham), Phthorimaea, 29-186
sphenoplecta Meyrick, Filinota, 21-103
sphragidopis (Meyrick), Cerconota, 21-898
sphyrocopa Meyrick, Cymotricha, 29-554
spilocrossa Meyrick, Doliotechna, 21-1558
spinifera Meyrick, Acrolophus, 15-172
spinigera Meyrick, Anomoxena, 29-698
spinignatha Becker, Timocratica, 21-1506
spinosum Walsingham, Spanioptila, 20-114
spintheropis Meyrick, Strobisia, 29-491
spiridoxa (Meyrick), Chionodes, 29-81
spitzi (Viette), Aepytus, 5-86
spodinopis Meyrick, Stenoma, 21-1414
sporimaea Meyrick, Lygronoma, 21-1685
sporomantis Meyrick, Rhabdocrates, 34-21
sporozona (Meyrick), Compsolechia, 29-456
spretella (Denis & Schiffermüller), Niditinea, 15-306
spudasma (Walsingham), Peleopoda, 21-417
spumescens Meyrick, Urodus, 36-78
spurca (Zeller), Antaeotricha, 21-773
spurcatella (Walker), Antaeotricha, 21-774
squalens Meyrick, Dichomeris, 29-597
squamigera Walsingham, Aristotelia, 29-45
squamosa (Walsingham), Chlamydastis, 21-981
stabilis (Butler), Stenoma, 21-1415
stactogramma Meyrick, Axiagasta, 15-215
stagmatophoria Walsingham, Holophysis, 29-484
stagnicolor (Meyrick), Chlamydastis, 21-982
stalagmitis Meyrick, Acrocercops, 20-102
staphylina Meyrick, Urodus, 36-79
staphylitis (Meyrick), Inga, 21-1668
stasichlora Meyrick, Glyphipterix, 34-63
stasigastra Meyrick, Compsolechia, 29-457
stativa (Meyrick), Inga, 21-1669
staudingerana Maassen), Stenoma, 21-1416
staudingeri (Wagner), Callipielus, 5-11
stauromacha Meyrick, Acartophila, 21-1517
staurota Meyrick, Antaeotricha, 21-775
stella (Busck), Stenoma, 21-1295
stellaris (R. Felder & Rogenhofer), Strobisia, 29-492
stellatella Busck, Onebala, 29-631

stelliferella (Walker), Compsolechia, 29-458
steloglypta (Meyrick), Chlamydastis, 21-983
stemonias Meyrick, Thaumatolita, 21-1714
stemonodes (Meyrick), Marmara, 20-120
stenobathra Meyrick, Antaeotricha, 21-483
stenomorpha Meyrick, Glyphidocera, 23-35
stenopa Walsingham, Tiquadra, 15-438
stenota Walsingham, Catarata, 21-841
stephanodes Meyrick, Stenoma, 21-1417
stephanopsis (Meyrick), Alucita, 31-19
stereodesma (Meyrick), Inga, 21-1670
stereogramma Meyrick, Battaristis, 29-359
stereopa Meyrick, Cranaodes, 15-241
sterrhomitra (Meyrick), Antaeotricha, 21-776
sticta Walsingham, Recurvaria, 29-222
stictopus Walsingham, Bucculatrix, 19-16
stigmaphyllae Busck, Phyllonorycter, 20-132
stigmaphylli (Walsingham), Psittacastis, 21-293
stigmatias (Walsingham), Antaeotricha, 21-777
stigmatica Pfitzner, "Dalaca", 5-123
stigmatophora (Walsingham), Pyroderces, 27-39
stillata Meyrick, Compsolechia, 29-459
stimulatrix Meyrick, Tinea, 15-403
stirodes (Meyrick), Scrobipalpula, 29-246
stomatocosma (Meyrick), Timocratica, 21-1507
strabo Meyrick, Macrocirca, 21-406
strabonia (Meyrick), Chlamydastis, 21-984
straminella (Walker), Stenoma, 21-1418
strangoides Viette, Aplatissa, 5-58
stratella (Walsingham), Dichomeris, 29-598
stratella Zeller, Prays, 35-27
stratigera Meyrick, Dichomeris, 29-599
strenuella (Walker), Stenoma, 21-1419
striata Clarke, Alynda, 21-1523
striatella Busck, Ethmia, 21-396
striatella Busck, Stenoma, 21-1420
striatella (Murtfeldt), Symmetrischema, 29-267
strigiplenella Walker, Vazugada, 29-680
strigivenata (Butler), Stenoma, 21-1421
strigosa (Cockerell), Ethmia, 21-324
strigosella (Walker), Ethmia, 21-324
stringens Meyrick, Antaeotricha, 21-778
striolata (Meyrick), Lethata, 21-1065
strophalodes (Meyrick), Cerconota, 21-845
stupefacta Meyrick, Stenoma, 21-1422
stygeropa (Meyrick), Antaeotricha, 21-779
stygia Meyrick, Commatica, 29-747
stygnota (Walsingham), Eupolella, 23-1
stylonota (Meyrick), Cerconota, 21-899
suavis Meyrick, Baeonoma, 21-838
subactella (Walker), Anacampsis, 29-337
subaequalis Walsingham, Polyhymno, 29-195
subagrestis Meyrick, Stasixena, 21-1708
subalbata Meyrick, Brachypsaltis, 29-709
subalbella Meyrick, Brachmia, 29-520
subalbella (Walsingham), Brachmia, 29-520
subalbusella (Chambers), Brachmia, 29-520
subannulata Zeller, Dysgnorima, 21-1562
subapicalis (Walker), Compsolechia, 29-460
subcaerulea (Walsingham), Ethmia, 21-337b
subcervinella (Walker), Opogona, 15-323
subcuprea Meyrick, Tinea, 15-414
subdentata (Meyrick), Cymotricha, 29-555
subditella (Walker), Endrosis, 21-1563
subdulcis (Meyrick), Antaeotricha, 21-780

subferreus (Herrich-Schäffer), Lactura, 36-22
subfervens Butler, Dalaca, 5-3
subfervens (Walker), Lactura, 36-22
subfusca Meyrick, Acrolophus, 15-173
subicula Clarke, Doina, 21-81
subita Meyrick, Stenoma, 21-1423
subjectella (Walker), Euzonomacha, 29-771
sublatella (Walker), Compsolechia, 29-466
sublimbata (Zeller), Stenoma, 21-1424
sublimis Meyrick, Choropleca, 15-230
sublunaris Meyrick, Stenoma, 21-1134
sublustricella (Walker), Hapalonoma, 29-610
submersa (Meyrick), Antaeotricha, 21-582
submetallica Meyrick, Copocentra, 40-8
submissa Busck, Ethmia, 21-397
subnigritaenia Powell, Ethmia, 21-398
subnitens Walsingham, Bucculatrix, 19-17
subnotatella (Walker), Stenoma, 21-1425
subolivacea Walsingham, Blastobasis, 23-80
subovalis (Meyrick), Timocratica, 21-1507
subrosea Meyrick, Aristotelia, 29-46
subrutila Meyrick, Cymotricha, 29-531
subscriptella (Walker), Compsolechia, 29-444
subsimilis Walsingham, Ethmia, 21-399
substratella Walsingham, Dichomeris, 29-600
substricta Meyrick, Antaeotricha, 21-781
subtigrina Meyrick, Crepidochares, 15-242
subtilis (Hübner), Atteva, 36-8
subtincta (Meyrick), Prochola, 26-37
subvectella (Walker), Parelectroides, 29-798
subversa Walsingham, Energia, 21-994
succincta (Walsingham), Compsolechia, 29-461
succinctella (Walker), Menesta, 21-1071
suffectella (Walker), Compsolechia, 29-462
suffumigata Walsingham, Antaeotricha, 21-782
suffusella Sattler, Deltophora, 29-90
suffusella (Walker), Compsolechia, 29-371
sulphurea (Busck), Pycnotarsa, 21-1700
sumptella (Walker), Cymotricha, 29-556
superatella (Walker), Psittacastis, 21-294
superba Gozmány, Cosmopterix, 27-9
superciliosa Meyrick, Antaeotricha, 21-783
superella (Walker), Compsolechia, 29-440
superfusella (Walker), Compsolechia, 29-463
superstes Walsingham, Acrolophus, 15-174
supletella (Zeller), Holcocera, 23-101
suppressella (Walker), Antaeotricha, 21-821
surinamella (Möschler), Stenoma, 21-1426
surinamensis (Möschler), Cryptothelea, 16-18
susceptella (Walker), Compsolechia, 29-464
suspectella (Walker), Compsolechia, 29-465
suspensa Meyrick, Gelechia, 29-150
suspensilis Meyrick, Acrolophus, 15-175
sustentata Meyrick, Stenoma, 21-1427
sylpharis (Butler), Atteva, 36-5
symbolica (Meyrick), Onebala, 29-632
symmeles (Meyrick), Tinea, 15-415
symmicta (Dyar), Cryptothelea, 16-23
symmicta Walsingham, Stenoma, 21-1428
sympasta Meyrick, Holcocera, 23-102
sympathetica (Meyrick), Xystrologa, 15-452
sympatrica Clarke, Homoeoprepes, 26-2
symphonica Meyrick, Stenoma, 21-1429
symphora (Walsingham), Battaristis, 29-360
symphracta Meyrick, Stilbosis, 27-104

sympiestis Meyrick, Urodus, 36-80
symposias (Meyrick), Stenoma, 21-1430
sympotica Meyrick, Mompha, 25-25
synaphrista Meyrick, Apothetoeca, 29-704
synapta Durrant, Acrolophus, 15-176
synchorda Meyrick, Oenoe, 15-311
synchorista Meyrick, Ptilopsaltis, 15-348.1
syncrata (Meyrick), Parochromolopis, 33-3
syndecta Meyrick, Glyphipterix, 34-64
syndicastis (Meyrick), Timocratica, 21-1480
synedra (Meyrick), Chlamydastis, 21-985
synercta Meyrick, Antaeotricha, 21-784
synestia Meyrick, Recurvaria, 29-223
syngraphopa Meyrick, Battaristis, 29-361
syngraphopis Meyrick, Stenoma, 21-1431
syngraphora Meyrick, Battaristis, 29-361
synocha Meyrick, Battaristis, 29-362
synorista Meyrick, Glyphipterix, 34-65
synthetica Walsingham, Gelechia, 29-151
syntoma (Walsingham), Hypercallia, 21-228
syntripta Meyrick, Tiquadra, 15-439
syringota Meyrick, Trichotaphe, 29-673
syrmeutis Meyrick, Borkhausenia, 21-1540
syrphacopis (Meyrick), Acrolepiopsis, 39-16
tabacella (Ragonot), Phthorimaea, 29-185
tabacillus Weyenbergh, Oiketicus, 16-50
tabida (Butler), Cerconota, 21-900
taboga (Busck), Inga, 21-1671
tactella (Walker), Cymotricha, 29-549
tactica Meyrick, Dichomeris, 29-601
taeniarcha Meyrick, Acrocercops, 20-103
taeniata (Meyrick), Plumana, 16-8
tandilensis Köhler, "Platoeceticus," 16-71
tandilensis (Pastrana), Alucita, 31-20
tangolias Gyen, Trichotaphe, 29-674
tantalota Meyrick, Dysoptus, 15-253
tanysta Meyrick, Antaeotricha, 21-455
tapetiella (Zeller), Trichotaphe, 15-441
tapetzella (Linnaeus), Trichotaphe, 15-441
taphrocopa Meyrick, Cryptolechia, 21-66
tapinota Walsingham, Catalexis, 29-711
tapuja Pfitzner, "Dalaca," 5-124
arachodes Walsingham, Valentinia, 23-108
atarcta Walsingham, Amydria, 15-211
tardella (Walker), Compsolechia, 29-466
tectella (Walker), Stenoma, 21-1291
tectoria (Meyrick), Antaeotricha, 21-785
teganitis Meyrick, Gonionota, 21-189
tegulella (Walsingham), Onebala, 29-633
tehuacana (Busck), Epilechia, 29-502
telegraphella (Walker), Plocamosaris, 29-643
teleosema Meyrick, Antaeotricha, 21-786
teligera Meyrick, Cosmopterix, 27-25
telluris (Clarke), Ypsolopha, 35-25
tempestiva (Meyrick), Antaeotricha, 21-787
tenax Meyrick, Cosmopterix, 27-26
tenebralis (Hampson), Gonionota, 21-190
tenera (Zeller), Stenoma, 21-1310
tentatella (Walker), Holophysis, 29-485
tenuicaudella (Walsingham), Phyllonorycter, 20-133
tephrodesma (Meyrick), Antaeotricha, 21-788
tephropis Meyrick, Homosetia, 15-270
terea (Schaus), Aepytus, 5-68
terminalis Clarke, Coptotelia, 21-45
terpnodes Meyrick, Phylopatris, 29-802

terpnota Walsingham, Ethmia, 21-400
terpsichorella (Busck), Choropleca, 15-231
terracocta (Walsingham), Cotyloscia, 29-526
terrella (Walker), Phthorimaea, 29-185
terrenella (Busck), Compsolechia, 29-467
tersectella Zeller, Elachista, 22-4
tesquatella Chambers, Stilbosis, 27-105
tesquella Clemens, Stilbosis, 27-105
tessellatella Blanchard, Lindera, 15-294
tessellatus (Herrich-Schäffer), Phassus, 5-49
tesselloides (Schaus), Aepytus, 5-71
testacea Meyrick, Compsosaris, 29-750
testudinea Meyrick, Pseudocentris, 21-272
tetrabola Meyrick, Stenoma, 21-1432
tetradyas Meyrick, Zonochares, 15-453
tetraglossa (Meyrick), Choropleca, 15-231
tetragonella (Walker), Stenoma, 21-1162
tetragramma Meyrick, Cosmopterix, 27-27
tetrancyla Meyrick, Acrolophus, 15-177
tetraonella (Walsingham), Protodarcia, 15-347
tetrapetra (Meyrick), Antaeotricha, 21-789
tetrapetra Meyrick, Eristhenodes, 29-765
tetraphyta Meyrick, Pyramidobela, 21-1705
tetraplecta Meyrick, Calliprora, 29-68
tetratoma (Meyrick), Diploschizia, 34-74
tetraxoa Meyrick, Anomoxena, 29-699
tetrortha Meyrick, Compsolechia, 29-468
texanella (Chambers), Lita, 29-176.1
texanellus (Chambers), Acrolophus, 15-178
textrina (Meyrick), Inga, 21-1672
thalamias Meyrick, Pessograptis, 29-509
thalamobathra Meyrick, Antaeotricha, 21-664
thalamopa (Meyrick), Cymotricha, 29-557
thaleropa Meyrick, Stenoma, 21-1433
thalpodes (Meyrick), Cymotricha, 29-558
thaminodes Meyrick, Acrolophus, 15-179
thammii Zeller, Antaeotricha, 21-790
thamnocephala Clarke, Eraina, 21-1575
thamnolopha (Meyrick), Pseuderotis, 21-419
thapsinopa Meyrick, Antaeotricha, 21-791
tharsales (Walsingham), Acrolepiopsis, 39-17
theca Clarke, Batrachedra, 24-31
thecophora (Walsingham), Praeacedes, 15-341
themelia (Meyrick), Cymotricha, 29-559
theobromae (Busck), Stenoma, 21-1150
theoretica Meyrick, Antaeotricha, 21-792
thermodryas Meyrick, Dichomeris, 29-602
thermophaea (Meyrick), Ilingiotis, 29-613
thermopsamma Meyrick, Gnathotona, 21-105
thermoxantha (Meyrick), Inga, 21-1673
thesmiopa (Meyrick), Cymotricha, 29-560
thesmophora Meyrick, Antaeotricha, 21-793
thespia Meyrick, Stenoma, 21-1434
thiobasis (Duckworth), Rectiostoma, 21-1111
thiodes Meyrick, Recurvaria, 29-224
thisbe (Druce), Aepytus, 5-107
tholodes (Meyrick), Cerconota, 21-901
thologramma Meyrick, Stenoma, 21-1435
tholomicta Meyrick, Acrolophus, 15-180
thoracica (Grote), Biopsyche, 16-46
thoracica (Schaus), Oiketicus, 16-50
thoristes Busck, Stenoma, 21-1436
thrasynta Meyrick, Ilingiotis, 29-614
thrasyzela Meyrick, Cosmopterix, 27-28
thriophora Meyrick, Syrmologa, 15-365

thrombodes (Meyrick), Meridorma, 29-781
thrypsandra Meyrick, Trichotaphe, 29-675
thurberiella Busck, Bucculatrix, 19-18
thylacandra Meyrick, Stenoma, 21-1437
thylacosaris (Meyrick), Antaeotricha, 21-794
thymiata (Meyrick), Gelechia, 29-152
thymiota Meyrick, Stenoma, 21-1345
thymora Meyrick, Imma, 41-35
thyridopa (Meyrick), Lucyna, 21-230
thyrsitis Meyrick, Himotica, 21-198
thyrsogastra Meyrick, Glyphidocera, 23-36
thysanodes (Meyrick), Antaeotricha, 21-795
thysanora (Meyrick), Compsolechia, 29-469
thysanota Walsingham, Recurvaria, 29-225
thysiarcha Meyrick, Eomichla, 21-1573
tibialis Zeller, Antaeotricha, 21-796
tibicina Meyrick, Scythris, 28-11
tinactis (Meyrick), Antaeotricha, 21-797
tinctipennis (Butler), Cerconota, 21-902
tineoides Dammerman, Setomorpha, 15-359
tischeriella (Walsingham), Protodarcia, 15-348
titanica Walsingham, Holcocera, 23-103
titanoleuca Becker, Timocratica, 21-1508
titanota (Walsingham), Compsolechia, 29-470
tolmeta Walsingham, Stenoma, 21-1438
tolmetes Walsingham, Anchimacheta, 36-26
tornogramma Meyrick, Antaeotricha, 21-798
tornoptila Meyrick, Compsolechia, 29-421
torophragma (Meyrick), Cerconota, 21-894
torta (Meyrick), Acrolophus, 15-181
tortricella (Staudinger), Stenoma, 21-1236
tortricella (Walker), Gonioterma, 21-1008
toumeyi Jones, Oiketicus, 16-52
townsendi Townsend, Oiketicus, 16-55
townsendi Townsend, Oiketicus, 16-55a
toxocosma Meyrick, Agathactis, 29-693
trabeata Meyrick, Erithyma, 21-90
trabeella (R. Felder & Rogenhofer), Erithyma, 21-90
trachyacma Meyrick, Glyphidocera, 23-37
trachycantha Clarke, Doina, 21-82
trachycnemis Meyrick, Compsolechia, 29-471
trachydesma (Meyrick), Alucita, 32-21
trachyxyla Meyrick, Triclonella, 27-61
tractrix Meyrick, Antaeotricha, 21-799
traditionis Clarke, Euchionodes, 21-92
traducella Busck, Gelechia, 29-153
traili (Meyrick), Inga, 21-1674
trailii (Butler), Inga, 21-1674
trajectella Walker), Compsolechia, 29-472
transennata Meyrick, Microcolona, 26-4
transformata Meyrick, Metallocrates, 23-105
transfossa Meyrick, Cryptolechia, 21-67
transjectella (Walker), Compsolechia, 29-473
translucens Meyrick, Tinea, 15-416
translucida (Walsingham), Telphusa, 29-298
transversa Clarke, Gonionota, 21-191
transverseguttata (Zeller), Urodus, 36-81
transversella Busck, Ethmia, 21-401
transversestrigella (Dietz), Setomorpha, 15-359
transversiguttata Walsingham, Urodus, 36-81
transversus Walker, Phassus, 5-55
trapezias Meyrick, Compsolechia, 29-474
trastices (Busck), Chlamydastis, 21-986
travestita (Gozmány), Erechthias, 15-260
tremulella (Walker), Antaeotricha, 21-800

tretus Kaye, Acrolophus, 15-182
triacmopa (Meyrick), Promenesta, 21-1097
triancycla Meyrick, Urodus, 36-82
triangularides Pfitzner, Phassus, 5-35
triangularis H. Edwards, Phassus, 5-35
triangularis Möschler, Yponomeuta, 36-87
triangularis Walsingham, Blastobasis, 23-81
triargyra Meyrick, Triclonella, 27-62
triatomella (Walsingham), Acrolophus, 15-183
tribomias (Meyrick), Antaeotricha, 21-801
tricapsis (Meyrick), Antaeotricha, 21-802
tricausta Meyrick, Acrolophus, 15-184
tricentrota Meyrick, Battaristis, 29-363
tricharacta (Meyrick), Cerconota, 21-903
trichella Busck, Parastega, 29-793
trichinaspis (Meyrick), Scrobipalpula, 29-247
trichocolpa Meyrick, Stenoma, 21-1439
trichocyma (Meyrick), Onebala, 29-634
trichoneura (Meyrick), Cerconota, 21-904
trichonota Meyrick, Antaeotricha, 21-803
trichophysa Meyrick, Parectopa, 20-34
trichorda Meyrick, Stenoma, 21-1440
trichosoma Durrant, Acrolophus, 15-185
tridentella (Walsingham), Anacampsis, 29-338
tridesma Meyrick, Cerconota, 21-905
trierica Meyrick, Psittacastis, 21-295
trifasciata Walsingham, Athrinacia, 21-15
triformalis Forbes, Acrolophus, 15-197
triformellus Forbes, Acrolophus, 15-197
trifurcata (Meyrick), Inga, 21-1675
trigama (Meyrick), Machimia, 21-252
trigonana (Walsingham), Lotisma, 30-3
trigonella (Walsingham), Cymotricha, 29-561
trigonophorella (Zeller), Recurvaria, 29-225
trigonota (Walsingham), Haplochela, 29-503
trigramma Meyrick, Calliprora, 29-69
trilinearides (Pfitzner), Aepytus, 5-69
trilinearis (Pfitzner), Aepytus, 5-69
trilineata (Butler), Stenoma, 21-1441
trimacula Clarke, Atha, 21-1530
trimetalla Meyrick, Acrocercops, 20-104
trimolybda (Meyrick), Compsolechia, 29-541
trinidadensis Davis, Metaxypsyche, 16-15
trinidadensis Davis, Tetrapalpus, 15-367
trinota (Clarke), Ephysteris, 29-91
triplacodes Meyrick, Doliotechna, 21-1559
triplagella (Walker), Cotyloscia, 29-527
triplectra (Meyrick), Antaeotricha, 21-804
triplintha (Meyrick), Antaeotricha, 21-805
tripudians Meyrick, Choropleca, 15-232
tripunctata (Walsingham), Mompha, 25-26
tripunctella Walsingham, Heliodines, 40-26
tripustulata (Zeller), Stenoma, 21-1442
tripustulella (Walker), Antaeotricha, 21-806
trirecta Meyrick, Stenoma, 21-1443
trisecta (Walsingham), Antaeotricha, 21-807
trisinuata Meyrick, Antaeotricha, 21-808
trissobathra Meyrick, Doliotechna, 21-1560
trissoxantha (Meyrick), Catoptristis, 29-712
tristicta (Busck), Cymotricha, 29-562
tristis (Schaus), Astala, 16-37
tristrigata (Zeller), Stenoma, 21-1444
trithalama Meyrick, Mompha, 25-27
tritogramma Meyrick, Antaeotricha, 21-809
tritypa (Meyrick), Chlamydastis, 21-987

trivallata Meyrick, Antaeotricha, 21-810
trizeucta (Meyrick), Cerconota, 21-906
trochalosticta (Walsingham), Lethata, 21-1066
trochilea (Walsingham), Compsolechia, 29-475
trochiloides Walsingham, Homoeoprepes, 26-3
trochilosticta (Busck), Lethata, 21-1066
trochistis (Meyrick), Cerconota, 21-907
trochoscia Meyrick, Antaeotricha, 21-811
troctis Meyrick, Pigritia, 23-112
trojesa (Schaus), Schausiana, 5-56
trossulella Walsingham, Aristotelia, 29-47
trunca Meyrick, Machimia, 21-253
truncata Clarke, Doina, 21-83
truncatula (Meyrick), Chlamydastis, 21-988
trygaula (Meyrick), Inga, 21-1676
trymalopa (Meyrick), Cerconota, 21-908
trypherantis Meyrick, Antispila, 14-8
tryphon (Busck), Chlamydastis, 21-989
tuberosella (Busck), Symmetrischema, 29-266
tubicen (Meyrick), Inga, 21-1677
tumens (Meyrick), Antaeotricha, 21-812
tumulata (Meyrick), Cerconota, 21-909
tunicata (Busck), Hypercallia, 21-220
tupi (Pfitzner), Aepytus, 5-99
turbinalis Meyrick, Triclonella, 27-63
turrita (Meyrick), Cymotricha, 29-563
tyrocopa Meyrick, Thrasydoxa, 40-38
tyrocrossa Meyrick, Stenoma, 21-1445
tyroxesta (Meyrick), Cerconota, 21-855
uberrima (Meyrick), Gonionota, 21-192
ulosema Meyrick, Stenoma, 21-1446
umbrata Walsingham, Glyphidocera, 23-38
umbratella Walker, Antaeotricha, 21-813
umbraticella (Busck), Exoncotis, 15-263
umbraticostella Walsingham, Tinea, 15-417
umbratipalpis (Walsingham), Acrolophus, 15-186
umbrifera (R. Felder), Aepytus, 5-102
umbriferella (Walker), Antaeotricha, 21-814
umbrigera Meyrick, Triclonella, 27-64
umbrinervis Meyrick, Stenoma, 21-1447
umbripennis (Walsingham), Pectinophora, 29-506
uncigera (Walsingham), Acrolophus, 15-187
uncispinis Walsingham, Acrolophus, 15-188
uncta Meyrick, Glyphipterix, 34-66
uncticoma Meyrick, Stenoma, 21-1448
underwoodi Druce, Acrolophus, 15-189
undifraga Meyrick, Acrocercops, 20-105
undosa (Walsingham), Parectopa, 20-35
unguentana Meyrick, Stenoma, 21-1449
unguifera Meyrick, Glyphipterix, 34-67
ungulatella Busck, Ethmia, 21-402
ungulifera (Meyrick), Chlamydastis, 21-990
unicolor Walsingham, Tischeria, 9-8
unimaculella Chambers, Ithome, 27-77
unipectinicornis Hasbrouck, Acrolophus, 15-98
unipuncta Walsingham, Bucculatrix, 19-19
unisecta (Meyrick), Antaeotricha, 21-815
unisignis Meyrick, Stenoma, 21-1454
unistrigella (Busck), Battaristis, 29-364
unomaculella auct., Ithome, 27-77
uranophanes (Meyrick), Parascaeas, 21-1081
urbana (Butler), Stenoma, 21-1421
urbanella (Zeller), Acrocercops, 20-106
urichi Busck, Heliodines, 40-27
urophora (Walsingham), Diploschizia, 34-75

urosema (Meyrick), Gnorimoschema, 29-166
ursula Walsingham, Anacampsis, 29-339
uruguayensis (Berg), Stenoma, 21-1450
usaque Pfitzner, "Dalaca", 5-125
ustimacula (Zeller), Callistenoma, 21-1542
uterella (Walsingham), Phereoeca, 15-337
vacans (Meyrick), Anadasmus, 21-442
vacata Meyrick, Antaeotricha, 21-816
vaccula Walsingham, Stenoma, 21-1451
vadata Meyrick, Ascalenia, 27-71
vaga (Butler), Stenoma, 21-1452
vagabundella Forbes, Aristotelia, 29-48
vagatioella (Chambers), Coleotechnites, 29-83
vagella (Walker), Chionodes, 29-77
valdiviae Davis & Nielsen, Apoplania, 3-3
vallifera Meyrick, Cryptolechia, 21-52
vanduzeei Hasbrouck, Acrolophus, 15-190
vanis (Busck), Cerconota, 21-881
vannifera (Meyrick), Antaeotricha, 21-817
vapida (Butler), Stenoma, 21-1453
variabilis (Busck), Lita, 29-177
variabilis (Viette), Dalaca, 5-10
variabilis (Walsingham), Acrolophus, 15-191
variata Meyrick, Glyphipterix, 34-68
variegella (Blanchard), Lindera, 15-294
variolata Walsingham, Auximobasis, 23-59
varipes (Walker), Imma, 41-36
varronia (Busck), Trichotaphe, 29-676
vasifera Meyrick, Stenoma, 21-1454
vauriei Hasbrouck, Acrolophus, 15-192
venatella (Busck), Urodus, 36-83
venatum (Busck), Antaeotricha, 21-818
venefica Meyrick, Cosmopterix, 27-29
veneranda Walsingham, Gelechia, 29-154
venezuelae Davis, Dendropsyche, 16-25
venezuelensis Amsel, Antaeotricha, 21-819
venezuelensis (Amsel), Lindera, 15-293
veniflua Meyrick, Cryptolechia, 21-68
venifurcata Becker, Timocratica, 21-1509
venosa Busck, Galtica, 29-773
venosa (Butler), Orsotricha, 29-180
venosella (Walker), Anadasmus, 21-443
venosus (Blanchard), Blanchardina, 5-21
ventilatrix Meyrick, Stenoma, 21-1455
ventratella (Zeller), Symmetrischema, 29-268
vergarai (Povolný), Ptycerata, 29-199
vernalis Jones, Thyridopteryx, 16-60
verresi Schaus, Aepytus, 5-96
verruculella (Zeller), Mompha, 25-28
versatella (Walker), Compsolechia, 29-476
versatilis (Meyrick), Inga, 21-1678
versicolor Meyrick, Glyphipterix, 34-69
versicolorella Walker, Tocmia, 29-826
vespertilio Meyrick, Acrolophus, 15-193
vestita Walsingham, Glyphidocera, 23-39
veteranella (Zeller), Aristotelia, 29-49
vetusta Meyrick, Symmoca, 23-43
vetustella (Walker), Trichotaphe, 29-677
vexata Meyrick, Stenoma, 21-1456
vexatalis (Walker), Moca, 41-15
vexillata (Meyrick), Gonionota, 21-193
v-flavum (Haworth), Oinophila, 15-312
vibicata Pfitzner, "Dalaca", 5-126
vicana Meyrick, Arsitotelia, 29-50
victrix Walsingham, Acrolophus, 15-194

vigasi (Schaus), Astala, 16-39
vigia Beutelspacher, Acrolophus, 15-195
vigilaciella Clemens, Plutella, 35-16
vigilans Meyrick, Ilingiotis, 29-615
vilis Meyrick, Tiquadra, 15-440
villosula Zeller, Clistothyris, 29-724
viminea Meyrick, Parectopa, 20-36
vinacea (Meyrick), Nanodacna, 26-8
vinifera Meyrick, Stenoma, 21-1457
vinitincta (Walsingham), Acompsia, 29-517
violacea (Busck), Eunomarcha, 29-770
violacea Butler, Dalaca, 5-3
violaria Meyrick, Trichotaphe, 29-666
violenta (Meyrick), Cosmopterix, 27-30
virens (Meyrick), Antaeotricha, 21-820
virescens Nielsen & Davis, Simacauda, 10-4
virescens Walsingham, Brachmia, 29-523
viretella (Zeller), Anacampsis, 29-340
virginalis (Butler), Stenoma, 21-1310
virginea Meyrick, Doliotechna, 21-1561
virginia (Busck), Inga, 21-1679
viridans Meyrick, Sorotacta, 29-810
viridiceps (R. Felder & Rogenhofer), Stenoma, 21-1458
viridis (Busck), Antaeotricha, 21-820
viridis Butler, Pisinidea, 21-270
viridisquamata Zeller, Dasycarea, 18-1
viridula (Zeller), Caloptilia, 20-17
viscida Meyrick, Ceromitia, 12-13
vita (Busck), Stenoma, 21-1459
vitellinella Staudinger, Pseudoecophora, 21-1698
vitellus Poey, Acrolophus, 15-196
vitrea Meyrick, Hyalopseustis, 21-1034
vitreola Meyrick, Stenoma, 21-1460
vivax Busck, Stenoma, 21-1461
vivida (Meyrick), Gonionota, 21-194
vividella (Busck), Chlamydastis, 21-914
vociferans Meyrick, Filinota, 21-104
vogli Amsel, Antaeotricha, 21-483
vogli Viette, Pfitzneriana, 5-60
volcanella Powell, Ethmia, 21-403
volcanica Walsingham, Arrhenophanes, 17-2
volitans Meyrick, Stenoma, 21-1462
volubilis Meyrick, Compsolechia, 29-477
voluptaria (Meyrick), Inga, 21-1680
voluptella R. Felder & Rogenhofer, Glyphipterix, 34-70
vorax (Meyrick), Fascista, 29-104
vulcanicola Meyrick, Lotisma, 30-4
vulgaris Nielsen & Robinson, Callipielus, 5-20
walchiana (Stoll), Antaeotricha, 21-821
walkeri (Walsingham), Charistica, 29-723
walsinghami (Busck), Phereoeca, 15-336
walsinghami Möschler, Acrolophus, 15-197
watsoni (Jones), Cryptothelea, 16-19
wellingi Powell, Ethmia, 21-404
westwoodii Berg, "Oiketicus", 16-72
whalleyi Duckworth, Thioscelis, 21-1472
whitelyi (Druce), Acrolophus, 15-198
willineri Pastrana, Hexeretmis, 31-23
wygodzinskyi E. M. Hering, Phyllocnistis, 20-147
xanthobasis (Zeller), Rectiostoma, 21-1112
xanthobyrsa (Meyrick), Cerconota, 21-871
xanthocarpa Meyrick, Commatica, 29-748
xanthocephala (Walsingham), Lamprolophus, 40-31
xanthochorda Meyrick, Hapalothyma, 19-11
xanthographa Walsingham, Athrinacia, 21-16

xantholitha Meyrick, Ethmia, 21-343
xanthopetala (Meyrick), Antaeotricha, 21-822
xanthophaeella (Walker), Stenoma, 21-1463
xanthophanes Meyrick, Anthinora, 29-700
xanthoplecta Meyrick, Glyphipterix, 34-71
xanthoptila Meyrick, Antaeotricha, 21-602
xanthorrhoa (Zeller), Ethmia, 21-370
xanthoselena Meyrick, Deoclona, 29-689
xanthoselene (Walsingham), Deoclona, 29-689
xanthosema (Meyrick), Zymrina, 21-1720
xanthosoma (Dognin), Timocratica, 21-1510
xanthosoma (Dognin), Timocratica, 21-1510a
xanthosomella (Maassen), Tinea, 15-418
xanthostoma Walsingham, Holophysis, 29-486
xanthota Walsingham, Triclonella, 27-65
xanthotarsa Becker, Timocratica, 21-1511
xanthotricha Meyrick, Recurvaria, 29-227
xanthura Walsingham, Cosmopterix, 27-31
xenica (Meyrick), Erechthias, 15-260
xeniella (Zeller), Acrocercops, 20-107
xenodes Meyrick, Tinea, 15-419
xenodroa Clarke, Odonna, 21-1690
xerodes (Walsingham), Trichotaphe, 29-678
xerospila Meyrick, Dasmophora, 15-244
xerota Walsingham, Opogona, 15-324
xiphias (Meyrick), Acrolepiopsis, 39-18
xiphodes Meyrick, Syntetrernis, 26-39
xiphura Meyrick, Urodus, 36-84
xuthobasis (Duckworth), Rectiostoma, 21-1113
xuthosaris Meyrick, Antaeotricha, 21-823
xuthostola (Walsingham), Trichotaphe, 29-679
xylinaspis (Meyrick), Chlamydastis, 21-991
xylinella (Walker), Acrolophus, 15-199
xylinopa Meyrick, Stenoma, 21-1464
xylobathra Meyrick, Gelechia, 29-155
xylochroa (Meyrick), Taygete, 29-278
xylocosma Meyrick, Antaeotricha, 21-824
xylodeta Meyrick, Stereodmeta, 29-814
xylograpta Meyrick, Stenoma, 21-1413
xylophragma Meyrick, Acrolepiopsis, 39-21
xylostella (Linnaeus), Plutella, 35-18
xylozona Meyrick, Pelocnistis, 29-800
xylurga (Meyrick), Antaeotricha, 21-825
xystidota Meyrick, Eomichla, 21-1574
xystrota Meyrick, Acrocercops, 20-108
yamorkinei Köhler, "Clania", 16-73
ybyrajuba Becker, Stenoma, 21-1465
yuccaella (Boll), Tegeticula, 13-2
yuccasella (Riley), Tegeticula, 13-2
yumaella Kearfott, Dyotopasta, 15-251
yungas (Viette), Aepytus, 5-94
yunquella Forbes, Mea, 15-298
zacharis Meyrick, Hybroma, 15-278
zachroa (Meyrick), Eunebristis, 29-607
zacualpania (Dyar), Astala, 16-38
zalodisca Meyrick, Glyphipterix, 34-72
zanclogramma (Meyrick), Antaeotricha, 21-826
zanclophora Meyrick, Acrolophus, 15-200
zebrata Powell, Ethmia, 21-405
zebrina (Butler), Erechthias, 15-260
zebrina (Walsingham), Compsolechia, 29-478
zebrulella Forbes, Acrocercops, 20-109
zelleri Butler, Callistenoma, 21-1542
zelleri Walsingham & Durrant, Antaeotricha, 21-827
zelotes (Walsingham), Antaeotricha, 21-828

zephyritis Meyrick, Stenoma, 21-1466
zeugmatica Meyrick, Scythris, 28-12
zihuatanejensis Vázquez, Oiketicus, 16-51
zingarella (Walsingham), Eunebristis, 29-608
zischkai Viette, Aepytus, 5-79
zobeida Meyrick, Stenoma, 21-1467
zomias Meyrick, Dichomeris, 29-603
zonaria Clarke, Palinora, 21-263
zonostoma (Meyrick), Vazugada, 29-680
zophocrossa Meyrick, Glyphidocera, 23-40
zophodes (Meyrick), Moca, 41-16
zostera Clarke, Muna, 21-256
xygoterma Meyrick, Choropleca, 15-233
zygotoma Meyrick, Stegasta, 29-257
zymotica Meyrick, Coleophora, 24-12

Addendum* [p. 58]

albicaudis Meyrick, Cosmopterix, 27-6.1
athymopa Meyrick, Blastobasis, 23-63.1
caryoplecta Meyrick, Myrophila, 29-617.1
cervinella (Walsingham), Zaratha, 26-39.1
chalcopeda Meyrick, Palaeomystis, 26-9.1
chalcothorax Meyrick, Prochola, 26-22.1
disposita (Meyrick), Hypatima, 29-503.1
formularis Meyrick, Untomia, 29-494.1
hilda Meyrick, Atmozostis, 22-0.2
hippurista Meyrick, Recurvaria, 29-204.1
invigorata Meyrick, Auximobasis, 23-53.1
iopyrrha Meyrick, Diacholotis, 26-0.1
nolckeni Zeller, Antispila, 14-3.1
petalistis Meyrick, Elachista, 22-3.1
sancticola Meyrick, Prochola, 26-34.1
smaragdophanes Meyrick, Aristoptila, 22-0.1
spermotoca Meyrick, Helcanthica, 26-0.1

INDEX TO GENERA

Abaraschia Căpuse, 42
Abebaea Hübner, 55
Acampsia [sic] Westwood, 49
Acartophila Meyrick, 38
Accompsia [sic] Bruand, 49
Acedes Hübner, 23
Achanodes Meyrick, 24
Acompsia Hübner, 49
Acraeologa Meyrick, 47
Acrocercops Wallengren, 26
Acrolepiopsis Gaedike, 57
Acrolophus Poey, 19
Acrophiletes Meyrick, 51
Acureuta Zeller, 23
Adela Latreille, 18
Adrasteia Chambers, 47
Adrastia Kirby, 47
Adricara Walker, 57
Adullamitis Meyrick, 51
Adullanitis [sic] Gaede, 51
Aeaea Chambers, 44
Aechmia Treitschke, 54
Aecimia [sic] Boisduval, 54
Aedemoses Walsingham, 31
Aedia Duponchel, 29
Aedilia Gistel, 18
Aepytus Herrich-Schäffer, 17
Aepytus Herrich-Schäffer (subgen.), 17
Aerotypia Walsingham, 51
Aesyla Chambers, 26
Aetola [sic] Frey, 57
Aetole Chambers, 57
Afdera Clarke, 27
Agapalsa Falkovitsh, 42
Agarica Sodoffsky, 23
Agathactis Meyrick, 51
Ageliarches Meyrick, 51
Agriastis Meyrick, 48
Agriastsi [sic] Busck, 48
Agriocoma Zeller, 28
Agrioscelis Meyrick, 40
Agriotorna Meyrick, 39
Alapa Kieffer & Jörgensen, 55
Alicadra Walker, 57
Aliciana Clarke, 38
Alinguata Fleming, 54
Alloaepytus Viette (subgen.), 17
Allocota Meyrick, 49
Allocotaniana Strand, 49
Alsodryas Meyrick, 51
Altiura Clarke, 38
Alucita Linnaeus, 54
Alucitina Heydenreich, 54
Alynda Clarke, 38
Amadria [sic] Chambers, 20
Amadrya [sic] Chambers, 20
Amaurogramma Braun, 44
Amaurosetia Stephens, 38
Amblothridia Wallengren, 55
Amblytenes Meyrick, 42
Amiantastis Meyrick, 24
Amphiclada Meyrick, 57
Amphisyncentris Meyrick, 21
Amselghia Căpuse, 42
Amseliphora Căpuse, 42

Amydria Clemens, 20
Amydrya [sic] Kearfott, 20
Anacampsis Curtis, 48
Anacampsoides Bruand, 54
Anacompsis [sic] Desmarest, 48
Anadasmus Walsingham, 30
Anadetia Hübner, 55
Anapatris Meyrick, 31
Anaphora Clemens, 19
Anaphorina Strand, 19
Anatrachyntis Meyrick, 44
Anchimacheta Walsingham, 56
Anchimompha Clarke, 42
Ancipita Busck, 27
Andeabatis Nielsen & Robinson, 16
Anemerarcha Meyrick, 21
Anesychia Hübner, 29
Animula Herrich-Schäffer, 24
Animula Herrich-Schäffer (subgen.), 24
Aniuta Clarke, 38
Ankistrophorus Walsingham, 19
Anomoxena Meyrick, 51
Anoncia Clarke, 43
Anophora [sic] Kearfott, 19
Anorcota Meyrick, 44
Anorthodisca [sic] Gaede, 49
Anorthosia Clemens, 49
Antaeotricha Zeller, 31
Antequera Clarke, 43
Anterethista Meyrick, 51
Antherethista [sic] Gaede, 51
Anthinora Meyrick, 51
Anthistarcha Meyrick, 51
Antiolopha Meyrick, 25
Antipolistes Forbes, 20
Antispastis Meyrick, 57
Antispila Hübner, 18
Antistarcha [sic] Costa Lima, 51
Antoeotricha [sic] Walsingham, 31
Anybia Stainton, 42
Apethistis Meyrick, 49
Aphanaula Meyrick, 47
Aphanosara Forbes, 43
Aphanoxena Meyrick, 31
Aphelosetia Stephens, 41
Aphigalia Dyar, 41
Apista Hübner, 42
Aplatissa Viette, 17
Apoclisis Walsingham, 19
Apopira Walsingham, 52
Apoplania Davis, 16
Aporiptura Falkovitsh, 42
Apotactis Meyrick, 51
Apothetoeca Meyrick, 51
Apotomia Dietz, 23
Arauzona Walker, 40
Arctopoda Butler, 38
Ardania Căpuse, 42
Ardrupia [sic] Busck, 53
Argiope Chambers, 57
Argyractinia Falkovitsh, 42
Argyresthia Hübner, 56
Argyrestia [sic] MacKay, 56
Argyromiges Curtis, 25
Argyromis Stephens, 25

Argyrosetia Stephens, 56
Aristotelia Hübner, 45
Arla Clarke, 45
Aroga Busck, 45
Arogalea Walsingham, 45
Arotromima Meyrick, 51
Arotrura Walsingham, 44
Arrhenophanes Walsingham, 25
Artipenna Davis, 24
Aruga [sic] Janse, 45
Asarista Meyrick, 41
Ascalenia Wocke, 44
Ascleriductia Căpuse, 42
Aspidisca Clemens, 19
Astala Davis, 24
Astoxena Meyrick, 39
Astyages Stephens, 42
Atabyria Snellen, 23
Atachia Wocke, 41
Atelosticha Meyrick, 38
Atemelia Herrich-Schäffer, 55
Atha Clarke, 38
Athleta Walsingham, 31
Athrinacia Walsingham, 27
Atinea Amsel, 21
Atopocera Walsingham, 19
Atoponeura Busck, 52
Atoposea Davis, 54
Atopotorna Meyrick, 38
Atteva Walker, 55
Atticonviva Busck, 20
Augolychna Meyrick, 20
Aulacomima Meyrick, 49
Aureliania Căpuse, 42
Autocnaptis Meyrick, 24
Autoneda Busck, 44
Autoses Hübner, 23
Auximobasis Walsingham, 41
Auxocrasia [sic] Walsingham, 35
Auxocrossa Zeller, 35
Auxotricha Meyrick, 27
Axiagasta Meyrick, 20
Azinis Walker, 29

Babaiaxa Busck, 29
Bacculatrix [sic] Flint, Noble & Shaw, 25
Bacescuia Căpuse, 42
Baeonoma Meyrick, 33
Baraschia Căpuse, 42
Barticeja Povolný, 45
Barymochtha Meyrick, 20
Basanasca Meyrick, 20
Basileura Nielsen & Davis, 18
Batrachedra Herrich-Schäffer, 42
Battaristis Meyrick, 48
Bazira Walker, 19
Bedellia Stainton, 25
Begoe Chambers, 51
Beltheca Busck, 51
Benanderpia Căpuse, 42
Besciva Busck, 51
Bima Falkovitsh, 42
Biopsyche Dyar, 24
Blabophanes Zeller, 22
Blanchardina Viette, 16

Blastobasis Zeller, 41
Blastotere Ratzeburg, 56
Boocara Butler, 40
Borkhausenia Hübner, 38
Bourgogneja Căpuse, 42
Brachiacma [sic] Common, 50
Brachicrossata [sic] Hartmann, 49
Brachiloma Clemens, 31
Brachmia Hübner, 49
Brachyacma Meyrick, 50
Brachyaema [sic] Povolný, 50
Brachycrossata Heinemann, 49
Brachyplatea Zeller, 28
Brachypsaltis Meyrick, 51
Brachysymbola Meyrick, 19
Braclunia [sic] Stephens, 49
Brithyceros Meyrick, 20
Brochometis Meyrick, 50
Bruchiana Jörgensen, 51
Brymblia Hodges, 38
Bucculatrix Zeller, 25
Busckia Dyar, 25
Butalis Treitschke, 44
Buxeta Walker, 56
Bythocrates Meyrick, 20

Cachura Walker, 22
Caenogenes Walsingham, 19
Calada Nielsen & Robinson, 16
Calaritania Mariani, 42
Calcomarginia Căpuse, 42
Caleophora [sic] Căpuse, 42
Callartona Hampson, 57
Calliathla Meyrick, 55
Callipielus Butler, 16
Calliprora Meyrick, 45
Callistenoma Butler, 38
Caloptilia Hübner, 25
Cameraria Chapman, 27
Capillaria Haworth, 18
Carna Walker, 50
Carpochena Falkovitsh, 42
Carposina Herrich-Schäffer, 54
Carposina Herrich-Schäffer (subgen.), 54
Carthara Walker, 55
Caryolestis Meyrick, 21
Casape Walker, 20
Casas Wallengren, 42
Casigneta Wallengren, 42
Casignetella Strand, 42
Catebrachmia Rebel, 45
Catacrypsis Walsingham, 41
Catalexis Walsingham, 51
Catarata Walsingham, 33
Cathegesis Walsingham, 49
Cathelotis Meyrick, 53
Catoptristis Meyrick, 51
Cecidolechia Kieffer & Jörgensen, 38
Cecidolechia Strand, 38
Cecidoses Curtis, 18
Cellaria [sic] Neave, 49
Cemiostoma Zeller, 25
Ceratophora Heinemann, 49
Ceratophysetis Meyrick, 29
Cerconata [sic] Busck, 33

Cerconota Meyrick, 33
Ceroclastis Zeller, 25
Ceromitia Zeller, 18
Cerostoma Latreille, 55
Cerostoma.- Stephens, 55
Cervitinea Amsel, 22
Cerycangela Meyrick, 52
Chaetochilus Stephens, 55
Chalcomima Meyrick, 52
Chalconympha Meyrick, 55
Chalybe Duponchel, 29
Characia Falkovitsh, 42
Chariphylla Meyrick, 27
Charistica Meyrick, 52
Chedra Hodges, 42
Chelaria Haworth, 49
Cheleria [sic] Lhomme, 49
Chezale Walker, 39
Chilocampyla Busck, 26
Chionoda [sic] Hübner, 45
Chionodes Hübner, 45
Chlamydastis Meyrick, 34
Chnoocera Falkovitsh, 42
Choropleca Durrant, 20
Chresotes Butler, 22
Chrysocorys Curtis, 57
Chrysoryctis Meyrick, 23
Cibyra Walker (subgen.), 17
Cirrha Chambers, 46
Cladodes Heinemann, 49
Clepticodes Meyrick, 21
Clinograptis Meyrick, 21
Clistoses Kieffer & Jörgensen, 18
Clistothyris Zeller, 52
Cnismorectis Meyrick, 21
Cnissostages Zeller, 25
Coenyphantes Hübner, 56
Coleophora Hübner, 42
Coleostoma Meyrick, 52
Coleotechnistes [sic] Riley, 45
Coleotechnites Chambers, 45
Colinita Busck, 44
Colonanthes Meyrick, 52
Colpocrita Meyrick, 21
Commatica Meyrick, 52
Comodica Meyrick, 21
Comotechna Meyrick, 27
Compsistis Meyrick, 27
Compsocrita Meyrick, 21
Compsolechia Meyrick, 48
Compsosaris Meyrick, 52
Compsoschema Walsingham, 25
Conchyliospila Wallengren, 22
Conquassata Gozmány, 41
Copocentra Meyrick, 57
Copocercia Zeller, 46
Copticostola Meyrick, 52
Coptodisca Walsingham, 19
Coptotelia Zeller, 27
Coptotriche Walsingham, 18
Corethropoea Falkovitsh, 42
Corinea Walker, 55
Coriscium Zeller, 25
Corita Clarke, 39
Cornulivalvulia [sic] Căpuse, 42

Corothropoea Căpuse, 42
Corythangela Meyrick, 42
Cosmopterix Hübner, 43
Cosmopteryx Zeller, 43
Costoma Busck, 27
Cotaena Walker, 54
Cotyloscia Meyrick, 50
Crambodoxa Meyrick, 52
Cranaodes Meyrick, 21
Crasimorpha Meyrick, 49
Creagria Sodoffsky, 55
Credemna [sic] Forbes, 55
Credemnon Wallengren, 55
Credemon [sic] Moriuti, 55
Cremastobombycia Braun, 26
Crembalastis Meyrick, 57
Crepidochares Meyrick, 21
Cricotechna Falkovitsh, 42
Cronicombra Meyrick, 54
Cryphiotechna Meyrick, 21
Crypsynarthra Lower, 38
Cryptolechia Zeller, 27
Cryptothelea Duncan, 24
Cryptotheles [sic] Costa Lima, 24
Cuphodes Meyrick, 26
Curtorama Davis, 24
Cyane Chambers, 20
Cycloplasis Clemens, 57
Cycnodia Herrich-Schäffer, 41
Cymatomorpha Meyrick, 49
Cymotricha Meyrick, 50
Cynestomorpha Meyrick, 49
Cynotes Walsingham, 41
Cyphophora Herrich-Schäffer, 42
Cyptasia Walker, 56
Cyrictodes Meyrick, 29

"**Dalaca**" auct., 17
Dalaca Walker, 16
Damophila Curtis, 42
Darlia Clarke, 52
Dasmophora Meyrick, 21
Dasycarea Zeller, 25
Daulia Walker, 19
Davendra Moore, 57
Decadarchis Meyrick, 21
Decantha Busck, 39
Dectobathra Meyrick, 50
Deia Clarke, 39
Deltophora Janse, 45
Demobrotis Meyrick, 21
Dendroneura Walsingham, 22
Dendropsyche Jones, 24
Deoclana [sic] Fletcher, 51
Deoclona Busck, 51
Depressariodes Turati, 27
Derchis Walker, 19
Desmaucha Meyrick, 53
Despina Clarke, 39
Deuteroptila Meyrick, 49
Diachalastis Meyrick, 20
Dialectica Walsingham, 26
Dianasa Walker, 56
Diaphthirusa Hübner, 22
Diastaltica Walsingham, 52

Diastoma Möschler, 34
Diataga Walsingham, 21
Dichomeris Hübner, 50
Dicranoctetes Braun, 41
Dicranoses Kieffer & Jörgensen, 18
Dicte Chambers, 18
Dinotropa Meyrick, 39
Diploschizia Heppner, 55
Dipremna Davis (subgen.), 54
Dissoptila Meyrick, 52
Disthymnia Hübner, 29
Dita Clarke, 39
Ditrigonophora Walsingham, 56
Doina Clarke, 27
Dolaca [sic] Druce, 16
Dolidiria Busck, 30
Doliotechna Meyrick, 39
Donacivola Busck, 41
Dorata Busck, 21
Dorota [sic] Kearfott, 21
Doshia Clarke, 27
Doxa Walsingham, 39
Drastea Walsingham, 21
Drepanoterma Walsingham, 52
Dromiaulis Meyrick, 43
Druceiella Viette, 16
Ductispira Căpuse, 42
Dumitrescumia Căpuse, 42
Durrantia Busck, 30
Duvita Busck, 48
Dyotopasta Busck, 21
Dyselachista Spuler, 19
Dysgnorima Zeller, 39
Dysnepticula Börner, 18
Dysoptus Walsingham, 21
Dystinea Börner, 23
Dystopasta McDunnough, 21

Ecballogonia Walsingham, 43
Ecebalia Căpuse, 42
Echinoglossa Clarke, 45
Ecliptoloma Zeller, 29
Ecpathophanes Bradley, 25
Ectaga Walsingham, 27
Ectinocampa Silvestri, 21
Eddara Walker, 19
Ederesa Curtis, 56
Edosa Walker, 23
Eidothea Chambers, 45
Eidothoa [sic] Chambers, 45
Elachista Treitschke, 41
Elasiprora Meyrick, 52
Embola Walsingham, 57
Empedaula Meyrick, 52
Enaemia Zeller, 56
Endothamna Meyrick, 53
Endrosis Hübner, 39
Energia Walsingham, 34
Enteucha Meyrick, 18
Eomichla Meyrick, 39
Ephedroxena Meyrick, 21
Ephystereris [sic] Janse, 45
Ephysteris Meyrick, 45
Epichnopterix Hübner, 24
Epichnopteryx [sic] Heylaerts, 24

Epidictica Turner, 56
Epilechia Busck, 49
Epilegis Dietz, 23
Epimoryctis Meyrick, 39
Epipremna Davis (subgen.), 54
Episacta Turner, 49
Epistetus Walsingham, 41
Episyrta Meyrick, 21
Eraina Clarke, 39
Erechthias Meyrick, 21
Erechtias [sic] Ghesquière, 21
Ereunetis Meyrick, 21
Ergatis Heinemann, 45
Erineda Busck, 40
Eriopyrrha Meyrick, 56
Eriphia Chambers, 44
Eripnura Meyrick, 52
Eristhenodes Meyrick, 52
Eritarbes Walsingham, 44
Erithyma Meyrick, 27
Erminea Haworth, 56
Erysiptila Meyrick, 29
Erythriastis Meyrick, 52
Ethirostoma Meyrick, 52
Ethmia Hübner, 29
Euarne Möschler, 56
Eucalliathla Clarke, 55
Eucatoptus Walsingham, 45
Eucecidoses Brèthes, 18
Euceratia Walsingham, 55
Eucestis Hübner, 26
Euchionodes Clarke, 45
Euchiradia Hübner, 54
Euclemensia Grote, 43
Eucordylea Dietz, 45
Eucosmophora Walsingham, 26
Eudactylota Walsingham, 45
Eudodacles Snellen, 49
Eudolichura Clarke, 55
Eulepiste Walsingham, 19
Eumimographe Dognin, 28
Eumiturga Meyrick, 31
Eunebristis Meyrick, 50
Eunomarcha Meyrick, 52
Euota Hübner, 55
Eupista Hübner, 42
Eupleuris Hübner, 40
Eupolella Fletcher, 41
Eupolis Meyrick, 41
Eupragia Walsingham, 28
Euprora Busck, 25
Euproteodes Viette, 41
Eurynome Chambers, 25
Eurysacca Povolný (subgen.), 47
Eurysara Turner, 50
Euryzancla Turner, 50
Euspilapteryx [sic] Spuler, 26
Euspilopteryx Zeller, 26
Eustaintonia Spuler, 42
Eusynopa Lower, 22
Eutheca Grote, 19
Eutrichocnemis Spuler, 26
Euzonomacha Meyrick, 52
Evagora Clemens, 45
Evexia Gistel, 18

Evippe Chambers, 45
Exaeretia Stainton, 27
Exala Meyrick, 22
Exoncotis Meyrick, 21
Exosphrantis Meyrick, 39
Exoteleia Wallengren, 45

Faculta Busck, 46
Falculina Zeller, 34
Falkovitshia Căpuse, 42
Fapua Kieffer & Jörgensen, 53
Fascista Busck, 46
Felderia Walsingham, 19
Filinota Busck, 28
Fortinea Busck, 52
Frederickoenigia Căpuse, 42
Friseria Busck, 46

Gaeza Walker, 50
Galechia [sic] Desmarest, 46
Galtica Busck, 52
Gelechia Hübner, 46
Gelecia [sic] Watt, 46
Gelschia [sic] Nowicki, 46
Geniadophora Walsingham, 47
Glaseria Căpuse, 42
Glaucacna Forbes, 52
Glaucagna [sic] Gaede, 52
Globulia Căpuse, 42
Glyphidocera Walsingham, 41
Glyphipterix Hübner, 54
Glyphipterys [sic] Christoph, 54
Glyphipteryx Zeller, 54
Glyphiptoryx [sic] Mann & Rogenhofer, 54
Glyphiteryx [sic] Fischer von Röslerstamm, 54
Glyphopteryx Herrich-Schäffer, 54
Glyphptieryx [sic] Turati, 54
Glyphteryx [sic] Watt, 54
Glyphyteryx [sic] Hampson, 54
Glypipteryx [sic] Stainton, 54
Gnathotona Meyrick, 28
Gnorimochema [sic] Dyar, 46
Gnorimoschema Busck, 46
Gnorrimoschema [sic] Hartig, 46
Gompsosaris [sic] Gaede, 52
Gonada Busck, 28
Gonionota Zeller, 28
Gonioterma Walsingham, 34
Gonorimoschema [sic] Deurs, 46
Guenea Bruand, 46
Gymelloxes Viette (subgen.), 17

Habrophylax Meyrick, 28
Hagno Chambers, 29
Halimarmara Meyrick, 39
Hamadera Busck, 28
Hamadryas Clemens, 43
Hampsoniella Viette (subgen.), 17
Hamuliella Căpuse, 42
Hapalonoma Meyrick, 50
Hapalosaris Meyrick, 46
Hapalothyma Meyrick, 25
Haplochela Meyrick, 49
Haploptilia Hübner, 42
Harmaclona Busck, 25

Harpagandra Meyrick, 41
Harpalyce Chambers, 31
Harpella Schrank, 39
Harpepteryxx Sodoffsky, 55
Harpipterix Hübner, 55
Harpipteryx Treitschke, 55
Harpograptis Meyrick, 43
Harpopteryx Agassiz, 55
Hasta Busck, 28
Hastamea Fletcher, 28
Hecista Wallengren, 41
Hedycharis Turner, 56
Helcystogramma Zeller, 50
Heliodines Stainton, 57
Heliodinides [sic] Turner, 57
Heliostibes Zeller, 39
Heliozela Herrich-Schäffer, 19
Helopharea Falkovitsh, 42
Helvalbia Căpuse, 42
Hepialyxodes Viette (subgen.), 17
Heribeia Stephens, 54
Heringia Spuler, 45
Heringiella Börner, 42
Heringiola Strand, 45
Heterobathmia Kristensen & Nielsen, 16
Hexeretmis Meyrick, 54
Hibita Walker, 19
Hieroxestis Meyrick, 22
Himotica Meyrick, 28
Hinnebergia Spuler, 47
Hofmannophila Spuler, 39
Holcocera Clemens, 41
Holophysis Walsingham, 49
Holoscolia Zeller, 40
Homeoprepes [sic] Hodges, 43
Homilostola Meyrick, 24
Homodoxus Walsingham, 21
Homoeoprepes Walsingham, 43
Homonymus Walsingham, 19
Homosetia Clemens, 21
Homostinea Dietz, 21
Homotinea Meyrick, 21
Hoplophysis [sic] McDunnough, 49
Hormantris Meyrick, 21
Horomeristis Meyrick, 39
Huapina Bryk, 16
Hyalopseustis Meyrick, 35
Hyalospila Herrich-Schäffer, 22
Hybroma Clemens, 21
Hymenopsyche Grote, 25
Hypatima Hübner, 49
Hypatina [sic] Stephens, 49
Hypercallia Stephens, 28
Hyperskeles Butler, 39
Hyphantes Hübner, 56
Hyphypena Warren, 27
Hypoclopus Walsingham, 19
Hypocolypus [sic] Dyar, 19
Hypolepia Guenée, 55
Hypomartyria Kristensen & Nielsen, 16
Hyponomenta [sic] Turner, 56
Hyponomeuta Billberg, 56
Hyponomeuta Sodoffsky, 56
Hypopremna Davis (subgen.), 54
Hypsilophus Agassiz, 55

Hypsolopha Billberg, 55

Iconisma Walsingham, 41
Ide Chambers, 31
Idiocrates Meyrick, 28
Idioptila Meyrick, 40
Ifeda Hodges, 39
Ilarches Meyrick, 52
Ilingiotis Meyrick, 50
Imma Walker, 57
Infurcitinea Spuler, 21
Inga Busck, 39
Inotica Meyrick, 47
Instica [sic] Sharp, 47
Ionescumia Căpuse, 42
Ionnemesia Căpuse, 42
Iphimachaera Meyrick, 52
Irenia Clarke, 40
Irenicodes Meyrick, 41
Isembola Meyrick, 52
Ismene Stephens, 56
Isochasta Meyrick, 45
Isocorypha Dietz, 21
Isophrictis Meyrick, 44
Ithome Chambers, 44
Ithutomus Butler, 56
Ithytomus Meyrick, 56

Jobula Walker, 57
Johanssonia Borkowski, 18

Kakivoria Nagano, 40
Kasyfia Căpuse, 42
Keiferia Busck, 46
Klimeschja Căpuse, 42
Klimeschjosefia Căpuse, 42
Klinzigedia Căpuse, 42
Kuznetzovvlia Căpuse, 42

Labdia Walker, 43
Lactura Walker, 56
Lamelliformia Viette (subgen.), 17
Lamprolophus Busck, 57
Lamprozela Meyrick, 19
Larupsia [sic] Soffner, 46
Lata Kieffer & Jörgensen, 53
Lathontogenes [sic] Hodges, 50
Lathontogenus Walsingham, 50
Lathontogonus [sic] Diakonoff, 50
Latisacculia Căpuse, 42
Laverna Curtis, 42
Leistogenes Meyrick, 52
Lelita Clarke, 40
Lepidobregma Zimmerman, 21
Leptochersa Meyrick, 21
Leptozestis Meyrick, 44
Lepyrotica Meyrick, 21
Lerupsia Riedl, 46
Lethata Duckworth, 35
Leucanthiza Clemens, 26
Leucophasma Walsingham, 22
Leucophryne Chambers, 42
Leucoptera Hübner, 25
Leuroperna Clarke, 55
Limnaecia Stainton, 44

Lindera Blanchard, 22
Lioclepta Meyrick, 51
Lipatia Busck, 50
Lipomerinx Walsingham, 22
Lita Treitschke, 46
Lithariapteryx Chambers, 57
Lithcolletes [sic] Matsumura, 26
Lithocolletis Hübner, 26
Lithopsaestis Meyrick, 22
Locharca Meyrick, 46
Logisis Walsingham, 52
Longibacillia Căpuse, 42
Lophaeola Meyrick, 52
Lophoptilus Sircom, 42
Lossbergiana Viette, 16
Lotisma Busck, 53
Loxotoma Zeller, 35
Loxotrochis Meyrick, 57
Lozostoma Stainton, 22
Lucidaesia Căpuse, 42
Lucyna Clarke, 29
Lumacra Davis, 24
Lupercalia Busck, 28
Luzulina Falkovitsh, 42
Lvaria Căpuse, 42
Lychnocrates Meyrick, 38
Lygronoma Meyrick, 40
Lymnaecia [sic] Kimball, 44
Lyoneta [sic] Matsumura, 25
Lyonetia Hübner, 25
Lysigrapha Meyrick, 39

Macarocosma Meyrick, 40
Machimia Clemens, 29
Machlotica Meyrick, 54
Maclotica [sic] Busck, 54
Macroceras Staudinger, 41
Macrocirca Meyrick, 30
Macrozancla Turner, 50
Maculella Viette, 16
Maesara Clarke, 29
Magnifascia Povolný (subgen.), 47
Malachotriche [sic] Busck, 51
Malacotricha Zeller, 51
Malacotriche [sic] Busck, 51
Manchana Walker, 23
Mapa Strand, 55
Marmara Clemens, 26
Martyrhilda Clarke, 27
Mathildana Clarke, 40
Mattea Duckworth, 40
Mea Busck, 22
Megacraspedas [sic] Barnes & McDunnough, 44
Megacraspedus Zeller, 44
Melaneulia Butler, 29
Melanoleuca Stephens, 29
Melitonympha Meyrick, 55
Melochrysis Meyrick, 40
Membrania Căpuse, 42
Menesta Clemens, 35
Meridorma Meyrick, 52
Mesoptycha Zeller, 31
Metabolaea Meyrick, 52
Metallitis Sodoffsky, 18
Metallocrates Meyrick, 41

Metallosetia Stephens, 42
Metapista Căpuse, 42
Metaxypsyche Davis, 24
Metopleura Busck, 52
Microcolona Meyrick, 43
Microcraspedus Janse, 45
Microlechia Turati, 47
Microscardia Amsel, 23
Microtinea Amsel, 21
Mieza Walker, 56
Mnesichara Walsingham, 28
Moca Walker, 57
Molopostola Meyrick, 52
Mompha Hübner, 42
Monachozela Meyrick, 19
Monochroa Heinemann, 45
Monopis Hübner, 22
Monotemachia Falkovitsh, 42
Moriloma Busck, 44
Morophaga Herrich-Schäffer, 23
Mothonica Walsingham, 35
Multicoloria Căpuse, 42
Muna Clarke, 29
Myrmecozela Zeller, 22
Myrophila Meyrick, 50
Myrsila Boisduval, 54
Mysaromima Meyrick, 35
Mythoplastis Meyrick, 22

Naevipenna Davis, 24
Nanodacna Clarke, 43
Neaera Chambers, 41
Nealyda Dietz, 45
Necedes Walsingham, 29
Neda Chambers, 44
Nedenia Clarke, 29
Nemapogon Schrank, 22
Nematochares Meyrick, 29
Nemesia Căpuse, 42
Neochrista Meyrick, 51
Neodactylota Busck, 46
Neodecadarchis Zimmerman, 21
Neolophus Walsingham, 19
Neomachlotica Heppner, 54
Neomeristis Meyrick, 20
Neophylarcha Meyrick, 53
Neoschema Povolný, 46
Neotheora Kristensen, 16
Nepticula Heyden, 18
Nesolechia Meyrick, 49
Nesoxena Meyrick, 21
Neugenvia Căpuse, 42
Neurobathra Ely, 26
Neurostrata [sic] Ely, 26
Neurostrota Ely, 26
Nicanthes Meyrick, 43
Niditinea Petersen, 22
Noeza Walker, 51
Nosyrislia Căpuse, 42
Nygmia Hübner, 56

Obithome Hodges, 44
Odonna Clarke, 40
Oeceticus [sic] Herrich-Schäffer, 24
Oecia Walsingham, 41

Oedicaula Falkovitsh, 42
Oegoconides [sic] Neave, 46
Oegoconiodes Matsumura, 46
Oeketicus [sic] Lefebre, 24
Oenoe Chambers, 22
Oenophila [sic] Zeller, 22
Oeseis Chambers, 46
Oestomorpha Walsingham, 52
Oeta Grote, 55
Oiceticus [sic] Lahille, 24
Oicocestis [sic] Köhler, 24
Oiketicus Guilding, 24
Oiketicus Guilding (subgen.), 25
Oiketikus [sic] Seitz, 24
Oinophila Stephens, 22
Oliera Brèthes, 18
Oligos Treitschke, 56
Onebala Walker, 50
Opogona Zeller, 22
Opostega Zeller, 18
Opsodoca Meyrick, 22
Ordrupia Busck, 53
Ordupia [sic] Busck, 53
Orghidania Căpuse, 42
Orneodes Latreille, 54
Ornix Treitschke, 25
Orothyntis Meyrick, 19
Orphnolechia Meyrick, 35
Orsimacha Meyrick, 39
Orsotricha Meyrick, 46
Orthenches Meyrick, 55
Orthographis Falkovitsh, 42
Ortholophus Walsingham, 19
Ortohgraphis [sic] Căpuse, 42
Oscella Walker, 23
Osmarina Clarke, 29
Osphretica Meyrick, 23
Osrhoes Druce, 17
Otochares Meyrick, 22
Oudejansia Căpuse, 42
Oxybelia Hübner, 50
Oxycryptis Meyrick, 53
Oxylechia Meyrick, 53
Oxysactis Meyrick, 50

Pachdyta [sic] Meyrick, 22
Pachydyta Meyrick, 22
Pachygeneia Meyrick, 53
Pachyphoenix Butler, 40
Pachysaris Meyrick, 51
Paelia Walker, 54
Paepia Walker, 29
Palaephatus Butler, 22
Palinorsa Meyrick, 29
Palpula Blanchard, 22
Pammeces Zeller, 43
Panclintis Meyrick, 43
Pantheus Zimmerman, 21
Panthytarcha Meyrick, 22
Paragorgopis Viette (subgen.), 17
Paralechia Busck, 45
Parana Viette, 17
Paraneura Dietz, 22
Paranoea Walsingham, 51
Paraoiketicus Davis (subgen.), 25

Parapielus Viette, 16
Parascaeas Meyrick, 35
Paraspastis Meyrick, 35
Paraspistes Meyrick, 50
Paraspistis [sic] Busck, 50
Parastega Meyrick, 53
Parasymmoca Rebel, 41
Parategeticula Davis, 18
Paratiquadra Walsingham, 56
Paravalvulia Căpuse, 42
Parectopa Clemens, 26
Parelectra Meyrick, 53
Parelectroides Clarke, 53
Parochromolopis Gaedike, 54
Parornix Spuler, 26
Patzakia Căpuse, 42
Paucivena Davis, 24
Pavolechia Busck, 53
Pectinophora Busck, 49
Pedaliotis Meyrick, 22
Peleopoda Zeller, 30
Pelocnistis Meyrick, 53
Pelomimas Meyrick, 39
Penica Walsingham, 26
Periclymenobius Wallengren, 55
Perileucoptera Silvestri, 25
Perilicmetis Meyrick, 22
Perimede Chambers, 44
Perioristica Walsingham, 53
Periploca Braun, 44
Perisceptis Meyrick, 24
Perygra Falkovitsh, 42
Perygridia Falkovitsh, 42
Perzelia Clarke, 29
Pessograptis Meyrick, 49
Petalothyrsa Meyrick, 35
Petasanthes Meyrick, 35
Pexicnemidia Möschler, 56
Pfitzneriana Viette, 17
Pfitzneriella Viette, 17
Phaeoses Forbes, 22
Phaetusa Chambers, 45
Phagolamia Falkovitsh, 42
Phalarotarsa Meyrick, 27
Phalerarcha Meyrick, 54
Phanerodoxa Meyrick, 39
Phanerozela Meyrick, 19
Pharmacoptis Meyrick, 44
Phassus Walker, 16
Phelotropa Meyrick, 35
Phereoeca Hinton & Bradley, 22
Phialuse Viette, 17
Phigalia Chambers, 41
Philaenia Kirby (subgen.), 17
Philaustera Meyrick, 55
Phillonome [sic] Chambers, 25
Philodoxa Gistel, 18
Philomusaea Meyrick, 40
Philomusea [sic] Clarke, 40
Philonome Chambers, 25
Philtronoma Meyrick, 29
Phlongia Walker, 19
Pholcobates Meyrick, 29
Phrixosceles Meyrick, 26
Phthorimaea Meyrick, 46

Phthorimea [sic] Diakonoff, 46
Phthorimoea [sic] Povolný & Zakopal, 46
Phtorimea [sic] Oei-Dharma, 46
Phtyorimaea [sic] Turner, 46
Phycomorpha Meyrick, 54
Phyllocnistis Zeller, 27
Phyllocnitis [sic] Busck, 27
Phylloenistis [sic] Chambers, 27
Phylloetis [sic] Chambers, 27
Phyllonorycter Hübner, 26
Phyllorycter [sic] Walsingham, 26
Phylloschema Falkovitsh, 42
Phylopatris Meyrick, 53
Phytomimia Walsingham, 29
Pigritia Clemens, 42
Pilanaphora Walsingham, 19
Pingrassa Walker, 57
Pisinidea Butler, 29
Placostola Meyrick, 40
Platoeceticus Packard, 24
Plegmidia Falkovitsh, 42
Pleurota Hübner, 40
Plocamosaris Meyrick, 51
Plumana Busck, 24
Plutella Schrank, 55
Plutelloptera [sic] Walsingham, 55
Pluteloptera Chambers, 55
Poeciloptera Clemens, 55
Poeciloptilia Hübner, 25
Polyhymno Chambers, 46
Polypsecta Meyrick, 22
Polypseustis Dognin, 38
Pomphocrita Meyrick, 33
Pompostola Meyrick, 22
Pompostolella Fletcher, 22
Porphyrosela Braun, 27
Porpodryas Meyrick, 49
Porrectaria Haworth, 42
Postvinculia Căpuse, 42
Praeacedes Amsel, 22
Prasolithites Meyrick, 31
Prays Hübner, 55
Proboloptila Meyrick, 23
Prochalia Barnes & McDunnough, 24
Prochola Meyrick, 43
Proclesis Walsingham, 51
Prodoxoides Nielsen & Davis, 18
Prodoxus Riley, 18
Profilinota Clarke, 29
Proglaseria Căpuse, 42
Progona Dietz, 22
Promasia Chrétien, 22
Promenesta Busck, 35
Promiba [sic] Kirby, 18
Promolopica Meyrick, 53
Pronuba Riley, 18
Prophoraula Meyrick, 51
Prosodica Walsingham, 41
Prostesis Walsingham, 41
Prostomeus Busck, 49
Protasis Herrich-Schäffer, 40
Protodarcia Forbes, 23
Protolophe Rebel, 22
Protonyctia Meyrick, 57
Psacaphora Herrich-Schäffer, 42

Psecadia Hübner, 29
Psecadioides Butler, 22
Psephocrita Meyrick, 19
Psephomeres Meyrick, 31
Pseudanaphora Walsingham, 19
Pseudarla Clarke, 46
Pseudastasia Walsingham, 57
Pseudatemelia Rebel, 38
Pseuderotis Clarke, 30
Pseudocaprima Walsingham, 56
Pseudocentris Meyrick, 29
Pseudoconchylis Walsingham, 19
Pseudodalaca Viette (subgen.), 17
Pseudoecophora Staudinger, 40
Pseudophassus Pfitzner (1914), 17
Pseudophassus Pfitzner (1938), 17
Pseudophilaenia Viette (subgen.), 17
Pseudotalara Druce, 56
Pseudotortrix Turner, 57
Pseudoxylesthia Walsingham, 21
Psilocorsis Clemens, 29
Psittacastis Meyrick, 29
Pterogyne Davis, 24
Pterolonche Walsingham, 21
Pteroxia Guenée, 55
Pthorimaea [sic] Issiki, 46
Ptilogenes Meyrick, 34
Ptilonostychia [sic] Fletcher, 53
Ptilopsaltis Meyrick, 23
Ptilostonychia Walsingham, 53
Ptycerata Ely, 46
Ptycerata Ely (subgen.), 46
Ptychoxena Meyrick, 25
Puermytrans Viette, 16
Pulicalvaria T. N. Freeman, 45
Pycnobathra, Lower, 44
Pycnopogon Chrétien, 55
Pycnotarsa Meyrick, 40
Pygmocrates Meyrick, 56
Pyramidobela Braun, 40
Pyroderces Herrich-Schäffer, 44

Quadratia Căpuse, 42

Razowskia Căpuse, 42
Rectiostoma Becker, 35
Recurvaria Haworth, 47
Retha Clarke, 40
Reuttia O. Hofmann, 47
Revonda Clarke, 40
Rhabdocrates Meyrick, 54
Rhamnia Căpuse, 42
Rhindoma Busck, 29
Rhinosia Treitschke, 50
Rhitia Walker, 22
Rhobonda Walker, 50
Rhodanassa Meyrick, 35
Rhopalosetia Meyrick, 54
Rhynchotona Meyrick, 53
Ridiaschinia Brèthes, 18
Roseala Viette, 17

Sacculia Căpuse, 42
Safra Walker, 22
Sagaritis Chambers, 49

Sapinella Kirby, 19
Sarbena Walker, 56
Saridacma Meyrick, 54
Sathrobrota Hodges, 44
Satrapodoxa Meyrick, 53
"Scaeosopha" Meyrick, 44
Scardia Treitschke, 23
Scelorthus Busck, 57
Sceptea Walsingham, 41
Sceptia [sic] McDunnough, 41
Schaefferiana Viette (subgen.), 17
Schausiana Viette, 17
Schistonoea Forbes, 30
Schistophila Chrétien, 47
Schreckensteinia Hübner, 57
Scintilla Guenée, 55
Scleriductia Căpuse, 42
Sclerograptis Meyrick, 53
Scoliographa Meyrick, 29
Scrobipalpopsis Povolný, 46
Scrobipalpula Povolný, 47
Scrobipalpula Povolný (subgen.), 47
Scrobischema Povolný (subgen.), 46
Scythris Hübner, 44
Sematoptis Meyrick, 44
Semiomeris Meyrick, 51
Semiota Dietz, 23
Semodictis Meyrick, 49
Semophylax Meyrick, 49
Sericostola Meyrick, 54
Setiarcha Meyrick, 23
Setiostoma R. Felder & Rogenhofer, 54
Setiostoma Zeller, 35
Setomorpha Zeller, 23
Siderograptis Meyrick, 39
Silotroga [sic] Kirby, 49
Simacauda Nielsen & Davis, 18
Simoca [sic] Weiler, 41
Simoneura Walsingham, 53
Sirogenes Meyrick, 50
Sititroga [sic] Costa Lima, 49
Sitotrega [sic] Borg, 49
Sitotroga Heinemann, 49
Sitotrogus [sic] Matsumura, 49
Snellenia Walsingham, 40
Socorypha [sic] Busck, 21
Sophronia Hübner, 47
Sorotacta Meyrick, 53
Spanioptila Walsingham, 26
Spiladarcha Meyrick, 56
Squamicornia Kristensen & Nielsen, 16
Stachyocera Ureta, 16
Stachyostoma Meyrick, 53
Stagmaturgis Meyrick, 53
Stasixena Meyrick, 40
Stathmopoda Herrich-Schäffer, 40
Stegasta Meyrick, 47
Stenoma Zeller, 35
Steremniodes Meyrick, 53
Stereodmeta Meyrick, 53
Stibarenches Meyrick, 53
Stigmalla [sic] Herrich-Schäffer, 18
Stigmella Schrank, 18
Stilbosis Clemens, 44
Stollia Căpuse, 42

Stomopterix [sic] Turati, 47
Stomopteryx Heinemann, 47
Strobisia Clemens, 49
Stromopteryx [sic] Pierce & Metcalfe, 47
Struthoscelis Meyrick, 40
Suireia Căpuse, 42
Syblis Guenée, 55
Symmetrischema Povolný, 47
Symmoca Hübner, 41
Symmoletria Gozmány, 41
Symphanactis Meyrick, 53
Symphypoda Falkovitsh, 42
Synactias Meyrick, 53
Synadia Walker, 55
Synallagma Busck, 43
Syncamaris Meyrick, 54
Syncerastis Meyrick, 56
Syncraternis Meyrick, 23
Synempora Davis & Nielsen, 16
Syngenomictis Meyrick, 49
Syntetrernis Meyrick, 43
Syrmologa Meyrick, 23
Syrrhoaula Meyrick, 24
Systasiota Walsingham, 49
Systrophoeca Falkovitsh, 42

Tabernillaea Meyrick, 53
Tabernillaia Walsingham, 53
Tachasara Walker, 19
Tachiptilia [sic] Chambers, 48
Tachoptilia [sic] Daltry, 48
Tachyptilia Heinemann, 48
Tachyptilix [sic] Hartmann, 48
Taeniodictys Forbes, 23
Taeniostola Meyrick, 54
Taeniostolella Fletcher, 54
Talitha Clarke, 27
Tamarrha Walker, 29
Taphrosaris Meyrick, 53
Taruda Walker, 29
Taygete Chambers, 47
Tecia Kieffer & Jörgensen, 53
Tegeticula Zeller, 18
Teinoptila Sauber, 56
Teladoma Busck, 44
Telea Stephens, 47
Telephusa [sic] Beirne, 47
Telphusa Chambers, 47
Teratomorpha Walsingham, 29
Teresita Clarke, 40
Tetrapalpus Davis, 23
Teuchophanes Meyrick, 51
Thalassonympha Meyrick, 55
Thanatopyche Butler, 24
Thaumatolita Walsingham, 40
Theama E. M. Hering, 40
Theatria Walsingham, 30
Thelethia Dyar, 18
Themiscyra Walker, 56
Theoxenia Walsingham, 29
Theristis Hübner, 55
Thia H. Edwards, 18
Thiastyx Viette (subgen.), 17
Thioscelis Meyrick, 38
Thiotricha [sic] Hartig, 47

Thiotrica [sic] Inoue, 47
Thiotricha Meyrick, 47
Tholerostola Meyrick, 45
Thomictis Meyrick, 25
Thrasydoxa Meyrick, 57
Thripsigenes [sic] Clarke, 53
Thrypsigenes Meyrick, 53
Thylacopleura Meyrick, 57
Thyridopteryx Stephens, 25
Thyrsomnestis Meyrick, 53
Thysanoscelis Walsingham, 19
Thysanosedes [sic] Druce, 19
Thysanoskelis Walsingham, 19
Tildenia Povolný, 47
Timocratica Meyrick, 38
Tinaegeria Walker, 40
Tinea Linnaeus, 23
Tineidia Zagulajev, 22
Tinexotaxa Gozmány, 21
Tiquadra Walker, 23
Tirasia Walker, 19
Tischeria Zeller, 18
Tisheria [sic] Busck, 18
Titaenoses Hinton & Bradley, 22
Tocasta Busck, 42
Tocmia Walker, 53
Toecorhychia Butler, 56
Toenga Tindale, 16
Tolleophora Căpuse, 42
Tollsia Căpuse, 42
Topaza Walker, 57
Tortricomorpha C. Felder, 57
Toxoceras Chrétien, 44
Trachoma Wallengren, 55
Trachyphilia [sic] Le Marchand, 48
Trachyptilia [sic] Le Marchand, 48
Trapeziophora Walsingham, 54
Trepsitypa Meyrick (subgen.), 54
Triadogona Meyrick, 21
Trichembola Meyrick, 53
Trichophaga Ragonot, 24
Trichophassus Le Cerf, 16
Trichorrhabda Meyrick, 18
Trichostibas Zeller, 56
Trichotaphe Clemens, 51
Tricladia R. Felder (subgen.), 17
Triclonella Busck, 44
Tricotaphe [sic] Riley, 51
Trierostola Meyrick, 24
Trisyntopa Lower, 23
Trycherodes Meyrick, 29
Tuberculia Căpuse, 42
Tuchyptilia [sic] Kirby, 48
Tuta Kieffer & Jörgensen, 46
Tyriomorpha Meyrick, 40

Ulna Căpuse, 42
Untomia Busck, 49
Urangela Busck, 44
Urbara Walker, 19
Urodus Herrich-Schäffer, 56
Usara [sic] Busck, 54
Ussara Walker, 54
Utilia Clarke, 40

Valentinia Coolidge, 18
Valentinia Walsingham, 42
Valvulongia Căpuse, 42
Vazugada Walker, 51
Ventia Walker, 23
Vinzela Walker, 57
Vladdelia Căpuse, 42

Walshia Clemens, 44
Wilsonia Clemens, 42
Wiltshireia Amsel, 29

Xerocausta Meyrick, 19
Xylesthia Clemens, 24
Xylestia [sic] Dyar, 24
Xyrosaria [sic] Kearfott, 56
Xyrosaris Meyrick, 56
Xystrologa Meyrick, 24
Xytrops Viette (subgen.), 17

Yleuxas Viette (subgen.), 17
Yponomeuta Latreille, 56
Ypsolopha Latreille, 55
Ypsolophus Fabricius, 55

Zagulajevia Căpuse, 42
Zangheriphora Căpuse, 42
Zaratha Walker, 43
Zelleria Stainton, 56
Zelosyne Walsingham, 53
Zemiocrita Meyrick, 29
Zetesima Walsingham, 38
Zonochares Meyrick, 24
Zymologa Meyrick, 24
Zymrina Clarke, 40

Addendum

Aristoptila Meyrick, 58
Atmozostis Meyrick, 58
Diacholotis Meyrick, 58
Helcanthica Meyrick, 58
Palaeomystis Meyrick, 58